After Phrenology

After Phrenology

Neural Reuse and the Interactive Brain

Michael L. Anderson

A Bradford Book
The MIT Press
Cambridge, Massachusetts
London, England

This book was set in Stone Sans and Stone Serif by Toppan Best-set Premedia Limited, Hong Kong.

Library of Congress Cataloging-in-Publication Data

Anderson, Michael L., 1968– author.
After phrenology : neural reuse and the interactive brain / Michael L. Anderson.
p. ; cm.
"A Bradford book."
Includes bibliographical references and index.
ISBN 978-0-262-02810-3 (hardcover : alk. paper), 978-0-262-54473-3 (pb)
I. Title.
[DNLM: 1. Brain—physiology. 2. Neuropsychology—methods.
3. Cognition—physiology. WL 300]
QP376
612.8′2—dc23

2014013237

To my father

Contents

Acknowledgments

Over the years I have reaped the benefits of many fruitful conversations with many different people; it would be impossible to thank them all, deserve it though they do. I am especially indebted to Luigi Agnati, Athena Aktipis, Matt Bateman, Tim Bayne, Tony Chemero, Morten Christiansen, Stephen Cowley, Rick Dale, Carrie Figdor, Barb Finlay, Kevin Gold, Josh Kinnison, David Landy, Tim Oates, Michael Silberstein, Sune Vork Steffensen, Marcie Penner-Wilger, Don Perlis, Luiz Pessoa, Mike Richardson, Norbert Schwarz, Nick Shea, Terry Stewart, and Lucina Uddin. Thanks, too, to my entire department at Franklin & Marshall College for creating an extraordinarily congenial and supportive environment.

The bulk of this book was written while I was a Fellow at the Center for Advanced Study in the Behavioral Sciences, at Stanford University. It was an amazing year in a remarkable place. I offer my gratitude to all the CASBS staff and especially to Cynthia Pilch for keeping the atmosphere vibrant and inviting. Special thanks to Naomi Baron, Bob Brulle, Carla Faini, Allan Horwitz, Tomás Jiménez, Dan Jurafsky, Stephen Kosslyn, Melissa Lane, Jon Levy, Michael Macovsky, Petra Moser, Craig Murphy, Ken Shultz, Deborah Tannen, Mark Vail, and Joanne Yates. Tim Schroeder was the best bartender—both qua mixologist and qua engaged philosophical interlocutor—that CASBS has ever seen. And the year would not have been as healthy, nor the book written as quickly (and chapter 7 could not have been written at all) without the perfect combination of escritorial competition, athletic inspiration, and expert guidance on all things sociolinguistic offered by Cynthia Gordon.

Over the past several years my work has been sustained by two leaves from Franklin & Marshall College and award number 0803739 from the National Science Foundation. I am very grateful for this support. Thanks are also due to my editor at MIT Press, Phil Laughlin, who has made the process remarkably smooth, and to two anonymous reviewers for the press, who provided very helpful suggestions.

Finally, I owe a large debt of gratitude to Begoña Aristimuño, who suffered through the many mental and physical absences that the writing of this book entailed. Thank you.

Introduction

When you start to think about it, phrenology wasn't such a bad idea. Certainly it is not deserving of the degree of scorn that is heaped on it, so much that a book about neuroimaging entitled *The New Phrenology* (Uttal 2001) will be immediately understood to be a trenchant critique. In fact, by introducing the doctrine of phrenology, Gall made a number of extremely important contributions to neuroscience. Consider some brief excerpts from his letter to von Retzer, where he convincingly established the brain as the organ of the mind:

> I adduce the following proofs:
>
> 1. The functions of the mind are deranged by the lesion of the brain: they are not immediately deranged by the lesion of other parts of the body.
> 2. The brain is not necessary to life; but as nature creates nothing in vain, it must be that the brain has another distinction; that is to say
> 3. The qualities of the mind; or, the faculties and propensities of men and animals, are multiplied and elevated in direct ratio to the increase of the mass of brain, proportionally to that of the body; and especially in proportion to the nervous mass. ... (Gall 1798, 1857, p. 146)

He defines faculties in terms of *observable individual differences*:

> 1. We can make the qualities of the mind alternately act and repose; so that one, after being fatigued, rests and refreshes itself, while another acts and becomes fatigued in turn.
> 2. The dispositions and propensities exist among themselves, in variable proportions in man, as also in animals of the same kind.
> 3. Different faculties and propensities exist separately in different animals.
> 4. The faculties and propensities develop themselves at different epochs; some cease, without the other diminishing, and even while the other increases. (Gall 1798, 1857, p. 146)

He argues for the functional differentiation of the brain:

> [E]ach external organ of sense is in communication by nerves with the brain; and at the commencement of the nerves is a proportionable mass of brain which constitutes the true internal organ of each sensitive function. ... The same mind which sees through the organ of sight, and which smells through the olfactory organ, *learns by heart* through the organ of memory, and does good through the organ of benevolence. It is the same spring which puts in motion fewer wheels for you and more for me. In this way the general functions of the brain are established. (Gall 1798, 1857, pp. 147–148, emphasis in original)

With which of these points do we nowadays part company? Naturally, such a selective review of Gall's thinking ignores his empirical focus on the shape of the skull as a way of studying the brain and the mind. But although that particular choice of dependent variable proved scientifically fruitless and led to an immense amount of social mischief besides, it must be remembered that in the sciences of the mind we are *always* limited to inference from indirect measures, be they behavioral or physiological, and our instruments are hardly more immune to misuse for nefarious social purposes (see, e.g., Gould 1996; Guthrie 2003).

And here again, the focus on differential brain size was not a terrible idea. If indeed the biological basis of our behavioral dispositions should be understood by analogy with bodily organs, and if indeed the relative power of those organs can be indexed by their size (since "our faculties ... are multiplied and elevated in direct ratio to the increase of the mass of brain"), and if, finally, the shape of the skull faithfully renders the contours of the underlying organs, then it is perfectly reasonable to expect to discover relationships between local skull shape and behavior. The research program failed not because Gall chose the wrong dependent variable but because he chose the wrong organizing frame, which dictated where to look, what to look for, and how to interpret what was seen.

So as my former teacher Sarah Broadie once said to me, "Let's be clear just where metaphysics is rearing its ugly head, here." It rears its ugly head in (the history of) neuroscience precisely in the choice of the guiding analogy for understanding brain function, a choice necessarily governed by a limited conception of what relevant kinds of (biological) things there are. What manner of thing, after all, can the brain be? It can't be a kind of rock, or a kind of wax (can it?), or a collection of birds (which would be ridiculous). But a collection of organs is *not* ridiculous just in so far as it stems from the thought that animals are composed of organs (although here one is led to wonder if "organ" really names a natural kind). Insofar as animals have organs, it stands to reason that this complex bit of flesh we call the brain could contain a subset of those organs. Having settled on this classification, the rest quickly follows.

Indeed, where phrenology did invite, or at least *attract* ridicule (putting aside the *ad hominem* charge of head hunting), the target was not the analogy with organs but the idea that the brain could be a *collection*. For through the imperfect microscopes of the day, the brain seemed rather to be *one* thing, an extended reticulum of "protoplasmic prolongations" (Gerlach 1872). Thus did the debate over the neuron doctrine—the hypothesis that the brain is composed of discrete cells with very special properties—become entwined with that over the localization of function (Finger 1994) despite the fact that the greatest champion of that doctrine, Santiago Ramón y Cajal, was decidedly not a supporter of either the definition of psychological "faculties" or their assignment to discrete, localized neural "organs."

Instead, Ramón y Cajal brought an evolutionary and computational perspective to the matter:

> Countless modifications during evolution have provided living matter with an instrument of unparalleled complexity and remarkable functions: the nervous system. ... [W]e may think of the nervous system as entrusted with several tasks: collecting a large number of external stimuli; classifying them as to kind; and communicating them with great speed, range and precision to motor systems." (Ramón y Cajal 1995, pp. 3–4)

Of the sense organs he writes:

> In essence these organs are *computational devices* ... that select in a very specific way from the middle range of the immensely broad energy spectrum those wavelengths for which they are adapted. ... [T]he cerebral cortex of vertebrates, and the cerebral ganglion of invertebrates, do not need to create images; complete images are formed by the sense organs and supplied instead to the cerebral cortex or cerebral ganglion in highly refined ways that actually reflect the intensity and all the subtle nuances inherent in the excitatory stimuli. In the final analysis, the marvelous structural organization of the eye and ear is the primary reason for the dominant position of the cerebral cortex. (Ramón y Cajal 1995, pp. 8–9)

In Ramón y Cajal's view, brain function is to be understood in terms of a hierarchy of reflexes, in the most sophisticated instances of which one responds not just to external but also to internal, and not just to current but also to stored stimuli. What differentiation of function exists is driven largely by the nature of the stimuli and the differential demands the stimuli thereby place on its processing and classification. In such a brain there can be no region for circumspection or poetic talent, for although a particular sensory experience or association may be stored in a particular place—and may be a matter of "the assembly of functional connections established between different categories of representational neuron" (Ramón y Cajal 1995, p. 17)—the behavioral characteristics of the organism are realized only by the fluid activity of the whole system in its environment. We would not see such sophistication return to neuroscientific debates for many decades and rarely see it even now, and so I hope the current volume can be seen in part as refining and developing the functionalist perspective first articulated for the neurosciences by Ramón y Cajal. As will become clear by the conclusion of this volume, his thinking about function largely dissolves the debate over localization, so deeply changing its terms that the debate essentially ceases to be a matter for serious contention.

Nevertheless, the attraction to the particular version of functional differentiation exemplified by faculty psychology was such that once it was established that the brain was *in fact* composed of discrete parts, assigning individual faculties to discrete brain parts proved irresistible. This point reminds us that the failure of metaphysical imagination in phrenology was twofold, for it encompassed not just the organ analogy but also Gall's taxonomy of faculties. However useful it might be to classify and compare individuals in terms of their acquisitiveness, impulse to propagation, and sagacity, what warrants the thought that these characteristics will be useful to structuring the neuroscience of behavior and divide the brain at its functional joints? The wholesale importation of categories from faculty psychology for such use ought to have struck Gall and his followers as a risky scientific bet in and of itself, even apart from the commitment to localization of function in discrete neural organs. We are perhaps a bit wiser now.

Or are we? That contemporary cognitive neuroscience must be seen as deeply continuous with phrenology is brought home forcefully by what can only be described as a brilliant (and bold!) *reductio ad absurdum* of current scientific practice, recently published by Russell Poldrack (2010).

> Imagine that fMRI had been invented in the late 1860s rather than the 1990s. Instead of being based on modern cognitive psychology, neuroimaging would instead be based on the faculty psychology of Thomas Reid and Dugald Steward, which provided the mental "faculties" that Gall and the phrenologists attempted to map onto the brain. Researchers would … almost certainly have found brain regions that were reliably engaged when a particular faculty was engaged, … [and] Gall and his contemporaries would have taken these neuroimaging results as evidence for the biological reality of his proposed faculties. (Poldrack 2010, p. 753)

Poldrack's point, one that I will be amplifying in this volume, is that we in the cognitive neurosciences have been equally captured and possibly limited by a specific taxonomy of mental function. That taxonomy has been inherited from cognitive psychology, which, noble and successful though it may be, is nevertheless likely motivated by concerns, desiderata, and, yes, even metaphysical assumptions quite different from—possibly even incompatible with—those that are most illuminating for an investigation into the functional organization of the brain. This is not to claim it is not worth knowing how the various parts of the brain act and interact to support the aspects of cognition as they are defined in psychology (chapter 1 of this volume is dedicated to just this question). It is rather that we need to be open to the development of ontologies that let the phenomena speak on something closer to their own terms (Thagard 1992).

Poldrack outlines three research strategies that have been variously applied in the investigation of the functional structure of the brain, and in each one can see to varying degrees the continuing influence of the phrenological mind-set. I discuss the two core strategies, below.

> The most obvious strategy within cognitive neuroscience is what one might call *the "where" strategy*:
>
> 1. Design a manipulation that is thought to modulate the engagement of some particular mental process.
> 2. Analyze neuroimaging data to identify regions whose activity is modulated by this manipulation.
> 3. Conclude that the active regions are involved in the manipulated process. (Poldrack 2010, p. 755)

The "where" strategy is the one most obviously continuous with the phrenological enterprise, depending as it does on finding correlations between mental processes and neural powers. (For a classic example of this approach see, e.g., Posner, Petersen, Fox, & Raichle 1988.) Indeed, it might fairly be said that the "where" strategy represents an advance over phrenology in (only) three respects. It is experimental rather than observational; by measuring changes in the blood oxygenation level-dependent (BOLD) signal, neuroimaging employs a more plausible (albeit equally indirect) index of neural engagement; and it generally relies on experiments developed within and mental processes identified by cognitive psychology, which might be fairly said to have better empirical support than Reid's faculty psychology. This is not to say that these advances are not crucial but is rather to emphasize the fact that this approach deserves the label "neophrenology" (Uttal 2001) if not the scorn that generally accompanies it.

The influence of the phrenological mind-set on the second common research strategy identified by Poldrack is a bit more subtle and more thoroughly mediated by contemporary cognitive psychology.

As neuroimaging has matured, the "where" strategy has given way to what one might call *the "what" strategy,* which focuses more directly on characterizing the function of a specific brain region:

1. Design a task that independently manipulates two or more different mental processes, one of which is hypothesized to be performed by some particular region.
2. Examine the imaging data to identify the relative response of the region in question to these manipulations.
3. Conclude that the region in question performs a particular one of the manipulated processes. (Poldrack 2010, p. 755)

There is no doubt that this is an improvement over the "where" strategy, not least because the functional analysis and experimental design will naturally lead to an increasing refinement not just of the specificity of the neural associations for specific mental operations but also of our understanding of the operations themselves. Given this, it is worth quoting Poldrack's analysis of the situation at length by way of bringing out the subtle phrenological thinking that still motivates the approach.

It is instructive to project forward and think about what the ultimate result would be from several decades of science using the current approach. It is tempting to conclude that this approach would help us learn what each brain area does, but the reality may be somewhat less informative. In particular, although this approach is likely to uncover a broad set of functions that rely upon each region, it is unlikely to identify a fundamental functional role in mental activity for a particular region (e.g., the basic computations that each region performs). As an analogy, imagine a group of people individually trying to understand the function of a knife blade. One person tests its ability to cut peaches. Upon finding that the blade cuts through peach flesh but not the pit, he or she concludes that the knife is specialized for peach flesh removal. Another person might test its ability to screw various types of screws; finding that the knife blade works well to screw flathead and Phillips screws, but not Allen screws, he or she might conclude that it is specialized for a subset of screwing functions. Although each of these is a valid description of the functions that the knife performs, neither seems to be an accurate description of the fundamental function of a knife blade, such as "cutting or manipulating objects depending on their hardness." (Poldrack 2010, p. 756)

I agree entirely with Poldrack's description of the current situation and what a steady-as-she-goes future would have in store for us (Anderson 2010b; Cabeza & Nyberg 2000). But in light of the extensive evidence recently compiled (e.g., Anderson 2010b; Anderson, Kinnison, & Pessoa 2013; Poldrack 2006) and reviewed in chapter 1 of this volume that individual neural assemblies are deployed in many very different circumstances, I have lately come around to a rather different diagnosis of the underlying disease. For exactly what sort of metaphysical stance would lead one to suppose that something as versatile as a knife blade has anything like a "fundamental function"? It has some fundamental physical characteristics that make it useful in a variety of circumstances. Knowing what those characteristics are is surely useful, but to search for the functional essence of a knife is to be in the grip of a deep philosophical, ontological error.

I have come to wonder if we are making a similar error in the brain sciences. For we must be ever mindful that, although our scientific devices cost many millions of dollars more than does a set of calipers, measuring the change in local brain blood oxygenation puts us little closer to *fundamental function* than does measuring cranial bumps. The BOLD signal is just another dependent variable, no different in kind from response time, error rates, or the numbers chosen on a Likert scale. What we are typically doing in cognitive neuroimaging is *measuring the response tendencies* of local neural assemblies, and from this we are *inferring computational functions* (Posner et al. 1988). And why do we make this particular inference? Credit (or blame) must be assigned to the particular, componential version of the computational theory of mind (CTM) that has been as widely adopted in the contemporary cognitive sciences as the organ analogy was by phrenology (Edelman 2008; Fodor 1981, 1987; Gallistel & King 2009; Haugeland 1978, 1981; Neisser 1967; Pylyshyn 1980, 1984). If the brain is a computational information-processing device—a transformer of representations—then we must ask what is represented, and where. Where are the transformations implemented, and what exactly are those transformations? Is this a hand representation, that a motion detector, and the other an action planner? With this version of CTM as our organizing frame, there *must* be an answer to such questions. And if so, then we must keep trying to find it; just as each organ must have its function, each brain region must have its fundamental computation.

Now, in the past I have been an advocate for the computer analogy of the brain, suitably modified (Anderson 2003; Anderson & Rosenberg 2008), and "information processing" is in many ways a perfectly apt description of what the brain does. Moreover, it must also be admitted that phrenological localization of function need not *follow* from a computational approach to the brain; parallel distributed processing models are, in part, an attempt to embrace computation without componential localization. But it is nevertheless true that computational approaches to the brain and the component-oriented practice of analysis, decomposition, and localization have been reliable allies over the years, and I have thus come to believe that if we are ever to truly get beyond phrenology, we need to deeply reconsider both.

This book is an attempt to provide a guide to such a fully postphrenological science of the brain. My specific criticisms of CTM (henceforth CCTM, to remind the reader that I will be reexamining both the *componential* and the *computational* assumptions underlying this approach) emerge over the course of this volume, but it is worth an initial if brief reflection on an important disanalogy between the brain and a computer: whereas a computer is typically understood as a device that carries out a specific instruction set on (and in response to) inputs, brain responses to stimuli are characterized instead by specific deviations from intrinsic dynamics. As we come to better understand the significance of the brain's continuous dynamics, this disanalogy seems likely to loom ever larger (Bechtel 2012; Chemero 2009; Fox et al. 2005; Honey et al. 2007, 2009; Kelso 1995; Spivey 2007). At the very least it ought to shake our confidence that we understand the nature of the computations that the brain is performing; it just generally seems unlikely that we have already identified the right basis set (to adapt a term from a somewhat different context)—the right set of fundamental operations—out of which cognitive processing is built. Thus, we need to focus more attention not just on figuring out *what* those operations are but on how to *figure out* what those operations are.

And the fact is, the cognitive roles played by even very small regions of the brain are highly varied. As I demonstrate in chapter 1, a typical brain region contributes to tasks as diverse as finger awareness, magnitude comparison, task switching, and response inhibition, among many other things, and coherent cognitive function is reflected largely in the different neural partnerships that these regions establish under different circumstances (Anderson & Penner-Wilger 2013). Given this, should the bulk of our scientific efforts really revolve around the attempt to discern which specific computational operator must characterize the contributions of individual neural regions? Of course, I have spent a great deal of time doing just that—as has anybody who works in the field. As I note above, it is a virtual requirement of the framing assumptions we work within. But it seems to me more scientifically prudent to devise alternate models for understanding structure-function relationships in the brain—models that do justice to the cross-domain activity profile of a typical region but do not require the assignment of a specific, individual cognitive operation. As I try to establish in this book, I believe this represents a promising road untaken.

It is not that I am sure our science can do without the speculative metaphysics that accompanies componential explanatory mechanistic models *forever*, although it is worth remembering that there *are* kinds of intelligibility other than that provided by the analyze-and-decompose approach valorized by seventeenth-century philosophy and science (Bechtel 1998; Chemero 2009; Kitcher 1989; Pickering 2010; Salmon 1971, 1989; van Fraassen 2002, 2008; van Gelder 1998). But *successful* explanatory models generally emerge from a background of descriptive and predictive models (I am thinking in particular of the history of astronomy and that of chemistry) of a sort that are largely lacking in the cognitive neurosciences. By adopting and then immediately experimenting under the thrall of CCTM, we may have short-circuited a necessary process of scientific data accumulation without which our explanatory models will remain unmoored. In this, cognitive neuroscience is apparently following the lead of psychology, which has a long tradition of (too) readily borrowing concepts and frameworks from sciences such as physics and information systems. As R. B. Cattell wrote in a somewhat different context:

> Anyone familiar with the history of psychology will recognize that it has been a problem child among the sciences, attempting from an early age to gain the privileges of adult stature without first submitting to the discipline of an exact descriptive stage. The evasion of a laborious apprenticeship arises, first, from psychology's unfortunate and traumatic experience of trying to become a descriptive science in the wrong medium. For it spent much time, with Titchener and James (in America) attempting to classify the elements of the stream of consciousness. When it turned from this cul-de-sac into a study of behavior, it fell foul of old semantic pitfalls and traded in such artifacts as "faculties" or, more commonly, became imprisoned in a mechanically rigid doctrinaire system which considered all personality traits as "reflexes." (Cattell 1946, p. 2)

Reflecting on such analyses, I realize that the divergent histories within the subfields of psychology may account for the fact that much of the very best behavioral neuroscience comes from the field of learning, where there has been a very long tradition of description and predictive modeling (cf. Anderson 2010a). But it is certainly not too late to remedy the situation for the cognitive neurosciences more generally, and the fact that thousands of neuroimaging experiments have recently been gathered into data repositories such as BrainMap (Laird, Lancaster, & Fox 2005) and Neurosynth (Yarkoni et al. 2011) offers the opportunity to efficiently jump-start a more descriptive and dispositional science of the brain.

My current approach to this problem, developed in some detail beginning with chapter 4 of this volume, is to quantify the functional properties of neural assemblies in a multidimensional manner, in terms of their tendency to respond across a range of circumstances—that is, in terms of their *dispositional tendencies*—rather than trying to characterize their activities in terms of computational or information-processing operations. As we will see, there is evidence that this sort of quantification of individual brain regions in terms of their activity profiles allows one to index the underlying causal properties of those regions—the exercise is not *merely* descriptive—as well as to determine some of their other task-relevant dynamic properties including functional partnerships. This approach is roughly analogous to the way researchers in personality psychology quantify individuals in terms of multiple characteristics such as openness, extroversion, conscientiousness, and the like and use these to predict a range of other properties including personal affiliations, mental health outcomes, career success, and more. As in the study of personality, such an effort in the neurosciences will involve a long process of trying to identify the most predictive and explanatory factors and to relate these to the cognitive and behavioral outcomes of greatest interest. That is, rather than adopt wholesale the psychological categories suggested by a particular theory of the mind, the process I am advocating tries explicitly to give the brain its scientific voice—to let it show us what aspects of its world it is in fact attuned to. An important part of the project I am describing—for which I am here merely laying some foundations—will thus involve significant revision to the vocabulary of cognition, the way we categorize and label experiments and mental activities (Poldrack 2010), and I believe it is important to let the brain provide guidance here.

Part II of the book (Bodies) takes a step back from these detailed analyses of the brain and its functions and explores the larger-scale issues of how animals interact with their environments and one another and why and how—in response to what—their behavior changes over time. In this part we focus a great deal of attention on the evidence coming from the embodied cognition movement (Anderson 2003; Barrett 2011; Chemero 2009) and what it tells us about the fundamental nature of perception, action, and cognition. Here I try to develop a picture of the brain as a complex causal mediator of the relationship between body and environment—this as a supplement to, and perhaps even in the end a replacement for CCTM. What emerges in the course of these chapters is that the brain is best understood as first and foremost an action controller, responsible for managing the values of salient organism-environment relationships. I argue that the multidimensional neural dispositions developed in part I should be understood as the brain's differential propensities to influence the organism's response to the various features or affordances in its environment. In a brain like ours, where each region is involved in multiple tasks and coherent function is a matter of establishing the appropriate neural partnerships, the multiple activation patterns elicited in complex situations naturally compete, and the dominant pattern determines the shape of our interaction with the environment. Our brains are architecturally oriented to action selection.

Finally, part III of the book (Beings) argues that the approach to the brain outlined in part I is a much better fit with the results detailed in part II and thus offers a more promising road forward for a unified science of minded organisms than many current paradigms. I focus in particular on language and argue that far from necessitating the postulation of the central, symbolic, computational resources envisioned by CCTM, our capacity for language in fact represents the highest achievement of—but also a natural development for—a brain evolved for managing action and interaction. Language is a cultural tool for managing our social interactions with one another, and our mastery of it is a sign not of a specialized, dedicated neural adaptation but rather of our general capacity to reuse and repurpose existing neural machinery for multiple purposes. The view being propounded here, this is to say, places neuroscience, embodied cognition, and what Bruner (1990) calls "cultural psychology" into a single coherent framework. The aim is not just to provide a more nuanced view of the functional architecture of the brain but to support the (eventual) elucidation of the biological foundations of a culturally situated psychology.

I end with an appendix listing 23 sets of open problems that we face after phrenology. I briefly describe each research question and suggest a few avenues for approaching it. But obviously, the hope is to enlist the efforts and creativity of a wide range of scientists and scholars.

The book is structured in a kind of trees-and-forest manner. The chapters form the backbone of the argument and march deliberately through the details of the framework and the empirical evidence for it. The interludes, in contrast, try to focus on one or two general themes and paint the big picture with broad strokes. A book composed of nothing but the chapters would be a self-contained academic treatise. A book composed of nothing but the interludes would be something closer to a (very short!) popular work on the same subject. The idea is that by combining the two in one it will be possible to appreciate the full weight of the evidence behind the framework while mitigating the risk that the details obscure the shape of the whole. We shall see. As they do not say but probably should, the proof of the writing is in the reading.

The long-term hope is that the research framework I offer here will allow the cognitive neurosciences to integrate more smoothly with the fields of learning, embodied cognition, ecological psychology, psychiatry, and even disciplines such as sociology and economics because results in those fields tend to be themselves dispositional in nature, being typically expressed, for instance, in terms of changing propensities to behave given changes in environmental conditions. I hope the book will provide the framework for a new synthesis in the cognitive and behavioral neurosciences.

Part I Brains

1 Neural Reuse and the Need for a New Approach to Understanding Brain Function

The notion that brain areas are highly selective and exhibit considerable specialization has been the dominant guiding idealization in the brain sciences for many decades. In the selective brain each neural region responds to a restricted class of inputs and contributes primarily to a single cognitive domain such as language or motor control. The rapid acceptance of this doctrine was spurred in part by Paul Broca's (1861) description of the patient "Tan," whose stroke-induced injury to a region in left frontal lobe left him unable to utter anything but that eponymous syllable yet did not impair his ability to *understand* language. A series of findings since that time have cemented neural selectivity in both the public and the scientific imagination as the fundamental principle governing the functional architecture of the brain.

As is well known, functional localization in the eighteenth and nineteenth centuries was heavily influenced by faculty psychology (Reid 2002), leading to the notion that individual capacities such as parental love or verbal memory might be supported by distinct, relatively circumscribed regions of the brain. In contemporary cognitive neuroscience the notion of neural selectivity has been combined instead with a particular computational approach to cognition that gained acceptance starting in the 1950s as part of the "cognitive revolution" (Broadbent 1958; Chomsky 1959; Miller 1956; Newell, Shaw, & Simon 1958; see Miller 2003 for a discussion). The core of this view is the notion that the brain is fundamentally an information-processing device, a system that operates by progressively transforming—compressing, warping, combining, categorizing—sensory (and other) representations in support of behavioral outcomes. The notion of localization, combined with the information processing approach, has led to the contemporary hypothesis that individual cognitive/computational operations might be strictly localized in individual regions of the brain. Indeed, much of the excitement over the emerging techniques in functional neuroimaging, including positron emission tomography (PET) and functional magnetic resonance imaging (fMRI), was rooted in the expectation that these tools would finally allow us to determine what operations the various parts of our brains actually performed. One set of early proponents of this approach put it this way:

> [C]urrent analysis of the operations involved in cognition (J. R. Anderson 1980) and new techniques for the imaging of brain function during cognitive tasks (Raichle 1983) have combined to provide support for a new hypothesis. The hypothesis is that elementary operations forming the basis of cognitive analyses of human tasks are strictly localized. Many such local operations are involved in any cognitive task. A set of distributed brain areas must be orchestrated in the performance of even simple cognitive tasks. The task itself is not performed by any single area of the brain, but the operations that underlie the performance are strictly localized. This idea fits generally with many network theories in neuroscience and cognition. However, most neuroscience network theories of higher processes (Goldman-Rakic 1988b; Mesulam 1981) provide little information on the specific computations performed at the nodes of the network, and most cognitive network models provide little or no information on the anatomy involved (McClelland & Rumelhart 1986). Our approach relates specific mental operations as developed from cognitive models to neural anatomical areas. (Posner et al. 1988, p. 1627)

This view of the fundamental functional organization of the brain remains widespread in both scientific circles and the popular imagination. And yet, over the past several years it has come under increasing critical scrutiny, largely as a result of the application of the very functional neuroimaging techniques that were meant to uncover and illuminate the specific functional contributions of each region of the brain. In this chapter I lay out the evidence for a different architecture, one based on the fundamental principle of neural reuse: the use of local regions of the brain for multiple tasks across multiple domains (Anderson 2010b). For instance, although Broca's area has been strongly associated with language processing, it turns out to also be involved in many different action- and imagery-related tasks, including movement preparation (Thoenissen, Zilles, & Toni 2002), action sequencing (Nishitani et al. 2005), action recognition (Decety et al. 1997; Hamzei et al. 2003; Nishitani et al. 2005), imagery of human motion (Binkofski et al. 2000), and action imitation (Nishitani et al. 2005; for reviews, see Grodzinsky & Santi 2008; Hagoort 2005; Tettamanti & Weniger 2006). Similarly, visual and motor areas—long presumed to be among the most highly specialized in the brain—have been shown to be active in various sorts of language processing and other "higher" cognitive tasks (Damasio & Tranel 1993; Damasio et al. 1996; Glenberg & Kaschak 2002; Hanakawa et al. 2002; Martin et al. 1995, 1996; Martin, Ungerleider, & Haxby 2000; Pulvermüller 2005; see Schiller 1996 for a related discussion). Excitement over the discovery of the fusiform face area (Kanwisher, McDermott, & Chun 1997) was quickly tempered when it was discovered that the area also responded to cars, birds, and other stimuli (Gauthier et al. 2000; Grill-Spector, Sayres, & Ress 2006; Hanson & Schmidt 2011; Rhodes et al. 2004). The ensuing debates over the "real" function of these areas have still not been resolved, and in light of these results researchers have started to question the boundaries between psychological domains once thought separate and distinct, such as perception and cognition (Anderson, Richardson, & Chemero 2012; Barsalou 1999, 2008) and cognition and emotion (Pessoa 2008, 2012).

This is just a short list of some highly studied regions for which the prospect of a clear-cut mapping of function to structure appears dim. Recent meta-analyses of imaging results have tended to support this emerging challenge. For example, Russell Poldrack (2006) estimated the selectivity of Broca's area by performing a Bayesian analysis of 3,222 imaging studies from the BrainMap database (Laird et al. 2005). He concludes that current evidence for the notion that Broca's area is a "language" region is fairly weak, in part because it was more frequently activated by nonlanguage tasks than by language-related ones. Similarly, several statistical analyses of experiments from large collections of neuroimaging results (Anderson et al. 2010; Anderson & Pessoa 2011; Anderson, Kinnison, & Pessoa 2013) demonstrate that most regions of the brain—even fairly small regions—appear to be activated by multiple tasks across diverse task categories (Anderson 2010b). These results, reviewed in some detail in section 1.1 below, also suggest that the brain achieves its variety of function by using the same regions in a variety of circumstances, putting them together in different patterns of functional cooperation.

The remainder of this chapter is spent detailing the evidence for neural reuse and why the brain is built this way. In the volume as a whole, I hope to establish not just that neural reuse is a fundamental feature of the functional architecture of the brain but also that this fact calls for a thorough rethinking of how we do brain science. We need to rethink the principles of brain evolution and development, the methods we use for function-structure mapping, and even the categories we use in the neural and behavioral sciences. But we get to those issues later. For now we turn to the evidence that will drive our consideration of them.

1.1 Neural Reuse in the Evolution of the Brain

Although as I noted above, the twin notions of functional localization and faculty psychology have long been powerful allies, largely dominating the scientific scene up to the present time, in science—as in history more generally—there are always parallel countercurrents. In this case the most important and relevant of those countercurrents has been a long tradition—going back at least to Spencer (1855) and Ramón y Cajal (1995), not to mention Darwin (1872) and James (1950)—of treating mind and brain in an evolutionary context (Anderson 2003; Anderson & Chemero in press; Barrett 2011; Chemero 2009; Clark 1997). This is a good place to start when trying to understand the significance of the emerging challenge to localization.

A long-standing guiding principle of both embodied cognitive science and evolutionary psychology (Barkow, Cosmides, & Tooby 1992) is that cognition was built within a system primarily fitted to situated action. The central nervous system—the neocortex most definitely included—is first and foremost a control system for an organism whose main job is managing the myriad challenges posed by its environment. "Higher" cognitive faculties such as language and abstract reasoning had to find their neural niche (Dehaene 2009; Iriki & Sakura 2008) within the constraints imposed (and the opportunities offered) by the way existing neural resources were deployed for this purpose, in a way mediated and guided by whatever continuing selection pressure there is to maintain fast, effective, and efficient solutions to pressing environmental challenges. Insofar as this is true, then—and this is the other guiding principle shared between evolutionary psychology and embodied cognitive science—this phylogenetic history should have left detectable traces on both brain and behavior. Where evolutionary psychology and embodied cognitive science part company is in their understanding of what those traces will look like and where to find them.

In particular, evolutionary psychology has adopted a methodological focus on the challenges posed by the environment of selection (Buss 2005), which has in turn led many researchers in this area to spotlight the efficiency of individual algorithmic and heuristic *solutions* to those problems. One result of this focus had been the acceptance of a version of faculty psychology exemplified in the "adaptive toolbox" model of mind (Gigerenzer & Selten 2002), a framework also sometimes known as "massive modularity" (Carruthers 2006). Two main considerations have been primarily responsible for the focus on independent, modular neural implementations of these tools. First, evolvability appears to require that these tools be separately targetable by selection pressures (Barrett & Kurzban 2006); and second, the demand for immediate and efficient real-time operation points to massive parallelism. Together, these considerations appear to require a functional architecture featuring modular, separately modifiable nearly decomposable subsystems. Below I recount some of the evidence that the brain is not built this way and so return to the important issues of functional efficiency and evolvability at the end of the chapter.

Unlike evolutionary psychology, embodied cognitive science (ECS) has been more interested in understanding the ways in which thinking is both influenced and partially constituted by emotional and physical states, bodily activity, and interactions among self, others, and environment (Ackerman, Nocera, & Bargh 2010; Chandler & Schwarz 2009; Chemero 2009; Kelso 1995; Lee & Schwarz 2010; Varela, Thompson, & Rosch 1990). When one is considering the neural supports for cognition, this perspective naturally places greater weight on the functional relations and interactions *between* neural structures than on the actions of individual regions. Moreover, this perspective has led ECS to focus less on the efficiency of individual processes and more on overall efficiency in the use of bodily, environmental, and social resources for cognitive ends.

This brings us to one of the initial points of contrast between theories of neural architecture rooted in evolutionary psychology (on this point a staunch ally of CCTM; see Barrett & Kurzban 2006 for discussion) and those rooted in ECS. Neural reuse theories (Anderson 2010b) generally accept the ECS insight that, rather than developing new structures de novo, resource constraints and efficiency considerations dictate that whenever possible neural, behavioral, and environmental resources should have been reused and redeployed in support of any newly emerging cognitive capacities. Functionally autonomous and dedicated neural modules just do not seem to make good design sense given the importance of efficient use of available resources and of ongoing interactions in shaping function. For ECS cognition is largely supported by "old wheels, springs, and pulleys only slightly altered" (Darwin 1862, p. 284) and reconfigured to serve present purposes.

A logical place to look for evidence of such a history is in the distribution of and relationships among the neural structures supporting various cognitive functions. ECS predicts that neural structures originally evolved or developed for one purpose will be reused in later emerging functionality. That is, rather than following an evolutionary/developmental pathway wherein we develop specialized, dedicated neural hardware to meet each new environmental/behavioral challenge, ECS suggests that much local

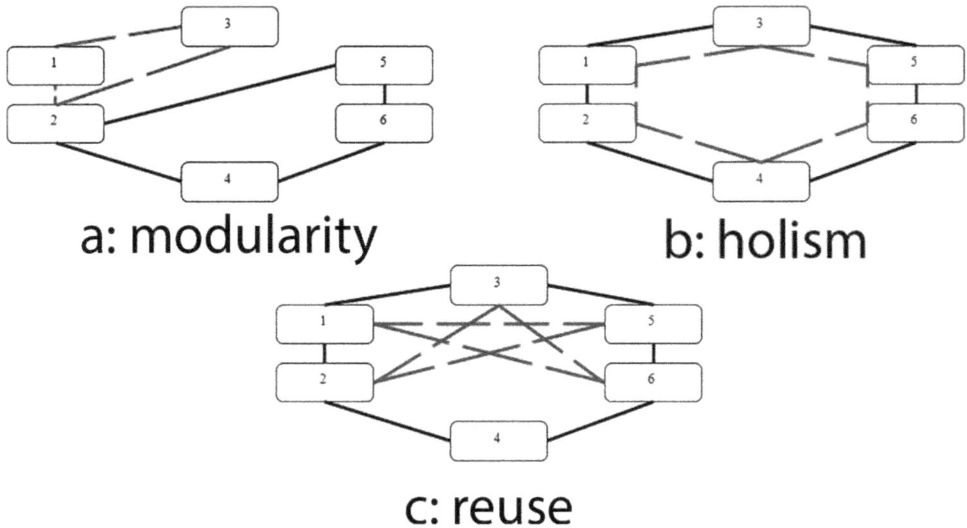

Figure 1.1
Three logical possibilities for the functional structure of the brain.

neural structure is conserved but is often combined and recombined by different organisms in different ways to achieve diverse purposes.

Imagine a simple brain consisting of six neural structures that could be combined in various ways to support two cognitive-behavioral tasks. Figure 1.1 illustrates three logical possibilities for how the neural structures could be functionally arranged to support the tasks in question. In the sort of modular brain predicted by evolutionary psychology (figure 1.1a), structures 1, 2, and 3 would combine to support one task (represented using dashed lines), and 2, 4, 5, and 6 would work together to support the other (represented with solid lines). Although there might be *some* neural and functional overlap and communication between the modules (structure 2 active in supporting both tasks), the neural underpinnings of different behaviors and abilities would be largely segregated. In contrast, if the brain is more holistically organized, all the structures might be engaged in supporting both tasks, with the behavioral differences possibly reflected in such things as different input characteristics and resulting oscillatory dynamics. Finally, it could be the case that many of the structures are used to support both tasks, but for each task they cooperate in different patterns and with different partners. So, for instance, in figure 1.1c, structure 1 cooperates with structures 2 and 3 in the "solid" task and with structures 5 and 6 in the "dashed" task.

If such reuse (an especially pure case of which is illustrated in figure 1.1c) obtains in the brain, then we should expect at least three things to be true of its functional structure. First, neural structures should be used and reused for diverse purposes in various task domains. That is, in contrast to what is illustrated by figure 1.1a, structures should not be classically selective in the sense of responding only to a highly restricted class of stimuli or tasks. Second, we should expect the functional differences between task domains to be reflected less in differences in what neural real estate is implicated in supporting the domains than in the different patterns of interaction between many of the same elements (in contrast to the brain illustrated in figure 1.1b). And third, we should expect later emerging (evolving or developing) behaviors/abilities to be supported by a greater number of different structures more broadly scattered in the brain. The reason is simple: the later something emerges, the more potentially useful existing elements there will be, with little reason to suppose they will be grouped locally. A more localist account of the evolution of the brain would instead expect the continual development of new, largely dedicated neural structures and would predict that the resulting functional complexes would remain tightly grouped, as this would minimize the metabolic cost of wiring the components together and communicating among them. In a number of recent publications (Anderson 2007a, 2007b, 2007c, 2008a, 2010b; Anderson & Pessoa 2011; Anderson, Kinnison, & Pessoa 2013; Anderson & Penner-Wilger 2013), I report evidence for all of these predictions. Some of this evidence is reviewed below.

Taking up the first prediction, Anderson and Pessoa (2011) examined the selectivity of 78 standard anatomical regions of the brain (based on the Freesurfer brain atlas: Fischl et al. 2004) by determining whether (and how often) each was active in 1,138 experimental tasks in 11 different BrainMap task domains: action execution, action observation, action inhibition, attention, audition, vision, emotion, language semantics, reasoning, explicit (semantic) memory, and working memory (Fox et al. 2005). The simple insights behind this work are that selectivity is the inverse of diversity and that we have various methods for measuring the diversity of, say, students in colleges (Chang 1999), housing prices in neighborhoods (Byrne & Flaherty 2004), or species in an ecosystem (Hill 1973; see Jost & Chao 2008; Schleuter et al. 2010 for reviews).

In the current case we wanted to see, for each region, what the diversity of activations was—how many neural activations fell into each of the 11 task categories. We used a measure of diversity variability (*DV*) based on standard deviation:

$$DV = \sqrt{\frac{\sum_{i=1}^{k}(Cat_i - mean)^2}{k}}$$

In this equation Cat_i refers to proportion of activations in each category, *mean* refers to the average proportion (always 0.091 with 11 categories), and *k* equals the number categories. Diversity is *(1 – DV)*, normalized such that the values range from 0 (all activations in one category) to 1 (activations spread equally across all 11 categories).

We determined that the overall average diversity of the 78 large anatomical regions of the brain was 0.70 (SD 0.12). The overall average diversity of cortical regions was 0.71 (SD 0.11) and of subcortical regions was 0.63 (SD 0.17). Put differently, the regions were active in an average of 95 tasks spread across nine cognitive domains. These results are represented graphically in figure 1.2 using a cool-to-hot scale (gray indicates no information).

The analysis was also performed in a brain divided into 1,054 neural regions. The overall average diversity of the 574 small cortical and 21 small subcortical regions activated by five or more experiments was 0.52 (SD 0.13). Those 595 regions were activated by an average of more than 10 experiments across five cognitive domains. The overall average diversity of the cortical regions was 0.52 (SD 0.13), and that of the subcortical regions was 0.59 (SD 0.12). These results were recently confirmed with different methods and metrics, for a range of brain region sizes, thoroughly measuring diversity in the brain voxel by voxel (Anderson et al. 2013). Indeed, my collaborators and I have investigated diversity in the brain using multiple metrics and ways of carving up the brain; no matter what we try, we cannot seem to disconfirm the finding. The upshot: local neural structures are *not* highly selective and typically contribute to multiple tasks across domain boundaries. Because the domains are highly varied, the observations cannot be explained by the similarity of the task domains. And because the brain activations being studied were generated using a subtraction-based methodology, such that the activations observed during the task of interest are compared to activations observed during a related control task, the finding is not explained by the fact that most experimental tasks have multiple cognitive aspects (e.g., viewing stimuli, recalling information, making responses). The control tasks would (mostly) ensure that the reported brain activity was supporting the particular cognitive function under investigation. Functional diversity appears to be a genuine feature of local brain organization, with important implications for understanding the brain's overall architecture.

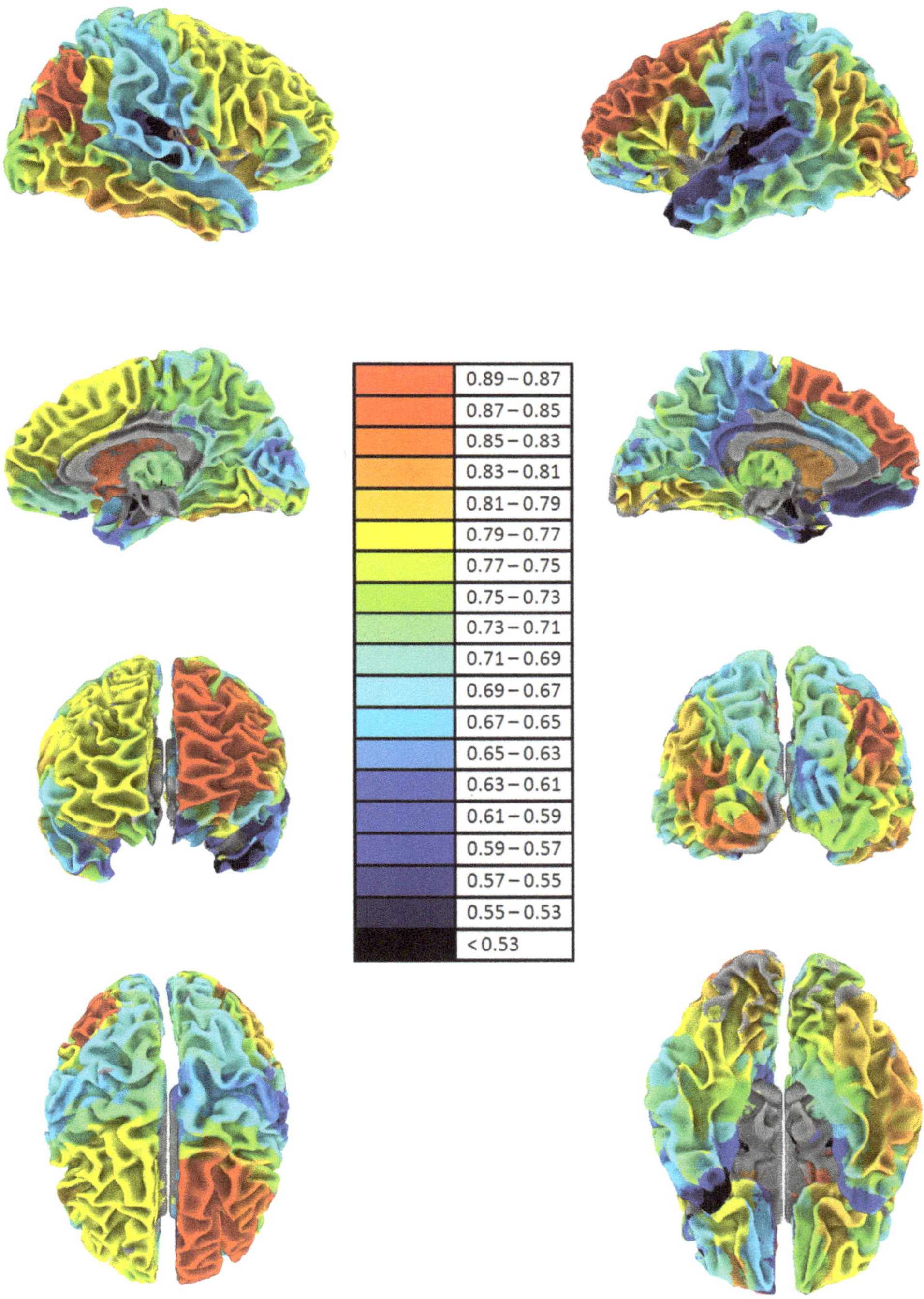

Figure 1.2
Task diversity of brain regions.

To examine the second prediction we performed a functional coactivation analysis of 1,127 experimental tasks from the data set (Anderson & Penner-Wilger 2013), falling into 10 of the BrainMap task domains (Fox et al. 2005; for this study we excluded action inhibition, as it contained too few experiments for this approach). In a functional connectivity analysis (Anderson, Brumbaugh, & Şuben 2010), one looks to see how often regions of the brain coactivate under various task conditions. If the regions coactivate more often than would be expected given the activation likelihood of the individual regions—that is, if the probability of region A and region B being active in the same task is significantly ($p < 0.01$) higher than would be predicted from the general probability of A being active and the general probability of B being active—then this indicates there is a functional connection between the regions.

The results of such a study can be represented as a graph. A graph is simply a set of nodes joined by edges, where the nodes and edges can represent various aspects of a modeled system. For instance, in an airline route map nodes are airports, and edges represent flights between them, and in a Facebook-style social network, nodes are people, and edges indicate "friendship." In a brain functional network like that depicted in figure 1.3, the nodes represent individual brain regions, plotted in a 3-D anatomical space, and the edges represent functional connections between them—that is, a higher-than-expected likelihood of coactivation during tasks in a given cognitive domain. By looking at the data in this format, it is easy to compare how often a given region is active in more than one domain and how often it has the same neural partners in more than one domain.

Figure 1.3 highlights the functional partners of the left precentral gyrus (the various functional roles of which are discussed further below) during semantics tasks, emotion tasks, and attention tasks. Visually, it is clear that although this neural region is active in supporting tasks in different domains, it rarely shares the same functional partners across domains.

We can make this individual visual result quantitative and general by comparing the average node overlap with the average edge overlap in a pairwise comparison of all the functional networks from the 10 cognitive domains analyzed. Referring back to figure 1.1, we can easily generate predictions for the three possible functional architectures. If the brain is largely modular, then we should expect both low node overlap and low edge overlap; because regions are largely dedicated to their individual function, there is by hypothesis low node overlap, and there can *not* be edge overlap between nonoverlapping nodes. If, however, the brain is holistically organized, we should expect high node and high edge overlap; as in

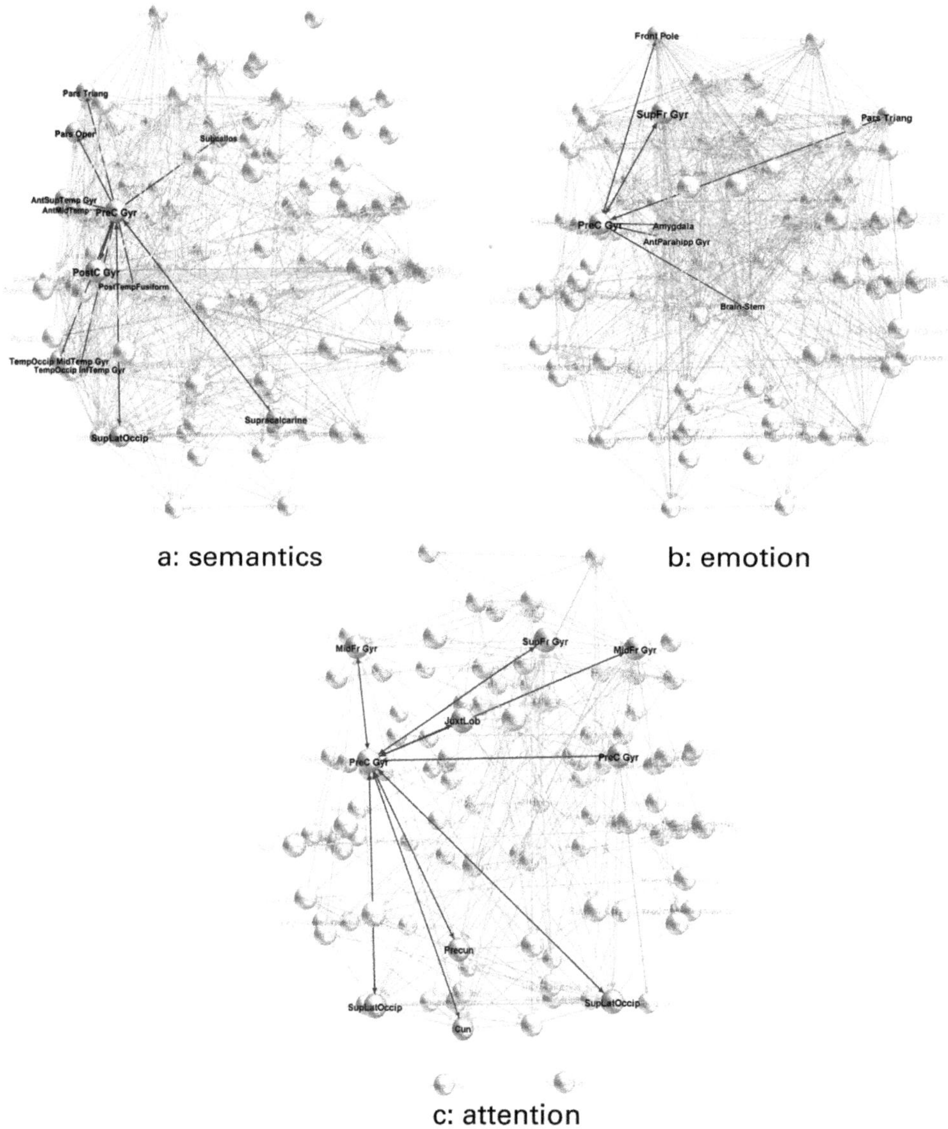

Figure 1.3
The functional partners of the left precentral gyrus under three different task conditions.

the diagram, both task domains use the same nodes, and these nodes communicate/cooperate with the same partners. Finally, if the brain developed by reusing individual neural structures for diverse purposes, then we should see high node overlap (because brain regions are used in both tasks) but low edge overlap (because they cooperate with different partners in each).

Using Dice's coefficient as our measure, $D = 2(o_{1,2})/(n_1 + n_2)$, where $o_{1,2}$ represents the number of shared components (edges or nodes) in the two networks, and n_x represents the total number of components in each network, we discover that the mean overlap for the nodes (D_N) = 0.60 (SD 0.13) while the mean overlap of the edges (D_E) = 0.09 (SD 0.07). Of course, one might worry that this result is simply an artifact of the fact that in networks there are many more possible edges than nodes, so one would expect to get this result just by chance. Thus, it is important to compare these averages with the expected overlaps between *random* networks with the same number of edges and nodes as our brain networks. Doing a pairwise comparison of random networks, mean (D_{rN}) = 0.50 (SD 0.11) and mean (D_{rE}) = 0.14 (SD 0.07). All differences are significant $p << 0.01$ (see figure 1.4). That is, the results indicate that between functional brain networks there is significantly *more* node overlap and significantly *less* edge overlap than would be expected by chance. These results replicate, with a much larger data set, those reported in Anderson (2008a) and strongly suggest that low

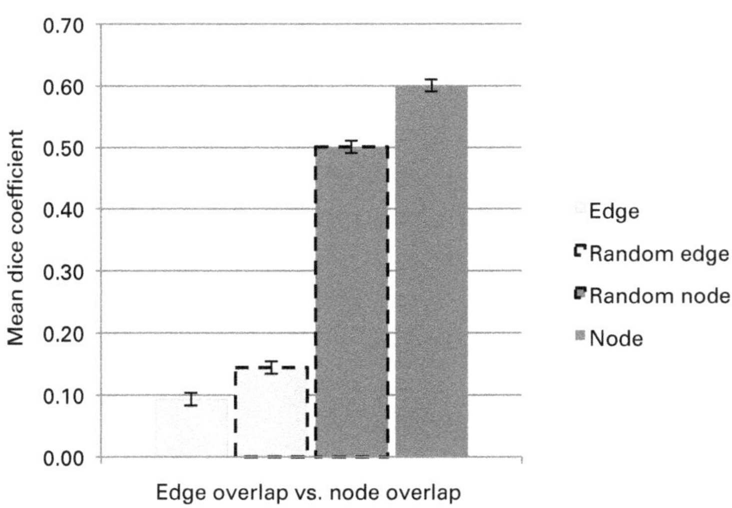

Figure 1.4
The average amount of overlap of nodes and edges in functional brain graphs as compared to random graphs.

edge overlap and high node overlap between task domains constitute a functionally significant feature of these brain networks.

As for the last prediction that more recently evolved functions will be supported by more broadly scattered regions of activation, I reported (Anderson 2007a) that language tasks activate more and more broadly scattered regions than do visual perception and attention. This finding was corroborated by a larger study (Anderson 2008a), which found that language was the most widely scattered domain of those tested, followed (in descending order) by reasoning, memory, emotion, mental imagery, visual perception, action, and attention. The significant differences in the degree of scatter were observed between attention and each of the following domains: language, reasoning, memory, emotion, and mental imagery; and between language and each of the following domains: visual perception, action, and attention. No other pairwise comparisons showed significant differences.

Note that, in addition to supporting the main predictions of neural reuse (and ECS), this last finding also corroborates one of the central assumptions of neural reuse, that cortical regions have specific biases that limit the uses to which they can be put without extensive rewiring. If neural structures could instead be easily put to almost any use, then given the increased metabolic costs of maintaining long-distance connections, we would expect the structures implementing functions to remain relatively localized. That this is not the observed pattern suggests that some functionally relevant aspect of local structure is relatively stable. In past work (Anderson 2007a; 2010b) I suggested that, given these findings, we should conceptualize the functional structure of the brain in terms of a set of fundamental computational operators—I called them "workings" (Bergeron 2007)—each of which has many different higher-level cognitive "uses." This differed from the sort of strict localization advocated by Posner, Kanwisher, and others (Kanwisher 2010; Posner et al. 1988) only in the expectation that these workings would be multimodal, multidomain operators capable of contributing to task processes across perceptual modalities and cognitive domains—a conceptual and architectural reform that suggested that the operators might *not* look like those typically postulated by contemporary cognitive theories.

Yet as I noted in that earlier work and wish to emphasize here, the evidence is equally compatible with a weaker (and thus an even less architecturally and functionally conservative) claim: local neural structures have a functional bias, a set of dispositional tendencies that capture the set of inputs to which the circuit will respond and govern the form of the resulting output. This notion is sufficient to account for the observations that local structures have multiple but limited uses without committing one to the idea that each circuit does *exactly one* specific thing. This is especially important because—as reviewed later in this chapter and throughout chapter 2—brain networks appear to have the capacity to change their effective connectivity over very short time scales. Although they cannot take on just *any* configuration, they appear to be able to take on more than one. Thus, the best current evidence seems to suggest that although some of the observed functional diversity results from the fact that a given circuit in a given configuration is often useful in multiple contexts, some of the observed diversity is likely *also* due to the fact that the local network can be in multiple different states. This matter is extensively discussed in this volume. For now the point to emphasize is that it would be worth our while to see how much science we can do with the weaker notion of functional *differentiation* in the brain before engaging in the kind of abduction to specific, local computational operations to which (I have lately come to think) the field is overprone. After all, we clearly have a great deal still to learn about how local structures cooperate to achieve cognitive function—how the activities of individual nodes contribute to or specify the functions of networks—and very likely the component operator model unduly limits the possibilities well in advance of the evidence.

1.2 Neural Reuse and Some of Its Cognitive Effects

One thing that is important to note is that the evidence for functional diversity in the brain becomes clear only when one looks across multiple experiments to characterize local function. For a modularist, this is not a natural scientific strategy to adopt, for if brain regions are dedicated to individual task domains or stimulus types, then once one has identified the stimulus or task that effectively causes a regional response, one need look no further to be in a position to characterize function. But here I want to draw attention to some of the persistent countercurrents to the dominant paradigm. In fact, some long-standing frameworks in cognitive science—most prominent among them conceptual metaphor theory and concept empiricism—seem to predict something very like neural reuse, and this provided the theoretical motivation to start looking at the activity of individual regions of the brain in multiple circumstances. So it is worth pausing to determine whether the observed functional diversity is already fully accounted for by these theories.

As is widely appreciated, conceptual metaphor theories comprise one of the most successful theoretical research programs in cognitive science. Originating with Lakoff and Johnson (1980, 1999) and extended by many others, perhaps most notably in the conceptual blending theory of Fauconnier and Turner (2002), conceptual metaphor theories suggest that cognition is dominated by metaphor-based thinking, whereby the structure and logical protocols of one or more domains, combined in various ways, guide or structure thinking in another. For a simple case, consider the *Love Is War* mapping taken from Lakoff and Johnson (1980, 1999). When employing this metaphorical mapping, people use their understanding of war—of how to interpret events and how to respond to them—to guide their thinking about love: one fights for a partner, makes advances, fends off suitors, or embarks on a series of conquests. Similarly, the *Life Is a Journey* mapping allows people to leverage their extensive experience and competence in navigating the physical world in order to facilitate planning for life more generally: we plan a route, overcome obstacles, set goals, and reach milestones. The theory has been widely discussed and tested and enjoys a raft of supporting evidence in linguistics and cognitive psychology.

A natural question that arises for such theories, however, is how the structured inheritance from one domain to another is actually achieved by the brain. Is it done abstractly, such that *mental models* (Gentner & Stevens 1983; Johnson-Laird 1983) of war or navigation are used as prototypes for building other models of love or life? Or is there a more basic biological grounding such that the very neural substrates used in supporting cognition in one domain are reused to support cognition in the other? Although some researchers favor the first possibility—notably Lera Boroditsky (e.g., Boroditsky & Ramscar 2002)—it seems fair to say that the greater effort has been focused on investigating the second.

This is at least in part because the debate over the biological basis of conceptual metaphors dovetails with another over the nature and content of cognitive representations—symbols, concepts, and (other) vehicles of thought—that has also played out over the last 20 years or so. At issue here is the degree to which the vehicles of thought—our mental carriers of meaning—are tied to sensory experience (Barsalou 1999; 2008). Concept empiricists (as they are called in philosophy) or supporters of modal theories of content (as they are called in psychology) are generally committed to some version of the thesis that "the vehicles of thought are reactivated perceptual representations" (Weiskopf 2007, p. 156). As one of the core statements of the modal position puts it, perceptual symbols, which "constitute the representations that underlie cognition," are "record[s] of the neural activation that arises during perception" (Barsalou 1999 pp. 578, 583; see Prinz 2002 for a general discussion). This position is meant to contrast with a rationalist or amodal one in which the vehicles of thought are inherently nonperceptual, abstract, logical, linguistic, or computational structures (see, e.g., Fodor 1975; Fodor & Pylyshyn 1988).

In the case of both debates it looked as if information about what neural resources were actually deployed to support cognitive tasks could provide evidence favoring one side or another. If planning a task used brain regions different from those used in planning (or imagining) a journey, then this would be prima facie evidence against the notion that the two were related via direct neural grounding. Similarly, if perceptual tasks and cognitive tasks appeared to be handled by distinct brain regions, this would appear to favor the amodal view.

In the event, a series of early findings bolstered the case for modal concepts, on the one hand, and for the idea that direct neural substrates supported metaphorical mappings, on the other. For example, a series of papers from the labs of Antonio Damasio and Alex Martin offered evidence that verb retrieval tasks activated brain areas involved in motor control functions, and naming colors and animals (that is, processing nouns) activated brain regions associated with visual processing (Damasio & Tranel 1993; Damasio et al. 1996; Martin et al. 1995, 1996, 2000). Similarly, it was discovered that perceiving manipulable artifacts, or even just seeing their names, activates brain regions associated with grasping (Chao & Martin 2000). All this suggested that class concepts such as HAMMER, RED, and DOG might be stored using a sensory and/or motor code and, more generally, that high-level, conceptual-linguistic understanding might involve the reactivation of sensorimotor experiences. This dovetailed nicely with the general idea behind direct neural support for the kinds of metaphorical mappings mentioned above: understanding in one domain would involve the reactivation of neural structures used for another, and the new domain would thereby inherit some of the semantic structure of the old. Thus, the finding that mental planning can activate motor areas even when the task to be planned itself involves no motor activity (Dagher, Owen, Boecker, & Brooks 1999) has long been taken to support the case that mappings such as *Life Is a Journey* are mediated by the direct sharing of neural resources by both domains.

It seems fair to say that these early discoveries prompted a much larger effort to uncover the neural underpinnings of high-level cognitive functions, one specifically focused on revealing the ways in which these underpinnings were shared with those of the sensorimotor system. The result is many hundreds of studies detailing the various ways in which neural substrates are shared among various cognitive functions. A representative sample of these studies is reviewed and discussed in sections 1.2.1 through 1.2.6. However, to preview the argument to follow: the effort to uncover instances of neural reuse has been *so* successful that even a cursory examination of the breadth and frequency of reuse suggests that there is much more reuse than can be accounted for by modal concepts or conceptual metaphor theory. That is to say, neural reuse is not just a new name for an old and well-known phenomenon but represents the operation of a more fundamental functional-structural principle than those articulated by existing frameworks. We review some of the evidence for this claim in the next subsections.

1.2.1 Reuse of Motor Control Structures for Language

A great deal of the effort to discover the specific neural underpinnings of higher cognitive functions has focused on the involvement of structures long associated with motor control functions. In a typical example of this sort of investigation, Pulvermüller (2005) reports that listening to the words "lick," "pick," and "kick" activates regions of primary motor cortex associated with mouth movements, hand movements, and leg movements, respectively. The finding is consistent both with the idea that comprehending these verbs relies on this motor activation, insofar as the concepts are stored in a motoric code or format (Goldman 2012), and also with the idea that understanding these verbs might involve (partial) simulations of the related actions. Either interpretation could easily be used as part of the case for concept empiricism.

Similarly, Glenberg and Kaschak (2002) uncover an interesting instance of the entanglement of language and action that they call the "action-sentence compatibility effect" (ACE). Participants are asked to judge whether a sentence makes sense or not and to respond by pressing a button, which requires a move either toward or away from their body. In one condition "yes" is away and "no" is toward; another condition reverses this. The sentences of interest describe various actions that would *also* require movement toward or away, as in "put a grape in your mouth," "close the drawer," or "you gave the paper to him." The main finding is of an interaction between the two conditions, such that it takes longer to respond that the sentence makes sense when the action described runs counter to the required response motion. More striking, this was true even when the sentences described abstract transfers, such as "he sold his house to you," which implies a direction without describing a directional motor action. Following the reasoning originally laid out by Sternberg (1969), an interaction between two manipulated factors implies at least one shared component between these two different processes—movement and comprehension. A likely candidate for this component would be a neural circuit involved in motor control, a supposition confirmed by Glenberg, Sato, and Cattaneo (2008a), who report that use-induced motor plasticity also affects language processing, and Glenberg et al. (2008b), who report that language processing modulates activity in the motor system. This connection is confirmed by the highly practical finding that one can improve reading comprehension by having children manipulate objects (Glenberg, Brown, & Levin 2007). Thus, this seems to be another clear case in which motor control structures are involved in, and perhaps even required for, language comprehension, whether via simulation (e.g., in the concrete transfer cases), metaphorical mapping (e.g., in the abstract transfer cases), or some other mechanism. Glenberg has suggested both that the effect could be explained by the activation of relevant action schemas (Glenberg et al. 2008b) and by the activation and combination of appropriate affordances (Glenberg & Kaschak 2002; Glenberg et al. 2009). Whatever precise mechanism is involved, the finding has been widely interpreted as support for both concept empiricism and conceptual metaphor theory (although see Anderson 2008b for a dissent).

1.2.2 Reuse of Motor Control Structures for Memory

Another interesting description of the motor system's involvement in a different cognitive domain comes from Casasanto and Dijkstra (2010), who describe bidirectional influence between motor control and autobiographical memory. In their experiment participants were asked to retell memories with either positive or negative valence while moving marbles either upward or downward from one container to another. Casasanto and Dijkstra found that participants retrieved more memories and moved marbles more quickly when the direction of movement was congruent with the valence of the memory (upward for positive memories, downward for negative memories). Similarly, when participants were asked simply to relate some memories without prompting for valence, they retrieved more positive memories when instructed to move marbles up and more negative memories when instructed to move them down. Because the effect is mediated by a mapping of emotional valence on a spatial schema, the finding seems to most naturally support conceptual metaphor theory. The fact that the effect was bidirectional—recounting memories affected movement, *and* movement affected memory retrieval—is a striking detail that seems to suggest direct neural support for the mapping.

1.2.3 Reuse of Structures Mediated by Spatial Cognition

Many of the apparent overlaps between higher-order cognition and sensorimotor systems appear to be mediated by spatial schemas in this way. For example, Richardson, Spivey, Barsalou, and McRae (2003) report that verbs are associated with meaning-specific spatial schemas. Verbs such as "hope" and "respect" activate vertical schemas, whereas verbs such as "push" and "argue" activate horizontal ones. As the authors put it, "language recruits spatial representations during real-time comprehension." In a similar vein Casasanto and Boroditsky (2008) suggest that our mental representations of time are built on the foundations of our experience with space. These findings appear to provide strong and relatively unproblematic support for conceptual metaphor theory and perhaps also for a generic theory of concept empiricism, according to which the content of our concepts is grounded in (but does not necessarily constitute a simulation or reactivation of) sensorimotor experiences.

On the other hand, even when simulation *is* an important aspect of the reuse of resources between different domains, it does not always play the functional role assigned it by concept empiricism or conceptual metaphor theory. For some time there has been growing evidence that doing actions, imagining actions, and watching actions done by others all activated similar networks of brain regions (Decety et al. 1990, 1997; Jeannerod 1994). This has suggested to many that social cognition—understanding the actions and intentions of others—could involve simulating our own behaviors, a notion that attracted even more widespread interest after the discovery of mirror neurons (Decety & Grèzes 1999; Gallese et al. 1996; Gallese & Goldman 1998; Rizzolati et al. 1996). The trouble for concept empiricism and conceptual metaphor theory is that the logic governing the reuse of resources for multiple purposes is quite different in this case. Here, the idea is that structures associated with behavioral control can be used to build predictive models of others by inputting information about another agent into the system that would normally be used to guide one's own actions (and reactions). Although it could be argued that using simulation in support of such "mind reading" (Gallese & Goldman 1998) requires a kind of metaphorical mapping (he is *like me* in relevant ways), in fact this is simply a necessary assumption to make the strategy sensible and is very unlike the kinds of metaphorical mappings envisioned by conceptual metaphor theory. In the case under discussion, the use of neural resources in behavioral control and in predictive modeling works because other people are *already* like me in the relevant sense and because the neural resources can provide services useful in both cases. The idea behind conceptual metaphor theory, in contrast, is that a new domain such as planning *acquires* its particular structure because it inherits it from a prestructured domain such as motor control. The logic is thus quite different in the two cases.

Even some of the evidence for the reuse of *spatial* operations in other cognitive domains—which has been a mainstay of research into concept empiricism and conceptual metaphor theory—suggests the existence of more kinds of reuse than can be accounted for by these theoretical frameworks. Consider just a few of the various manifestations of the spatial-numerical association of response codes (SNARC) effect (Dehaene, Bossini, & Giraux 1993): (1) When participants are asked to judge whether numbers are even or odd, responses are quicker for large numbers when made on the right side of space (canonically with the right hand, although the effect remains if responses are made while hands are crossed) and quicker for smaller numbers when responses are made on the left side of space. (2) Participants can accurately indicate the midpoint of a line segment when it is composed of neutral stimuli (e.g., *XXXXX*) but are biased to the left when the line is composed of small numbers (e.g., *22222* or *twotwotwo*) and to the right when the line is composed of large numbers (e.g., *99999* or *ninenine-nine*). (3) The presentation of a number at the fixation point prior to a target detection task will speed detection on the right for large numbers and to the left for small numbers. Hubbard et al. (2005b) hypothesized that the SNARC effect can be accounted for by the observed reuse in numerical cognition of a particular circuit in left inferior parietal sulcus that plays a role in shifting spatial attention. Briefly, the idea is that, among the representational formats we make use of in numerical cognition, there is a mental "number line" on which magnitudes are arrayed from left to right in order of increasing size (at least for speakers of left-right languages) (Zebian 2005). Once numerals are arrayed in this format, it is natural to reuse the circuit responsible for shifting spatial attention for the purpose of shifting attention between positions on this line. The resulting magnitude-influenced attentional bias can explain the SNARC effect.

This redeployment of visuospatial resources in support of alternate cognitive uses is somewhat difficult to explain from the standpoint of either concept empiricism or conceptual metaphor theory. In these examples the effects would not be accounted for by the fact that the meanings of numbers might be grounded in basic sensorimotor experience; nor is it immediately obvious what metaphorical mapping might be implicated here. In fact, if the reuse of spatial schemas *were* in support of some semantically grounding structural inheritance from one domain to the other, we would expect the numbers to be arrayed *vertically*, with magnitude increasing with height, given general associations of *more* with *higher*. Instead, the reuse in this case appears driven by more abstract functional considerations: when one is doing certain numerical tasks, a number line is a useful representational format, and something like the visuospatial sketchpad (Baddeley 1986) offers a convenient and functionally adequate storage medium. Similarly, reusing the spatial shifting mechanism is a sensible choice for meeting the functional requirements of the task and need not ground any semantic or structural inheritance between the domains.

1.2.4 Reuse of Structures for Numerical Cognition

In fact, several such examples can be found in the domain of numerical cognition. There is strong evidence that some of the brain regions in the left precentral gyrus and left angular gyrus associated with finger gnosis (finger awareness) are activated during tasks requiring the representation of number (Dehaene et al. 1996; de Jong et al. 1996; Göbel et al. 2004; Jancke et al. 2000; Kuhtz-Bushbeck et al. 2003; Liu et al. 2006; Numminen et al. 2004; Pesenti et al. 2000; Pinel et al. 2004; Venkatraman, Ansari, & Chee 2005). Zago et al. (2001) found increased activation in the premotor strip in a region implicated in finger awareness during multiplication performance compared to a digit reading condition, and similar findings were reported by Andres, Seron, and Oliver (2007), who found that hand motor structures were activated during adults' number processing in a dot-counting task. Moreover, repetitive transcranial magnetic stimulation (rTMS) (Rusconi, Walsh, & Butterworth 2005) and direct cortical stimulation (Roux et al. 2003) have been found to disrupt both finger gnosis and tasks requiring the representation of number. These findings are consistent with neural reuse; one of the neural structures originally evolved or developed for finger awareness has been reused in the (presumably) later-emerging function of number representation, and it now serves both uses.

Here again, this reuse of a basic sensorimotor function in an alternate cognitive domain does not seem to follow the logic of conceptual metaphor theory or concept empiricism. The most prominent explanations for the overlap do not make the claim that magnitudes inherit their meanings from finger representations, nor that there is some mathematical metaphor built on our finger sense. Rather, the idea is that this neural circuit, originally developed to support finger awareness, is offering some functionally relevant *resource* in the domain of numerical cognition. For instance, Butterworth (1999) suggests that the fingers provide children a useful *physical* resource for counting, with the neural result that the supporting structures for finger awareness and magnitude representation now overlap, and

Penner-Wilger and Anderson (2008; 2013) suggest instead that the circuit in question might *itself* offer useful resources (such as a storage array).

Interestingly, in this particular case there *does* appear to be a functionally relevant inheritance resulting from the overlap of structures—just not one that plays a role in grounding conceptual content. One landmark phenomenon in number representation is the *distance effect*—it is harder to differentiate numbers that are closer together in magnitude than those that are farther apart (e.g., 3 vs. 4 is harder than 1 vs. 9) (Dehaene, Dehaene-Lambertz, & Cohen 1998). The same phenomenon is also found in finger gnosis—it is harder to differentiate fingers that are closer together physically than those that are farther apart (Benton et al. 1983). Given that this effect in finger gnosis is likely a result of the underlying physical/spatial distribution of the nerves in the hand and the way the information is processed in the brain (and is therefore likely to obtain in many sensory systems), the fact that it is also observed in an apparently abstract functional domain such as magnitude comparison is all the more striking. In light of the independent evidence for the circuit sharing between part of the finger-sense system and part of the number-processing system, the persistence of a distance effect across these domains appears to reflect the influence of stable functional characteristics of the shared circuit. But it is not the sort of inheritance predicted by concept empiricism or conceptual metaphor theory. Rather than inheriting semantic structure, what we apparently have here is the persistence of performance characteristic imposed by the underlying implementation of the capacity.

This is not to question the notion that numbers and other mathematical concepts and procedures are in some way grounded in sensorimotor experience (Lakoff & Núñez 2000) nor to argue that sensorimotor experience might not be a necessary precursor to mathematical development. But this specific overlap in neural circuitry is not straightforward to explain in the context of such grounding, nor is it anything that would have been *predicted* on the basis of either conceptual metaphor theory or concept empiricism. In fact, proponents of conceptual metaphor theory in mathematics tend to focus on relatively higher-level concepts such as sets and to investigate how our understanding of them is informed by such things as our experience with physical containers.

A similar argument can be made when one is considering the interrelations of speech and gesture and the cognitive importance of the latter (see, e.g., Goldin-Meadow 2003). According to Goldin-Meadow (2003), gesture not only is typically used to signal different moments in the learning process (e.g., to index moments of decision or reconsideration in a problem-solving routine) but also appears to have utility in *advancing* the learning process by providing another representational format that might facilitate the expression of ideas currently unsuited (for whatever reason) to verbal expression. The motor control system is here being used for a specific cognitive purpose *not* because it is performing semantic grounding or providing metaphorically guided domain structuring but because it offers an appropriate physical (and spatiotemporal) resource for the task.

1.2.5 Reuse of Perceptual Structures to Support Other Types of Higher-Order Cognition

There are examples of the reuse of structures typically associated with perception that also make the same point. Although there are certainly studies that appear to unproblematically support concept empiricism—for example, Simmons et al. (2007) report the discovery of a common neural substrate for seeing colors and for knowing about (having concepts for) color—it appears that such studies represent only a small subset of a much broader phenomenon. Consider one of the earliest and most discussed cases of the reuse of neural structures for a new purpose, the Baddeley and Hitch model of working memory (Baddeley 1986; 1995; Baddeley & Hitch 1974, 1994). One strategy for remembering the items on a grocery list or the individual numbers in a phone number involves (silently) saying them to one's self (producing a "phonological loop"), which engages brain areas typically used both in speech production and in audition.

A pattern of findings supports the existence of a phonological loop and the engagement of both inner "speaking" and inner "hearing" to support working memory (see Wilson 2001 for a review). First, there is poor recall of similar sounding terms; second, there is poor recall of longer words; third, there is poor recall if the subject is made to speak during the maintenance period; and fourth, there is poor recall when the subject is exposed to irrelevant speech during the maintenance period. Moreover, imaging studies have found that such memory tasks cause activation in areas typically involved in speech production (Broca's area, left premotor cortex, left supplementary motor cortex, and right cerebellum) and in phonological storage (left posterior parietal cortex) (Awh et al. 1996).

In this interesting and complicated case we have something of a triple borrowing of resources. First is the use of a culturally specific, acquired representational system—language—as a coding resource, and second is the application of a particular learned skill—silent inner speech—as a storage medium. These two together imply the third borrowing—of the neural resources used to support the first two functions. And note that all of this borrowing is done in support of what is likely an enhancement of a basic evolved function for storing small amounts of information over short periods. This raises the obvious question of whether and to what degree evolutionary pressures might have shaped the language system so that it was capable of just this sort of more general cognitive enhancement (Carruthers 2002). In any case it seems clear that this sort of borrowing is very hard to explain in terms of concept empiricism or conceptual metaphor theory. In the case of sensorimotor coding in working memory, the phonological loop is not metaphorically *like* speech; rather, it *is* a form of speech. In this it is another instance of a straightforward functional redeployment—the reuse of a system for something other than its (apparent) primary purpose because it happens to have an appropriate functional structure.

1.3 Reuse Is Not Always Explained by Conceptual Metaphor Theory or Concept Empiricism

These various examples suggest something along the following lines. One of the fundamental principles guiding reuse is the presence of a sufficient degree of functional relatedness between existing and newly developing purposes. When these functional matches result in the reuse of resources for both purposes, this history *sometimes*—but not always—reveals itself in the form of a metaphorical mapping between the two task domains and sometimes, but not always, results in the inheritance or grounding of some semantic content. This way of thinking makes conceptual metaphors and "grounded" symbols into two possible side effects of the larger process of reuse in cognition. It also muddies the causal story a bit. Planning is like locomotion because it inherits the structure of the existing domain via neural overlap, but planning also overlaps with the neural implementation base of locomotion to the degree that it has functional requirements similar to those of locomotion.

The suggestion here is not that planning or communication or any other cognitive function has some predetermined Platonic structure that entirely reverses the causal direction typically supposed by conceptual metaphor theory. Rather, the idea is to point out the need to be open to a more iterative story, whereby a cognitive function finds its "neural niche" (Iriki & Sakura 2008) in a process codetermined by the functional characteristics of existing resources and the unfolding functional requirements of the emerging capacity (Deacon 1997; see also chapters 2 and 3).

Consider, in this regard, the particular articulatory character of human speech. Speech is produced by moving the vocal apparatus between various postures while making some noise (Fowler et al. 1980). Why should speech production be this way? In an article outlining their discoveries regarding the organization of the motor-control system, Graziano et al. (2002b) wrote:

> One possibility is that the mechanisms for speech were built on a preexisting mechanism for motor control, one that emphasized the specification of complex, behaviorally useful postures. When we stimulated the ventral part of the precentral gyrus in the mouth and face representation, we often caused the lips and tongue to move toward specific postures (Graziano, Taylor, & Moore 2002a). For example, at one site, stimulation caused the mouth to open about 2 cm and the tongue to move to a particular location in the mouth. Regardless of the starting posture of the tongue or jaw, stimulation evoked a movement toward this final configuration. This type of posture may be useful to a monkey for eating, but could also be an evolutionary precursor to the phoneme. (Graziano et al. 2002b, p. 355)

There are certainly functional characteristics that a unit of acoustic communication must have in order to adequately perform its communicative purpose, and not just any neural substrate would have had the required characteristics. But there remain degrees of freedom in how those characteristics are implemented. Speech, then, developed its *specific* articulatory character as the result of the structures on which it was built. Had the motor control system been oriented instead around (for example) simple, repeatable contractions of individual muscles—or had there been some other system with these functional characteristics available for reuse as acoustic communication was evolving—the result of the inheritance might have been a communication code built of more purely temporal elements, something closer to Morse code.

Finally, consider what may be a case *not* of the reuse of a basic sensorimotor area for higher cognitive functions but rather the reverse. As was already discussed above, Broca's area has long been associated with language processing, responsible for phonological processing and language production, but what has recently begun to emerge is its functional complexity (Grodzinsky & Santi 2008; Hagoort 2005; Tettamanti & Weniger 2006). The assumption under which I am working is that these functional overlaps should not be understood as the later reuse of a *linguistic* area for other purposes but are rather evidence that Broca's already performed some sensorimotor functions that were prerequisites for language acquisition and that made it a candidate for one of the neural building blocks of language when it emerged (Müller & Basho 2004; see chapter 7 for discussion). Note, however, that Broca's area is also activated in domains such as music perception (Maess et al. 2001; Tettamanti & Weniger 2006). Although it is possible that this is because processing music requires some of the same basic sensorimotor capacities as processing language, it seems also possible that this reuse was driven by functional features that Broca's acquired as the *result* of its reuse in the language system, and thus by some more specific structural similarity between language and music (Fedorenko et al. 2009). Whatever the correct history, this clearly represents another set of cases of functional reuse *not* explained by conceptual metaphor theory or concept empiricism.

If we assume the foregoing is sufficient to establish the existence of at least some cases of neural reuse that cannot be accounted for by these theoretical frameworks alone, the question naturally arises as to whether these anomalous cases should be dealt with by post-hoc elaborations of these theories (and/or by generating one or a few similarly specific theories) or whether this is a situation that calls for a global theory of reuse that supersedes and at least partially subsumes these existing frameworks. Far be it from me to argue a priori that one tack *must* be the correct one to take—science works best when we pursue multiple competing research paths—but the prevalence of reuse just outlined in section 1.2, and especially the *range* of domains that seem to share neural resources along with the particular *patterns* of sharing that we reviewed in section 1.1, would seem to provide a fairly strong justification for treating neural reuse as a fundamental and global architectural fact about the brain. The remainder of this volume takes this point to have been established.

Let me be clear, however, about what it is I *do not* take the evidence to have established. I was an early proponent of ECS (O'Donovan-Anderson 1996, 1997), continue to be a staunch advocate (Anderson 2003, 2008b, 2009; Anderson & Rosenberg 2008; Anderson et al. 2012), and do not think that any of the arguments made in this chapter falsify any of the claims made on its behalf. Quite the contrary! As I have already indicated, I think that neural reuse follows naturally from, and coheres perfectly with, the general expectations of ECS. What I do think is that conceptual metaphors and modal concepts as well as other kinds of cognitively relevant simulations are all examples of a much larger phenomenon of *borrowed* cognition, driven largely by neural reuse. The borrowing of spatial resources for number storage revealed by the SNARC effect (Dehaene et al. 1993), the use of gesturing in learning (Goldin-Meadow 2003), and the use of a common brain region for both finger and magnitude representation (Anderson & Penner-Wilger 2013; Penner-Wilger & Anderson 2008; Zago et al. 2001) can all be explained by positing that both later and earlier use share some functional requirements, such that one or more of the brain regions underlying the earlier use can also be of service in supporting the later use. Similarly, one does not need to posit any inheritance of semantic structure from one domain to another to explain the finding that finger differentiation exercises improve math performance (Gracia-Bafalluy & Noël, 2008). In fact, this same finding suggests that although sensorimotor experience can be crucial to the development of numerical cognition, insofar as it helps establish the functional structure of the brain regions used in both domains, the crucial experience need not be of using the fingers to do mathematics. Exercises that improve the sensory acuity of finger representations could be expected to improve performance on certain numerical tasks without the further requirement that the actual fingers be used in a mathematical context. In such cases there need not be, and we in fact see no evidence in these particular cases for, any conceptual grounding or semantic inheritance between these domains as a result of these overlaps.

These findings most certainly do not explain away the kinds of semantic inheritance from sensorimotor to higher-order cognitive systems so important to many ECS accounts of higher cognition. Quite the contrary: they urgently need to be explained. The worst effect my arguments will have on advocates of these theories is to strip away the illusion that the semantic inheritance observed in so many domains was ever actually explained by the discovery of neural overlaps. But such disillusionment should be welcomed by any scientist, as it lays down the direction of future research.

1.4 Neural Reuse Does Not Go Away, No Matter How Small the Brain Region

Before we return to a consideration of brain evolution in light of these findings, it is worth pausing to critically evaluate the evidence presented so far. As was pointed out by several of the commentators on my *Behavioral and Brain Sciences* article (Anderson 2010b), most of the evidence presented above rests on brain-imaging data, and, given the relatively poor spatial resolution of that technique, it might be premature (or at least overly bold) to draw definitive conclusions based on the overlap between neural activations measured using those techniques. The imaging data, that is, are *also* consistent with there being multiple quite different neural assemblies in close physical proximity such that the supporting circuitry that shows up in different cognitive context only *appears* to overlap because of the poor spatial resolution of current functional imaging techniques. Moreover, because neural activation may spread around the brain network, this can lead to false positives: regions that are activated only as a side effect of their connectivity and not because they are making a functional contribution to the task under investigation (see, e.g., Klein 2010; Mahon & Caramazza 2008; Ritchie & Carruthers 2010).

These are important points, and it is certainly the case that when one is looking at neuroimaging data alone, the smaller the regions, the fewer activations the region will contain, and the less diverse each region can appear. But imaging results demonstrating neural overlaps are not the only data cited above, and I think the broader picture sits uneasily with the assumption that if we only parcellate our cortex into small enough pieces we will eventually reveal the true selectivity that lies hidden. Consider, for instance, the data from cortical stimulation, cognitive interference and cross-domain interactions cited above (e.g., Glenberg & Kashak 2002; Glenberg et al. 2008a; Roux et al. 2003; Rusconi et al. 2005). Data like these, demonstrating that using neurons in one task makes them less available for another, seem to indicate that the *very same neurons* are being engaged by both tasks. This appears to weigh against the possibility that the poor spatial resolution of neuroimaging techniques is masking the fact that multiple neighboring structures are making separate contributions to cognition. Second—and more telling in my view—are the cases in which there *does* appear to be a functional or semantic inheritance that results from the sharing of components. This suggests that at least sometimes the functional contributions of the shared local neural structures are stable and detectible across multiple uses, a possibility that seems once again to strongly favor the use of the *very same structure* in two different cognitive contexts.

Moreover, there is some very intriguing evidence from comparative studies of neural reuse *at the level of single neurons and small networks* in various species. Such findings suggest that reuse is a fundamental evolutionary strategy for getting maximal functional flexibility out of scarce and expensive neural resources—and thus something we should expect in the human brain at all levels of description. Cisek and Kalaska (2010) review a large body of evidence from the nonhuman primate literature and find a great deal of evidence for the functional complexity of single neurons and networks. For instance, during tasks in which different movements must be made depending on the particular content of tactile or visual stimuli (e.g., whether successive stimuli are the same or not), both the sensory encoding and the decision-making processes appear to rely on the same neurons in premotor areas. They conclude a review of such findings:

> In all of these cases, the same neurons appear to first reflect decision-related variables such as the quality of evidence in favor of a given choice and then later encode the metrics of the action used to report the decision (Cisek & Kalaska 2005; Kim & Basso

2008; Roitman & Shadlen 2002; Schall & Bichot 1998; Yang & Shadlen 2007). (Cisek & Kalaska 2010, p. 274)

Similarly, neurons in the posterior parietel cortex (PPC) appear to be involved in processing information about object location and action intentions, among other things, and are known to be modulated by attention, behavioral context, expected utility, and other decision-related variables. Reviewing evidence for the functional complexity of neurons in PPC, Cisek and Kalaska write:

In short, the PPC does not appear to fit neatly into any of the categories of perception, cognition, or action; or alternatively, the PPC reflects all categories at once without respecting these theoretical distinctions. Indeed, it is difficult to see how neural activity in the PPC can be interpreted using any of the concepts of classical cognitive psychology (Culham & Kanwisher 2001). (Cisek & Kalaska 2010, p. 274)

The same could be said for species much deeper in the evolutionary tree. Consider recent work with *C. elegans*, the soil-dwelling roundworm (nematode) for which the first neuron network wiring diagram was published in 1986 (White et al. 1986) and updated in 2011 (Varshney et al. 2011). The *C. elegans* nervous system contains 302 neurons (282 in the somatosensory system and 20 in the small pharyngeal system) making over 8,000 connections, including chemical synapses, electrical junctions, and neuromuscular junctions. Interestingly, between individual animals the chemical synapses are about 75% identical (Durbin 1987), making this a very nice model system for studying brain-behavior relationships. Although some ablation studies have suggested that individual neurons could be assigned single, discrete functions (de Bono & Maricq 2005), the body of evidence paints a rather more complex picture.

Some *C. elegans* neurons perform more than one type of circuit function, including both motor and sensory functions or interneuron–motor neuron or interneuron–sensory neuron functions (White et al. 1986). Polymodal neurons are much more common in the male tail circuitry than anywhere else (S. W. Emmons et al., unpublished data; Sulston, Albertson, & Thomson 1980). M3 neurons of the pharynx have both motor and sensory functions. NSM neurons are both neurosecretory and motor neurons, and they may also have proprioceptive function (Albertson & Thomson 1984). A- and B-type VNC motor neurons are suggested to be proprioceptive. IL1 neurons in the head perform mechanosensory, motor, and interneuron functions, whereas OLQ neurons are both mechanosensory and interneurons. RIM, SMB, SMD, RMD, RMH, and RMF classes of head neurons seem to be both motor and interneurons. AVL is a motor neuron with additional interneuron-type synapses. DVA is an interneuron that also functions as a stretch-sensitive sensory neuron. DVB is a motor neuron for enteric muscles and is also an interneuron. Alternatively, some neurons have multiple functions within one modality. ASH sensory neurons, for example, function as mechanosensory, osmosensory, odorsensory, and nociceptive, and ADLs are chemosensory, odorsensory, and nociceptive. (Altun & Hall 2011)

There even seem to be examples of single neurons participating in generating completely opposite behavior as a result of alteration of the neuron's sensitivity, physical connections, and functional connectivity by various chemicals and genes (effects known collectively as neuromodulation; see Katz 1999; Katz & Calin-Jageman 2008; Marder & Thirumalai 2002 for discussion). For instance, the olfactory neuron AWCON can direct both attraction and repulsion to the same odor, depending on the presence of specific neuromodulators (Tsunozaki, Chalasani, & Bargmann 2008); and the nocioceptive ASH neurons can cause both social aggregation and avoidance, depending on whether the gap junction with RMG neurons and the associated aggregation circuit has been decoupled by the expression of the *npr-1* gene, which encodes a G-protein-coupled receptor (Bargmann 2012).

A profound violation of the one neuron–one behavior rule was uncovered by characterizing behaviors under different conditions. For example, avoidance of the repulsive odor octanol at particular concentrations can be generated by two different sets of sensory neurons. In well-fed animals, octanol avoidance is almost entirely mediated by the ASH nocioceptive neurons, but after an hour of starvation, octanol avoidance is distributed [among] ASH, AWB, and ADL nocioceptive neurons, revealing a change in circuit composition (see figure 1.5A). ... Food changes the compo-

Figure 1.5

Alternative, overlapping circuits for *C. elegans* sensory behaviors. Triangles are sensory neurons (arbitrary colors are used to highlight the neurons that appear in several panels); hexagons are integrating neurons. In each panel pathways for food regulation appear in dark blue. (A) Sensory neurons and neuromodulators that affect octanol avoidance in starved and well-fed animals. Blue, modulators or neuropeptides that enhance or accelerate avoidance; black, modulators that delay avoidance. Neuropeptides within this subcircuit are shown adjacent to the neuron that produces them; biogenic amines derive from external sources (Komuniecki et al. 2012). (B) Alternate sensory circuit for aerotaxis behavior. Aerotaxis is inhibited by food through inhibitory peptide signaling onto oxygen-sensing and modulatory neurons (Chang & Bargmann 2008; Chang et al. 2006; Cheung et al. 2005; Rogers et al. 2006). After growth in hypoxia, transcriptional regulation via HIF-1 collapses aerotaxis into a smaller circuit that is resistant to food regulation. (C) ASH chemical synapses are required for acute avoidance behavior, whereas ASH gap junctions in a hub-and-spoke circuit regulate aggregation; these two classes of ASH synapses are differentially regulated by the NPR-1 neuropeptide receptor (Hart, Sims, & Kaplan 1995; Macosko et al. 2009; Maricq et al. 1995). Reprinted from Bargmann (2012) with permission.

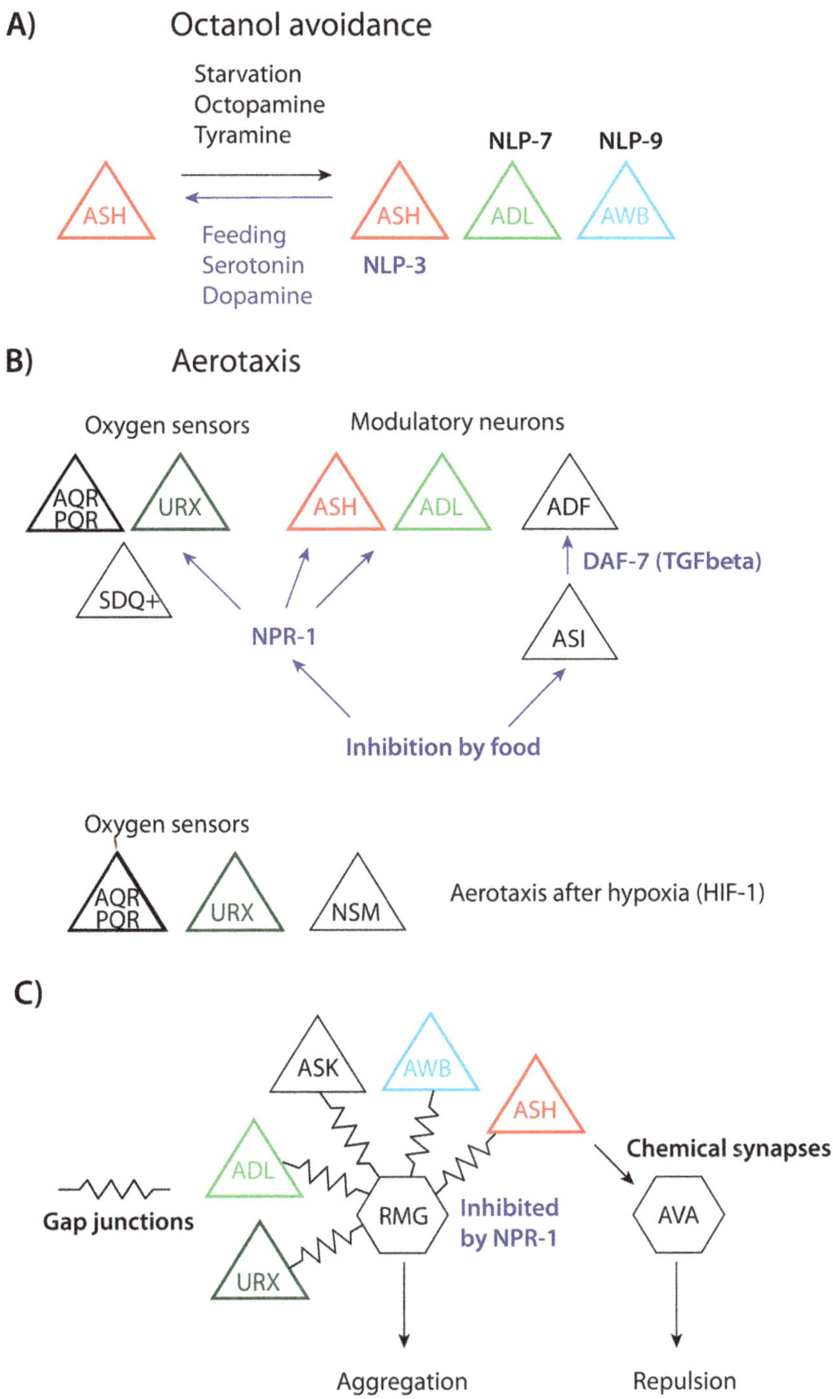
A)
Octanol avoidance
Starvation
Octopamine
Tyramine
NLP-7
NLP-9
ASH
ASH
ADL
AWB
Feeding
Serotonin
Dopamine
NLP-3
B)
Aerotaxis
Oxygen sensors
Modulatory neurons
AQR
PQR
URX
ASH
ADL
ADF
SDQ+
DAF-7 (TGFbeta)
NPR-1
ASI
Inhibition by food
Oxygen sensors
AQR
PQR
URX
NSM
Aerotaxis after hypoxia (HIF-1)
C)
ASK
AWB
ASH
ADL
Chemical synapses
Gap junctions
RMG
Inhibited
by NPR-1
AVA
URX
Aggregation
Repulsion

sition of a circuit for oxygen preference behavior (aerotaxis) as well. … Aerotaxis is more robust in starved than in well-fed animals due to the activity of multiple neuromodulators (Chang et al. 2006; Cheung et al. 2005; Rogers et al. 2006; see figure 1.5B). … The *npr-1* neuropeptide receptor that affects aerotaxis also regulates a second behavior, the aggregation of animals into feeding groups. Aggregation is triggered by a number of sensory neurons including the nocioceptive ASH neurons and oxygen-sensing URX neurons (Coates & de Bono 2002; de Bono et al. 2002) … integrated by one pair of npr-1-expressing neurons called RMGs (Macosko et al. 2009). … [The] npr-1 action in RMG uncouples the aggregation circuitry but leaves the avoidance circuitry intact. This allows ASH to generate different behaviors in two neuromodulatory states. (See figure 1.5C; Bargmann 2012: 460–461)

Examples of the redeployment of individual neurons to support multiple behaviors are not restricted to *C. elegans* (nor even, we shall see, to invertebrates).

Reuse may also be found in neurons involved in learning and memory. In the pond snail (*Lymnea stagnalis*), the breathing rhythm is generated by three synaptically connected neurons that form a central pattern generator. One of these neurons, RPeD1, is also necessary for many aspects of learning and memory, and removing the RPeD1 cell body can prevent the formation or reconsolidation of long-term memories (Sangha, Scheibenstock, & Lukowiak 2003). In honeybees (*Apis mellifera*), a single identified neuron (VUMmx1) in the suboesophageal ganglion mediates the reward pathway in associative olfactory learning, but this neuron has also been implicated in learning phenomena as diverse as second-order conditioning and blocking (Menzel 2009). (Niven & Chittka 2010, p. 285)

In fact, such examples of reuse enabled by neuromodulation can be found across the animal kingdom, suggesting it is a vitally important evolutionary strategy for deploying scarce neural resources to the greatest behavioral and adaptive effect. Neuromodulation comes in many guises, but two common types involve adjustment of sensory gain and gating of sensory inputs. Consider, for instance, the stress-induced analgesia seen in both rodents and humans (Akil et al. 1984; Bargmann 2012). This apparently involves the expression of G-protein-coupled opioid receptors that diminish neurotransmitter release in nocioceptive neurons, thereby reducing the sensation of pain. Stress can also change the configuration of large-scale brain networks across a number of species including humans (Hermans et al. 2011).

Sensory gating that changes the configuration of local information-processing structures is seen in the retina. Information from rods and cones converges on cone bipolar cells. However, the information from rods is mediated by the rod bipolar cells and AII amacrine cells, and high light levels cause the AII amacrine cells and the cone bipolar cell to decouple.

Finally, there is even evidence for state-related modulation of neural response properties; for instance, "one class of retinal ganglion cells in the salamander responds to light offset (OFF) under baseline conditions but light onset (ON) immediately after a stimulus that mimics an eye movement or saccade" (Bargmann 2012, p. 463).

Interestingly, Bargmann (2012) suggests that, given the ubiquity of neuromodulation, we should expect most neural circuitry to be structurally overconnected. Any given circuit will have a number of possible uses, only some of which are available at any given moment depending on the neuromodulatory state of the organism. The other uses will be "latent"—part of the set of functional possibilities afforded by the physical structure of the circuit but not currently expressed in its functional state. In general, the possible uses of any part of the nervous system will depend on the selection of "a set of functional synapses among a greater number of anatomically specified possibilities" (Bargmann 2012, p. 461). This fact becomes especially important to us in chapter 2, where we consider the question of how these multiple overlapping neural structures emerge in development.

Interpreting the implications of this work for human brains must of course be approached with caution. Consider, for instance, the fact that the simplicity of the organisms where neuromodulation is generally studied means that inactivating even a single synapse can have profound behavioral effects. This will not be generally true for humans, and so more work needs to be done to study mechanisms whereby the modulation of multiple synapses can be precisely coordinated to result in specific, reproducible changes in functional structure. Nevertheless it does seem likely that neuromodulation will turn out to be an important part of the functional operation of the human brain. And indeed there is evidence in the primate (including human) literature consistent with large-scale modulation of neural partnerships in support of cognitive function. For instance, there is evidence relating changes in the oscillatory coherence between brain regions (local and long-distance) to sensory binding, modulation of attention, and other cognitive functions (Steinmetz et al. 2000; Varela et al. 2001). Friston (1997) demonstrated that whether a given region of inferotemporal cortex was face selective depended on the level of activity in posterior parietal cortex. And McIntosh et al. (1994) investigated a region of inferotemporal cortex and a region of prefrontal cortex that both support face identification and spatial attention. They showed that during the face-processing task the inferotemporal region cooperated strongly with a region of superior parietal cortex, whereas during the attention task, that same region of parietal cortex cooperated more strongly with the prefrontal area. More recently, Cole et al. (2013) report evidence for "flexible hubs" in the brain, "regions that flexibly and rapidly shift their brain-wide functional connectivity patterns" depending on the task. These experimental findings are consistent both with the expectation of large-scale modulation of neural partnerships and with the data reviewed above that the differences between cognitive domains come down to differences in patterns of neural cooperation and not to differences in which regions are supporting the different tasks.

Clearly, all of the foregoing raises some profound challenges for structure-function mapping in the brain, both for determining the function of a given neural unit and even for easy *identification* of that unit. Some of these issues are taken up in chapters 3 and 4. But the evidence presented also cements the fundamental architectural principle behind neural reuse: individual pieces of the brain, from cells to regions to networks, are used and reused in a variety of circumstances, as determined by social, environmental, neurochemical, and genetic contexts. Insofar as this is so, the apparent ubiquity of neuromodulation seems to indicate the ubiquity of neural reuse, and the apparent ubiquity of neural reuse suggests the ubiquity of neuromodulation. But here we need to reflect on a complication. It *might* seem that the role of neuromodulation should instead lead us to strongly distinguish neural *reuse* from neural *multiuse*, reserving "reuse" for the case when a single neural element (neuron or network) is reused in the same state for multiple purposes and "multiuse" for the case when the element moves into a different functional configuration. But not only is our ability to *tell* when this has occurred severely limited, at least in humans (which itself must affect how we do good science, as I discuss in chapter 4), these situations are somewhat difficult to disentangle even in principle. Because there is reuse at multiple spatial scales, what is reuse at a one level of organization can be multiuse at another. Using the same neurons in a different configuration or when modulated by genetic or chemical factors is reuse of neurons, but multiuse in the local network. Reuse of a local region that cooperates with different partners is reuse of the region but multiuse at the level of the large-scale network. Thus, I think it is best to stick with the single term *reuse* with the realization that the *causes* of reuse in a particular case can include various kinds of neuromodulation and functional reconfiguration.

1.5 Neural Reuse, Evolution, and Modularity

Taken as a whole, the findings reported above appear to cast serious doubt on the long-standing belief in the brain's anatomical modularity. Although there are few defenders of a strong anatomical modularity hypothesis of the sort that predicts strict 1:1 mappings between regions and functions, Max Coltheart (2001) goes so far as to list anatomical modularity as one of the fundamental assumptions guiding cognitive neuropsychology. His argument is that neuropsychological research both requires and, in turn, supports the assumption that the brain is organized into anatomical modules. For if it were not, we would not observe the focal deficits characteristic of some brain injuries, nor would we be able to gather evidentiary support for double dissociations between tasks.

If this argument were sound, then the success of neuropsychology as a discipline would itself be prima facie evidence against neural reuse. In fact, the inference is fairly weak. First, it is possible for focal lesions to cause specific functional deficits and double dissociations in provably nonmodular systems (Chater & Ganis 1991; Plaut 1995; Van Orden, Jansen op de Haar, & Bosman 1997), and double dissociations do not by themselves support any inference about the underlying functional architecture of the brain (Van Orden, Pennington, & Stone 2001). In any event such deficits are the exception rather than the rule in human brain injuries. Even some of the patients most celebrated for having specific behavioral deficits often have multiple problems, even when one problem is the most obvious or debilitating (see Bergeron 2007; Prinz 2006 for discussions). The evidence coming from neuropsychology, then, is quite compatible with neural reuse. But is neural reuse compatible with the methodological assumptions of cognitive neuropsychology? Chapter 4 discusses some of the specific methodological changes that will be needed in the cognitive neurosciences in light of widespread neural reuse.

This discussion returns us to an important issue we put aside early on. Although the evidence presented strongly favors the kind of deployment of neural resources posited by ECS over that posited by evolutionary psychology and massive modularity, these latter positions nevertheless appear to have some strong considerations in their favor. In particular, advocates of these views argue that no nonmodular system could possibly evolve, and moreover that an organism with a nonmodular brain would be subject to debilitating amounts of processing interference. We consider these points in turn.

The argument about evolvability in fact has two prongs. The first focuses on the issue of incrementalism in the evolution of complex structures; the second on the evident fact of functional differentiation in the brain.

Carruthers (2006) follows Simon (1962) in making an argument of the first type:

> Simon (1962) uses the famous analogy of the two watchmakers to illustrate the point. One watchmaker assembles one watch at a time, attempting to construct the whole finished product at once from a given set of micro components. This makes it easy for him to forget the proper ordering of parts, and if he is interrupted he may have to start again from the beginning. The second watchmaker first builds a set of subcomponents out of given micro component parts and then combines those into larger subcomponent assemblies, until eventually the watches are complete. ... Simon's argument is really an argument from *design*, then, whether the designer is natural selection (in the case of biological systems) or human engineers (in the case of computer programs). It predicts that, in general, each element added incrementally to the design should be realized in a functionally distinct subsystem, whose properties can be varied independently of the others (to a significant degree, modulated by the extent to which component parts are shared between them). It should be possible for these elements to be added to the design without necessitating changes within the other systems, and their functionality might be lost altogether without destroying the functioning of the whole arrangement. (Carruthers 2006, pp. 13, 25; emphasis in original)

The argument from design set forth here is more convincing when it is applied to the original emergence of a complex system than when it is applied to its subsequent evolutionary development. What the argument says is that it must be possible for development to be *gradual*, with functional milestones, rather than all-or-nothing; but neural reuse hardly weakens the prospect of a gradual emergence of new functions. And the possibility that new functionality can be achieved by combining existing parts in new ways—which undermines independent variation and separate modifiability, as Carruthers (2006) admits here—suggests that a modular architecture is only one possible outcome from such gradualism.

Moreover, the strong analogy between natural selection and a designer may not be the most helpful conceptual tool in this case. When one thinks about the brain the way a human designer would, the problem that neural reuse presents is one of taking a given concrete network with a known function and imagining novel uses for it. That this process can be *very* difficult appears to place a heavy burden on reuse theories: How could such new uses ever be successfully designed? But suppose instead that, in building a given capacity, one is offered a plethora of components with *unknown* functions. Now the task is quite different: find a few components that do something useful and can be arranged so as to support the current task—*whatever* their original purpose. Thus, it is a problem of *design imagination* turned into a problem of *search*. Evolution is known to be quite good at solving problems of the latter sort (Newell & Simon 1976), and we review evidence in chapter 2 that search may be an important part of development as well. Thus, it is useful to keep this alternate analogy for the evolutionary process in mind here.

The other evolutionary argument for separate modifiability rests on the observation that we do indeed see stereotyped, heritable functional differentiation in animal nervous systems. *My* brain networks for vision and attention look a lot like *your* brain networks for vision and attention. Separately modifiable subsystems individually targeted by selection pressures provide a very natural way to explain this observation. Consider the following from Barrett and Kurzban (2006):

> Our position, then, is that functionally specialized mechanisms with formally definable informational inputs are characteristic of human (and nonhuman) cognition and that these features should be identified as the signal properties of "modularity." By this we intend an explicitly evolutionary reading of the concepts of function and specialization: modules evolved through a process of descent with modification, due to the effects they had on organisms' fitness. … As a direct and inseparable result of this evolutionary process of specialization, modules will become *domain specific*: Because they handle information in specialized ways, they will have specific *input criteria*. (Barrett & Kurzban 2006, p. 630)

As should be clear from the quote, it is a central part of massive modularity that each module should be separately modifiable, both in theory and in the course of evolutionary development. Indeed, here functional specialization and domain specificity are *results* of the fact that the modules are separately targeted by evolutionary pressures. So, how can we explain heritable functional differentiation in a brain without separately modifiable parts?

First, we should clearly establish that neural reuse does indeed militate against separate modifiability, for some authors, including Carruthers (2006), have suggested that these positions may in fact be compatible. Consider a brain with a set of relatively discrete neural units or regions $\{r_1, r_2, r_3 \ldots r_n\}$ and a set of functional complexes $\{f_1, f_2, f_3 \ldots f_n\}$ built from the regions and implementing a set of cognitive uses $\{u_1, u_2, u_3 \ldots u_n\}$. Neural reuse is the claim that any given region r_a is likely used in multiple complexes $\{f_a, f_b, f_c, \ldots\}$ and thus supports multiple uses $\{u_a, u_b, u_c, \ldots\}$. For now we are putting aside the finding that at least some regions can take on multiple configurations, which may make identifying functionally discrete units quite difficult; these issues are taken up in chapter 3. The point here is that modifying the shared region r_a seems likely to affect many if not all of the functional complexes in the set $\{f_a, f_b, f_c, \ldots\}$—specifically the subset that relied on any of the underlying causal features of r_a that were changed in the modification—and will thereby affect some nonempty subset of cognitive uses $\{u_a, u_b, u_c, \ldots\}$. How the physical configuration of a region determines its causal features and how individual neural regions contribute to the overall function of a complex are not well understood, so at this stage it is not possible to know whether functional complexes can rely on only some but not all of a region's causal features, nor whether a region can have its causal features independently modified. But these details change only the *scope* and not the *existence* of the problem here: there seems to be a basic conflict between neural reuse and separate modifiability.

Now, it is certainly true (and not in conflict with neural reuse) that there will be pairs of functional complexes $\{f_a, f_b\}$ that are functionally dissociable because they share no parts. And it is also true that even functional complexes that share *some* parts can be functionally distinguished by their reactions to disruptions of the parts not shared, and thus *some* overlap in the implementation of modules is compatible with modularity. That is, no modularist need claim that for *all* pairs of functional complexes $\{f_a, f_b\}$ it will be possible to functionally dissociate them by disrupting *any* of their respective parts because a few of the regions implementing the modules could be the same. But neural reuse does not claim that there is occasional overlap; it expects that it will *typically* (overwhelmingly) be the case that a given r_a is used in multiple complexes; that therefore, the members of the implementing set for any given complex f_a: $\{r_a, r_b, r_c, \ldots\}$ will be implicated in many functional complexes and that therefore, although not *every* complex f_b would be impacted by a modification to some member(s) of the implementing set of f_a, many others would. Insofar as this is so, then even though specific pairs of functional complexes will be functionally separable, it would appear that separate modifiability will not be a general characteristic of the brain.

Given this analysis, I want to insist that the evidence reviewed above (as well as the additional evidence broached in chapter 2) shows the brain is *not* a nearly decomposable system consisting of separately modifiable parts. Insofar as massive modularity rests on this feature of the brain, it would appear to be incompatible with the emerging facts. In the brain, function depends much more on the *interactions between* parts than on the *actions of* parts (Anderson et al. 2012; Kelso 2009; Sporns 2011). This crucial idea, and the reasons it must move us away from the component-dominated thinking of modularity in particular and typical CCTM models in general, comprise a main subject of this book.

But for now we need a candidate explanation for how heritable, species-typical neural assemblies can emerge in a brain that does *not* appear to be composed of separately modifiable subsystems that can be differentially targeted by selection pressures. In fact we will not be in a position to fully appreciate the solution until the end of chapter 2. But the key comes from seeing how evolutionary and developmental processes can work together. In short: stereotyped, genetically driven early neural projections, combined with conservation in the sensory properties of the environment and the set of tasks needing to be mastered, may be sufficient to induce species-typical functional biases in the brain, which would then be put together in consistent functional partnerships, without the need for these to be directly genetically encoded.

This brings us to the second large objection to neural reuse, that nonmodular systems would suffer from disabling degrees of interference and processing inefficiency. Here the key insight is to remember that the *environment* has features that organisms routinely exploit—for example, by developing perceptual and learning strategies that reflect some of the environment's structural invariants (Gigerenzer, Todd, & The ABC Research Group 1999; Gilovitch, Griffin, & Kahneman 2002). (It is perhaps worth flagging the fact that one of the recurring themes of this volume is that properties of the environment often play a crucial role in explaining facts about organisms, whether about their evolution, development, cognition, or behavior; for instance, explanations of the evolvability of brains need not reference only brain properties.) One such useful feature of most environments is that they do not pose all their problems all at once—inclement weather rarely comes along with predator abundance, pressing mating opportunities, and food shortages, for instance. And often when there *are* competing opportunities or challenges, there will be a clear priority (chapter 6 is largely dedicated to a discussion of the mechanisms of biased pattern competition that resolve such conflicts in the brain). Thus, an organism with massively redeployed brain regions can generally rely on the temporal structure of events in its environment to minimize interference. Were this environment-organism relationship different—or if it were to change—then neural reuse *does* predict that increased interference will be one likely result.

Interestingly, contemporary humans encounter just such a changed organism-environment relationship in at least two arenas, and the effect of reused regions can often be seen as a result—first, in the labs of some cognitive scientists who carefully engineer their experiments to exploit cognitive interference of various sorts (Glenberg & Kaschak 2002; Glenberg et al. 2008a); and, second, at the controls of sophisticated machinery, where the overwhelming attentional demands have been observed to cause massive processing bottlenecks, often with dangerous or even deadly results (Fries 2006; Hopkin 1995). It is no coincidence that, in addition to designing better human-machine interfaces, one important way of minimizing the problems caused by processing bottlenecks is to engineer the *environment* (Rasmussen & Vicente 1989), including, especially, changing its task configuration and social structure, for instance by designing more efficient teams (Hutchins 1995).

So the evidence seems to weigh heavily *against* the massively modular architecture posited by evolutionary psychology. But as the reader is very likely aware, there have been numerous empirical analyses of both anatomical and functional connectivity in the brain that show that the brain is in fact composed of numerous interacting modules (Achard et al. 2006; Buckner et al. 2009; Fox et al. 2005; Meunier et al. 2009; Sporns 2011; Valencia et al. 2009). Are these findings compatible with neural reuse? Yes, because in these studies the term "module" has an entirely different meaning. As noted briefly above, connectivity in the brain is typically represented as a graph—sets of nodes (neurons or brain regions) connected by edges indicating anatomical or functional connections between the nodes—and analyzed using graph theory. In graph theory the term module is used to refer to a set of nodes that are highly connected to one another and less connected with other parts of the graph. This is a *structural* rather than a *functional* characterization—even when one is analyzing a functional connectivity graph! That is, graph-theoretical modules are defined in terms of features of the abstract topology of the graph. Such modules have well-defined abstract structural characteristics but flexible functional characteristics—and the primary functional upshot of these studies is that brain function emerges from the dynamic interactions between the modules and *not* from the mere activation of them (see Sporns 2011 for extensive discussion). Moreover, the evidence suggests that the modular structure is *itself* dynamic (Cole et al. 2013; Honey et al. 2007). That is, the brain is continually, dynamically organizing into different functional coalitions; each brain region cooperates now with this set of partners and later with that set. This is *exactly* what neural reuse predicts, but it sits very uneasily with the massive modularity hypothesis. So as tidy as it would be for neuroscience if the modules in neural coactivation graphs identified brain structures with the functional features of mental modules, that is not the way the brain is organized.

In summary, the empirical evidence points to non(massively)modular, interaction-dominant brains in which neural resources are used and reused in varying patterns across a variety of circumstances. Moreover, there do not appear to be deeply compelling reasons why such brains could not evolve or serve the functional requirements of surviving in a difficult and complex environment. For the remainder of this book, then, I take these facts to have been established and turn to their implications. Chapter 2 addresses the developmental mechanisms that drive neural reuse and reliably produce the adult functional architecture that we typically observe. Chapter 3 takes up the issue of function-structure mapping in the brain by reviewing some related work on cognitive architectures and reflecting on how we need to rethink the very notion of function-structure relationships in light of widespread reuse. Finally, rounding out part I, chapter 4 takes up the issue of how to reform cognitive neuroscience itself, both its empirical methods and its interpretive practices.

Interlude 1 On the Importance of Neural Teamwork

Since the original invention of functional magnetic resonance imaging (fMRI) in the early 1990s, both the field of neuroscience and the public at large have become nearly obsessed with the technique. The result has been literally tens of thousands of functional brain-imaging studies investigating the neural underpinnings of everything from addition to addiction. Indeed, the beautiful pictures that fMRI produces have become so ubiquitous in both the public and the scientific imagination that we might fairly call the last 20 years of cognitive neuroscience "the age of the image."

That was then; this is now. Although we should not expect to see any less brain imaging in the coming years, we are undergoing a sea change in the way those images are being interpreted. Instead of reading these images individually, researchers over the past few years have begun pioneering efforts to interpret these images *collectively*. Because no human being can hope to assimilate and make visual sense out of thousands of brain images, investigating all manner of psychological phenomena, these researchers use various computational methods to analyze and find hidden patterns in all of those data.

This movement—recently dubbed "Cognitive Neuroscience 2.0" by Tal Yarkoni and colleagues (2010)—promises to radically alter our understanding both of the brain and of brain science. For a simple example of the sort of surprising insight that collective data interpretation can produce, consider the principle of selectivity. A guiding ideal of brain science for at least 50 years, selectivity is the notion that individual neurons, as well as larger networks, respond to only a narrow class of stimuli—straight lines but not curved ones; faces but not houses; nouns but not verbs. The principle of selectivity is what is behind the popular understanding of a brain composed of neural specialists clustered together rather like ethnic neighborhoods in Boston—vision in the Back Bay, language up on the North End, executive control somewhere in Southie. Certainly individual fMRI studies can reinforce this impression; for any given investigation, the brain will "light up" in only a few places, apparently highlighting the specialists responsible for the task under investigation.

But when you look at brain activity across many, many such individual studies, things do not really look that way. One early study (Anderson 2007a) investigated 135 experiments in four different cognitive domains—language, vision, attention, and memory—and colored the regions that were activated by tasks in each using standard four-color printing techniques, such that the amount of ink from each color added to a region was determined by the frequency with which it was activated by tasks in each category (figure 1a.1). Instead of seeing large regions of the brain painted in cyan or magenta or yellow or black, which would indicate dedication to tasks in a single domain, each brain region took on its own mixed hue, reflecting its contribution to many different tasks across the four domains. As detailed in chapter 1, this finding can be confirmed in various ways. Bottom line: the neighborhoods of the brain are highly integrated and functionally diverse.

This probably comes as a surprise to many readers. But when you think about it from an evolutionary perspective, it makes some sense. If the brain were a collection of regionally segregated specialists, then that would mean that new cognitive abilities would emerge only via the development of new dedicated brain tissue, like adding a new specialized organ or appendage. That is one possible pathway, of course, but evolution is also known to repurpose existing resources to meet emerging challenges. If that is the way the brain evolved—and it does seem to be a more efficient use of metabolically expensive brain matter—then what we would expect to see is what we indeed see: regions used and reused for a variety of purposes in different circumstances.

One important upshot of this is that there typically is not a "brain area for X." The brain does not operate by differentially activating one or a few local regions to achieve some task. Instead, the brain dynamically assembles different coalitions of partners. Achieving a task is *not* a matter of finding a single neural specialist but is rather about putting together the right neural team for the job.

"But," I hear you ask, "What about brain injury patients?" Neuropsychology in general, and the neuropsychological assessment of brain injury patients in particular, has long been associated with support for the localization of individual functions because when focal (i.e., highly localized) brain injuries appear to result in very specific cognitive deficits, it can be hard to resist the conclusion that the patient has damaged (and the scientist has

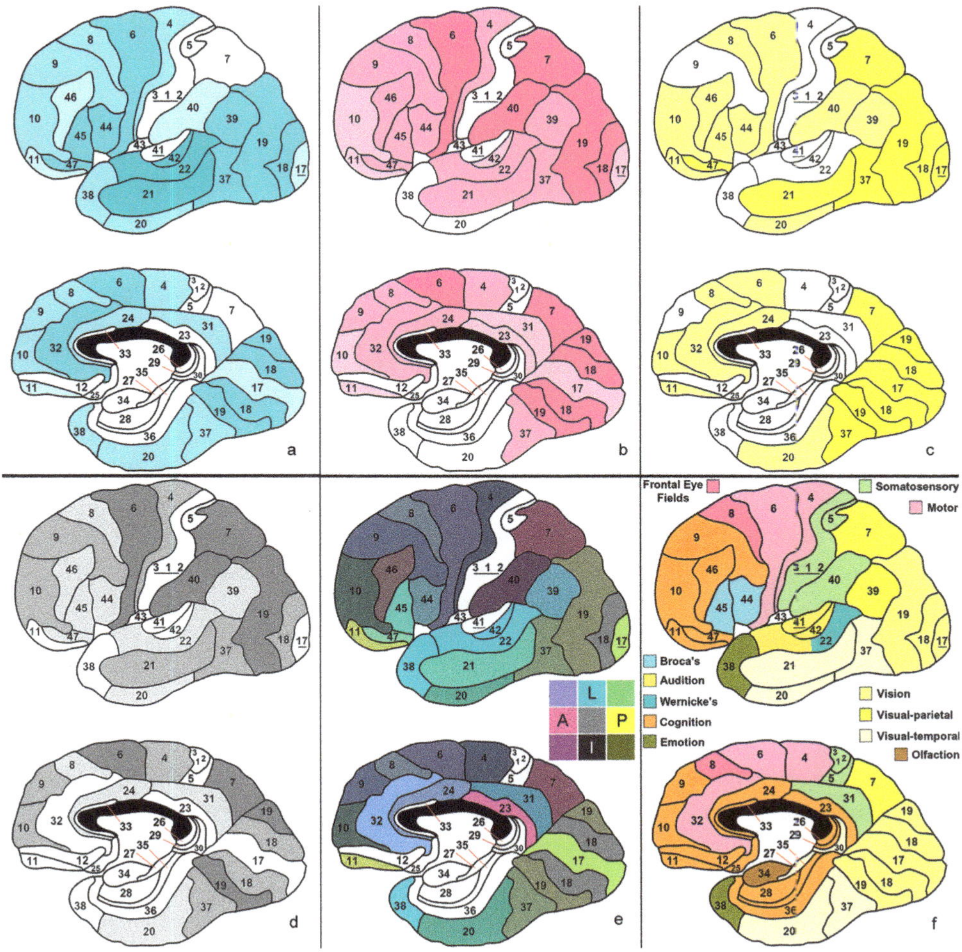

Figure 1a.1

Color-coded activations of left cortex. The figure illustrates the activations of Brodmann areas in the left hemisphere according to color and intensity, where color represents the cognitive domain and intensity the raw number of tasks in the domain activating the area. In this figure cyan represents language, magenta represents attention, yellow represents perception, and black represents imagery. Overlaying the single-color images (panels a–d) gives the four-color image in the bottom center (panel e). This image contrasts sharply with the common notion that cognitive domains are regionally localized, illustrated by panel f. Panel f image compiled by, and reprinted by permission of, Prof. Mark Wm. Dubin, MCD Biology, University of Colorado-Boulder.

thereby discovered) "the" brain region for the impaired aspect of cognition. Of course, one *should* resist, if for no other reason (and there are so many other reasons!) than the injury might merely have cut off communication between the brain regions *actually* responsible for the impaired ability.

But here again, thanks to the emergence of what might be called Neurology 2.0, the field (although not always the media) is beginning to put aside such oversimplified inferences, as is beautifully illustrated in an article by Dr. Aron Barbey and colleagues (2012). In this study of 182 patients with focal brain injuries, the authors use a simple but elegant method known as voxel-based lesion-symptom mapping (Bates et al. 2003) to identify the many regions of the brain that are causally related to general intelligence. The idea is straightforward: for each few millimeters of brain tissue, compare the test scores of patients *with* an injury there to the scores of patients *without* an injury there. If there is a significant difference in the average scores, that looks like pretty good evidence that those millimeters of brain tissue play some role supporting the abilities measured by the test.

Although the amount of information gained at each brain location is thus somewhat limited, the method nevertheless allows researchers to thoroughly map the causal network implicated in the cognitive functions of interest. In the current case the authors report finding an extensive, largely left-lateralized network of regions implicated in supporting general intelligence. The regions include those involved in supporting verbal reasoning, working memory, cognitive flexibility, and executive control, among other things. Their results specifically highlight importance of the white-matter communication pathways between these regions, suggesting that the communication between and the integration of information *from* these regions are at least as important to intelligence as the activities *of* these regions.

Of course, that should not be at all surprising, and it is true that the results are not necessarily earth-shattering by themselves. But the study is nevertheless important, because it represents a serious and successful attempt to do simultaneous justice to both functional differentiation and functional integration in the brain. We need more such work.

2 Interactive Differentiation and the Search for Neural Coalitions: Neural Reuse in the Functional Development of the Brain

In the first chapter I laid out some of the evidence that individual regions of the adult human brain are functionally diverse—used and reused in many different tasks across various cognitive domains. Moreover, we saw evidence that achieving functional specificity is a matter of assembling the right coalition of neural partners to accomplish the task in question. And finally, we reviewed some of the reasons that such a functional structure makes evolutionary sense, in terms of the efficient use of metabolically expensive and relatively scarce neural resources. But one question that was left largely unexplored related to the biological mechanisms that confer on the brain its ability to assemble the necessary functional partnerships to achieve diverse function. This is the question of *functional development*. In what follows I first outline one important and useful framework for understanding neural development, the *interactive specialization* framework (Johnson 2001, 2011), which shares with neural reuse an emphasis on the importance of interregional interaction to neurofunctional development. I argue, however, that although this framework offers a good starting point, it needs to be modified in light of the findings discussed in chapter 1 so as to place less emphasis on *specialization*; functional *differentiation* will turn out to be a better fit for the data. I then proceed to argue that, in addition to recognizing the importance of neural *interaction* to achieving local functional differentiation, we need to postulate the existence of a process of active searching for the neural partnerships that will support full adult function. Section 2.2 offers some reflections on the possible role of neural search in functional development, and it ends with a statement of the developmental framework I am advocating here: *interactive differentiation and search*. Section 2.3 offers some evidence for the existence of a neural search process, 2.4 speculates on the underlying biological mechanisms, and 2.5 reinterprets some well-established findings from the perspective of the interactive differentiation and search (IDS) framework.

2.1 From Interactive Specialization to Interactive Differentiation

Johnson elucidates the interactive specialization (IS) framework for understanding brain development by contrasting it with two competing approaches: the maturational viewpoint (Atkinson 1984; Kanwisher 2010), and the skill-learning perspective (Gauthier & Nelson 2001; Spencer, Thomas, & McClelland 2009). I follow his lead in my exposition, below.

The maturational viewpoint is closely allied with localization and modularity on the one hand and nativism on the other. According to this way of understanding functional development, behavioral change is largely the result of the maturation during normal development of individual brain regions, which naturally become increasingly specialized as they realize their "cortical fate" (the term comes from Pallas 2001, p. 418, although she would not be a supporter of the maturational viewpoint). In at least some cases this development proceeds according to a predetermined timetable, driven by genetics rather than by experience. For instance, Kanwisher (2010) argues that "genes may be largely responsible for wiring up the face system, with little or no role [for] experience with faces" (p. 5).

In contrast, the skill-learning perspective emphasizes the role of experience in neural development and thus foregrounds the possibility that different brain regions may support a given task at different stages in development, as experience shapes one region over another to better execute the task. For instance, it has been observed that the location of peak brain activity shifts from frontal to parietal areas as proficiency improves in a visuomotor sequence learning task (Sakai et al. 1998). In addition, Gauthier and colleagues (e.g., Gauthier et al. 1999) have shown that the ability to recognize individual objects and to categorize families of novel objects ("greebles") involves activation in the fusiform face area. Both of these findings appear to fit more closely with the notion that a region acquires its functional role as a result of experiential shaping and not solely as the result of genetic prespecification. The findings also suggest that brain regions may often support more than one specific function—not just over developmental time but at any given point in development.

Despite the clear contrast between these views, it must be emphasized that they come into direct conflict *only* when competing claims are made about the very same brain region—as in the debate among Kanwisher, Gauthier, and others over the function of the fusiform face area (Gauthier et al. 2000; Grill-Spector et al. 2006; Hanson & Schmidt 2011; Kanwisher et al. 1997; Rhodes et al. 2004)—because it could well be the case that different regions of the brain are relatively more or less specialized and relatively more or less driven by genetic versus experiential factors. Nevertheless, the tensions between regional specialization and diversity in functional architecture, and between nativism and empiricism in functional development, have shaped much of the debate over brain organization for the past few decades (Goldman-Rakic 1988a, 1988b; Kanwisher 2010; Pallas 2001; Schiller 1996; Sur, Garraghty, & Roe 1988; Sur, Pallas, & Roe 1990). The interactive specialization framework represents an explicit attempt to unify these various camps and, more importantly, to account for all of the data supporting each. For instance, in adjudicating the debate over the function of the fusiform face area, Johnson writes, "The IS framework potentially provides a midway account that involves both intrinsic and extrinsic factors, and predicts changes in both the degree of specialization and localization of face-evoked activity in cortex during development" (Johnson 2011, p. 12).

According to IS, we should understand functional development in terms of "changes in the response properties of cortical regions during ontogeny as regions interact and compete with each other to acquire their roles in new computational abilities" (Johnson 2001, p. 480). IS sides with skill learning on the importance of experience in development and (largely, although not entirely) with the maturational viewpoint on the issue of local specialization. In contrast to both, IS emphasizes the role of interregional interactions in shaping local function.

> [T]wo key ideas underlie this approach. The first is the notion that during postnatal development changes in the response properties of some cortical regions occur as they interact and compete with each other to acquire their role in new computational abilities. From this perspective, some cortical regions begin with poorly defined and broad functionality and consequently are partially activated in a wide range of different stimuli and task contexts. During development, activity-dependent interactions between regions sharpen up the functions and response properties of cortical regions such that their activity becomes restricted to a narrower set of circumstances (e.g., a region originally activated by a wide variety of visual objects may come to confine its response to upright human faces). In other words, some cortical regions become more specialized during development. This process mirrors that known to occur at a cellular level during development, in which receptive or activity fields become increasingly fine-grained during development (e.g., Wills et al. 2010). The second aspect to Interactive Specialization is that postnatal functional brain development, at least within cerebral cortex, involves a process of organizing patterns of interregional interactions. According to this view, the response properties of a specific cortical region are partly determined by its patterns of connectivity to other regions and their patterns of activity. The onset of new behavioral competencies during infancy will therefore be associated with changes in activity over several regions

(networks) and not just by the onset of activity in one or more additional region(s). (Johnson 2011, pp. 9–10)

Clearly, IS offers a sympathetic framework within which to specify the developmental processes that result in the adult functional architecture described in chapter 1. Most crucially, the recognition that the "process of organizing interregional interactions" is central to functional brain development fits in perfectly with the view of mature brain function urged by neural reuse. However, to be fully compatible with neural reuse, the IS framework requires two important modifications: first, the notion of regional specialization, largely inherited from the maturational viewpoint, must be reconsidered; and second, a more fleshed-out explanation for "why and how ... particular anatomically distant brain regions begin to cooperate in a functional network" (Johnson 2011, p. 16) must be provided. Although largely adopting the IS framework for understanding functional brain development, this chapter aims to address these two issues.

The easiest way to see the difficulty with the notion of specialization adopted by the IS framework is to recall that, at least according to the evidence presented in chapter 1, most regions of the mature, adult brain simply do not appear to realize that sort of specificity. Instead, local neural assemblies persist in supporting many tasks across ostensibly quite different task categories (such as attention, semantics, emotion, and working memory) and thus retain a complex response profile quite unlike that expected by the maturational viewpoint. Note this does *not* mean that regions of the cortex—much less subcortical structures—remain functionally undifferentiated. Although proponents of functional specialization often portray the choice as being between a brain composed of a collection of functionally specialized regions and "a bowl of porridge" (Kandel 2006), in fact we need to distinguish between *differentiation* and *specialization*. For given the apparent functional organization of the mature brain, brain development appears to be a process of functional differentiation that need not result in functional specialization.

In light of these findings I believe it would be better for a satisfactory developmental framework to characterize the matter in the following way: over the course of development, local neural assemblies will come to have particular, distinctive response profiles, as determined by a combination of intrinsic local cortical biases and extrinsic factors including experience and the influence of functional interactions with other regions of the brain. A region's response profile will certainly reflect its underlying functional capacities and determine the role(s) it can play in various functional coalitions. But although this might therefore be considered a kind of functional selectivity, it is quite different from the notion that brain regions will come to specialize in such tasks as "face perception" or "mind reading" and furthermore suggests a developmental pathway relatively unconstrained by (or simply in some sense insensitive to) the traditional categories of cognitive psychology. (These issues are treated in much more detail in chapters 3 and 4.) Besides, as Johnson (2011) points out, naming the specific function that a region performs (or even supposing it *has* some single specific function) involves a kind of abduction that is inherently underconstrained and uncertain. It would be best, then, for IS to replace the notion of functional specialization with that of functional differentiation. Although as currently articulated IS clearly revolves around the notion of functional specialization, it does not follow that IS must be *committed* to it, and Johnson explicitly leaves other possibilities open, noting, for instance, that individual neural regions may come to participate in "one *or more* functional networks" (Johnson 2011, p. 12, emphasis mine). We can call this modified posit *interactive differentiation.*

2.2 The Role of "Search" in Functional Development

Another illustration of the importance of adopting interactive differentiation over the pure IS framework comes from considering the case of sensory substitution. This also leads us—eventually, after a good deal of spadework—to the matter of the specific developmental mechanisms that may underlie the formation of functional networks. Sensory substitution refers to a family of phenomena wherein one takes information in a form suited to one sensory modality (e.g., light waves for vision) and transforms it into a form suitable to another sensory modality (e.g., tactile stimulation). Introduced by Bach-y-Rita and colleagues in the late 1960s (Bach-y-Rita et al. 1969), the first application of the principle was in a visual aid to the blind. The system consisted of a television camera connected to a large 20 × 20 array of 1-mm tactile stimulators placed 12 mm apart on the subject's lower back (the device was attached to the back of a dental chair; see figure 2.1, left).

The camera would generate a 20 × 20, 400-pixel black-and-white image (see figure 2.1, right), sent both to a monitor and to the tactile array, each element of which would vibrate with an intensity corresponding to the light falling on that pixel. After training with this device, blind subjects were able to identify block letters and common three-dimensional objects such as a teacup and a telephone and discriminate the angles of lines and the tilt of a checkerboard with very high accuracy (in some cases indistinguishable

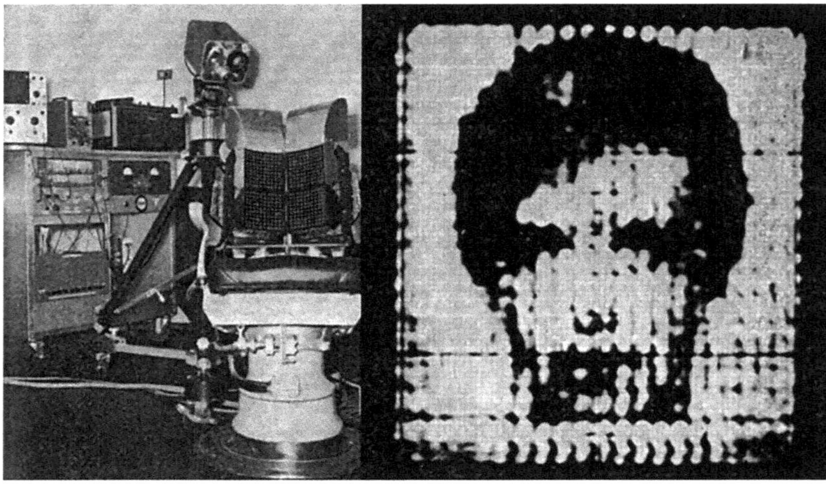

Figure 2.1
The sensory substitution device (left) and an example of the images it produces (right).

from the performance of sighted subjects looking at the monitor) and relatively low latency (on the order of seconds). The subjects spontaneously reported that the sensory information seemed to be coming from in front of them—that is, from the location of the objects—and not from the tactile array on their back, indicating that "the vision substitution system seems to become an extension of the sensory apparatus" (Bach-y-Rita et al. 1969, p. 964).

Follow-up work has shown that it is possible to "see" with auditory as well as tactile stimulation and, more interesting for our present purposes, that in doing so subjects are not using just "auditory" or "somatosensory" regions of the brain but also "visual" areas (Bach-y-Rita & Kercel 2003). This capacity of regions of the brain to become involved in processing nonstandard stimuli—called crossmodal plasticity—has been extensively studied using sensory substitution devices as well as with simpler paradigms including tactile discrimination tasks such as Braille reading. For instance, it has been shown that the visual cortex of early-blind subjects is activated by Braille reading (Sadato et al. 1996) and that the application of repetitive transcranial magnetic stimulation (rTMS) to disrupt processing in the activated visual areas disrupts Braille character recognition (Cohen at al. 1997; Kupers et al. 2007). This indicates that the activity in the occipital cortex is functionally relevant to the discrimination tasks and is not simply an artifact of neural activation spreading to an underused area. More recent work with sighted patients temporarily deprived of visual input demonstrates the same effects. It is worth reviewing this evidence in some detail.

Merabet and colleagues (2008) subjected sighted participants to 5 days of complete visual deprivation. During visual deprivation the participants underwent intensive training in Braille character recognition and showed significant increases in performance over the training period. Early in the training process participants exhibited a general increase in cortical excitability (Pitskel et al. 2007) and in some cases reported instances of visual hallucinations (Merabet et al. 2004). Blindfolded participants showed greater activation of occipital cortex during Braille reading when compared with nonblindfolded controls, and rTMS applied over the activated occipital areas disrupted performance on the Braille character discrimination task. After the removal of the blindfold, however, this difference between groups disappeared. That is, the previously blindfolded participants no longer showed significantly increased occipital activation, nor did rTMS disrupt character discrimination. Interestingly, task performance did not significantly change after blindfold removal.

Overall, this pattern of results does indeed seem to show "relatively rapid crossmodal recruitment of occipital cortex to process tactile information" (Merabet et al. 2008, p. 8). More striking is the apparent reversibility of the recruitment, with no accompanying reversal of the learning gains. How are such results to be explained? What functional mechanisms could be involved?

Pascual-Leone and Hamilton (2001) suggest that the body of evidence points to a "metamodal" functional organization for the brain. Although the brain generally *appears* to be modally organized—having different regions dedicated to processing visual, tactile, olfactory, and auditory stimuli—this misleading appearance is the result of a mechanism of functional suppression that engages in the course of normal development. Interestingly, the developmental process that Pascual-Leone and Hamilton outline is quite similar in spirit to the interactive specialization framework:

> Could it be that in fact, in normal development, multimodal sensory inputs feed into all cortical regions and that external influences and predefined functional superiority of given cortical areas shape the brain? Could it be the case that the brain is actually metamodal and that the impression of hierarchically structured unimodal systems is only the consequence of selection of functions by groups of neural networks that compete with each other in order to acquire specific processes? One might envision a cortical region with a functional role in special discrimination that might be predisposed to perform the kinds of processes that vision requires. In this

> setting, from early development onward, sight would be progressively selected as the input signal for such an "operator." Eventually, such an operator might appear to be "visual cortex" by virtue of its dominant input, when in fact, under certain conditions, the presence of metamodal inputs could be unmasked. (Pascual-Leone & Hamilton 2001, p. 432)

I say it is similar in spirit to the IS framework because it imagines a set of specialized "operators" refined over developmental time to preferentially respond to a select set of inputs and eventually achieving a relatively stable, mature functional profile. The hypothesis holds that there is, however, some hidden complexity underneath the surface that reveals itself in the case of injury, sensory deprivation, or extraordinary task demands that disrupt the status quo and "unmask" the outcompeted, nonselected inputs. The basic idea of unmasking is fairly simple: local regions may be inherently capable of dealing with more than one sort of input (hence the idea of "multi-" or "metamodal" operators), but during the course of development one sort of input comes to dominate and drive the work of the region. Thus, it is only when that source of dominant input is disrupted in some way that the underlying complexity is revealed ("unmasked") and the region can go to work on the nonstandard inputs. Now, insofar as this hypothesis *is* similar to IS, it will need similar modifications in light of neural reuse. In particular, given the evidence (of which there was but a glimmer in 2001; e.g., Cabeza & Nyberg 2000) for significant complexity in response profiles even in the mature brain, we do not need to posit the significant suppression of "nonstandard" inputs, which will become "unmasked" only in extraordinary circumstances. That is, there appears to be ample evidence for multi- (or meta) modality even under more typical experimental conditions such as those employed in fMRI studies (more on this in the paragraph, below). In addition, although the notion of a specialized "operator" is of course a natural one, we need to remember that, strictly speaking, the evidence points only to a sharpening and differentiation of response profiles over developmental time. Although such evidence is *consistent* with the existence of a single refined operator for each region of the brain, it is also consistent with the looser notion of a refined, functionally relevant cortical bias. In any case, I think it is clear that Pascual-Leone's notion of the metamodal organization of the brain is, suitably amended, completely consistent with the notion of interactive differentiation being sketched here and offers an adequate framework for understanding the kinds of crossmodal plasticity under consideration.

This being said, the notion of cortical "unmasking" needs some further examination, for although both Pascual-Leone (Pascual-Leone & Hamilton

2001) and Bach-y-Rita (1995) suggest it may be an important part of the developmental mechanism, I have two related worries about its misleading implications. First, although the disruption of normal input that characterizes cortical "unmasking" appears to *facilitate* crossmodal plasticity, it does not appear *necessary*. In addition to the general evidence reviewed in chapter 1 for the recruitment of neural regions by multiple diverse tasks, there is also evidence especially relevant to the current case. For instance, the occipital cortex appears to make a functional contribution to certain tactile discrimination tasks such as determining the orientation of tactile gratings (Zangaladze et al. 1999) and judging the distance between Braille dots (Merabet et al. 2004), even in sighted, nonblindfolded participants. Moreover, Amedi and colleagues (2002) observed neural activation in the occipital cortex of sighted participants for objects both when seen and when touched (see also Pietrini et al. 2004). So although unmasking might make the acquisition of new functions easier, perhaps by (temporarily) relieving certain regions of their normal functional pressures and thus making them more available for alternate uses, the process of establishing functional networks for new tasks can apparently persist without it.

Second, and more important, the metaphor of "unmasking" suggests a degree of passivity in the developmental process that does not appear to fit the evidence. In particular, the notion that the function will come to occupy the region of the cortex passively "offered up" by the sensory deprivation cannot explain the fact that in the study by Merabet et al. (2008), when the blindfold was removed, the occipital cortex ceased to play a functional role: the experimenters no longer observed occipital activation during the task, nor did rTMS disrupt performance. Yet there was *no loss of behavioral competence* on the part of the participants. Merabet and colleagues suggest that this may be because the disruptive effect of rTMS in blindfolded participants was nonlocal, and once the relevant pathways are once again masked by visual input, this effect is no longer seen:

> That is, the effect of TMS disruption, while delivered to occipital cortex, may be exerted at a distance along cortical networks that have undergone functional changes in connectivity due to visual deprivation itself. Such a hypothesis would provide a sensible explanation to the lack of impact of rTMS on tactile Braille character discrimination on day 6, when occipital cortex is no longer significantly activated by tactile stimuli on fMRI, but tactile Braille character recognition remains better than at baseline, presumably supported by activity at brain regions other than the occipital cortex. (Merabet et al. 2008, p. 8)

This is possible, of course, but the explanation undermines the claim that occipital cortex was *ever* functionally relevant to the task; it suggests that the occipital cortex merely passively carried the disruptive signal to the site of true functional relevance.

Given the depth and variety of evidence for the functional relevance of occipital cortex not just for tactile discrimination but also for other nonvisual tasks, this does not strike me as probable. And in any case it still leaves unexplained how it is that a different, *not* unmasked region of the brain would come to play a functional role in the Braille reading task. It appears that the explanation that best fits *all* of the evidence is that alternate neural partnerships were automatically established as part of the original learning. In the neurally "unmasked"—that is, blindfolded—condition, visual cortex was relatively more available to do the relevant processing, but backup processors, identified during functional development of the new skill, were ready to be brought online.

All of this suggests to me that the mechanisms of interactive differentiation underlying functional development (early as well as late skill acquisition) must also include a process of active search: the rapid testing of multiple neural partnerships to identify functionally adequate options. Although conditions might favor one neural option over another (and in some cases there might *be* only one viable option), as conditions change, different neural coalitions can be called on to take up the burden of function. Let us call this hypothesis about the course and mechanisms of functional brain development the *interactive differentiation and search (IDS) framework;* it is meant to incorporate the interactivity of IS and the underlying complexity of local neural function posited by the metamodal framework and to combine these with the notion of neural search that appears necessary for achieving adult functional architectures. (Table 2.1 offers a brief summary of the maturational viewpoint, IS, and IDS as a handy reference.) Obviously, this is highly speculative, and establishing the existence

Table 2.1
Brief schematic comparison of three views of development

Viewpoint	Type of local function	Mechanism of development
Maturational viewpoint	Single mental operation	Local Hebbian plasticity
Interactive specialization	Single mental operation	Local Hebbian plasticity + activity-dependent interactions between regions
Interactive differentiation and search	Complex functional bias	Local Hebbian plasticity + activity-dependent interactions + neural search

of some such process will take significant research effort, but it is far from clear to me how a more passive process (such as the "unmasking" of latent inputs) could account for both the variety of things that can be learned (what manner and degree of latent connections would need to remain in place?) and the apparent ability (at least in some cases) to rapidly shift the functionally relevant neural bases for a given skill. Moreover, as discussed in chapter 1, there does seem to be evidence that would lead one to suspect (1) that there is typically more than one neural option for supporting a given behavior ("degeneracy"; Sporns 2011) and (2) that functional neural networks are typically structurally overconnected and possessed of sets of both "available" and "latent" uses that can be triggered by circumstance (Bargmann 2012). Together these neurofunctional principles observed in mature brains seem able to adequately account for the observations being discussed; what remains is to suggest a mechanism for how the various options are discovered in the first place.

But before we discuss search in detail, it is worth pausing to offer a couple of crude but hopefully somewhat helpful analogies for how overall function might be supported according to each of these three frameworks. Imagine a large family of siblings who decide to learn to cook. According to the maturational approach, each sibling decides in advance what he or she will specialize in—pastry, sauté, grill, salad, and so forth. Each individually attends a rigorous and regimented cooking school, acquires the necessary skills, and meals in this household are the result of each person separately applying his or her skill and making a contribution to the table.

In contrast, the interactive specialization approach would be for the siblings to learn together, and, depending on the nature of their interactions and observations about who was best at learning each skill, each would eventually come to specialize in a different single kitchen skill. With this approach, there is room for complex collaborations to emerge, such that although it will *sometimes* be the case that an individual dish can be traced directly to a particular sibling, there will also be instances in which the cooking function is joint and interactive. For instance, they may discover while making a mayonnaise or similar sauce that it works best when one sibling holds the pot and whisks and another slowly pours in the oil. The right rate and amount of both whisking and pouring will have been learned over time based on the interactions between the siblings involved and the *outcome* of that interaction as it is iterated over multiple meals. Here although each sibling exercised a particular skill (specialty), the outcome depends on the interactions between those skills, and one might say that the function of mayonnaise making is in fact distributed over the pair.

Finally, the IDS approach imagines a similarly collaborative learning environment where each sibling trains with the others and acquires a set of skills determined both by his or her own intrinsic aptitude and preferences as well as by the aptitude and preferences of the others, until they come as a group to have both diversity and a degree of redundancy of skill. As in the IS example, many kitchen functions will be themselves interactive and distributed across siblings. But in this family, each sibling can do multiple things and adopts a role depending on the nature of the meal and the roles adapted for that meal by the others; if someone is already chopping, someone else will sauté. Thus, when this family is learning a new dish, there is an interactive "search" of the person-role configuration space as each sibling chooses a role and enters into the collaborations that will lead to the best meal.

Of course cooking skills like these are easy to identify and localize in individuals and so retain some of the feel of low-level components that I am hoping we can get beyond. Perhaps, then, a more abstract analogy is preferable. Consider the following from Lashley (1951):

> I can best illustrate this conception of nervous action by picturing the brain as the surface of a lake. The prevailing breeze carries small ripples in its direction, the basic polarity of the system. Varying gusts set up crossing systems of waves, which do not destroy the first ripples, but modify their form, a second level in the system of space coordinates. A tossing log with its own period of submersion sends out periodic bursts of ripples, a temporal rhythm. The bow wake of a speeding boat momentarily sweeps over the surface, seems to obliterate the smaller waves yet leaves them unchanged by its passing, the transient effect of a strong stimulus. Wave motion is not an adequate analogy, because the medium which conveys the waves is uniform, whereas the nerve cells have their individual characteristics of transmission which at every point may alter the character of the transmitted pattern. (Lashley 1951, p. 133)

Let us follow Lashley's lead here, taking his caveat to heart, and suppose it is the basic job of some mechanism to transport objects to and from various locations on a lake by placing bubblers on the bottom that create ripples and thus currents at the surface. According to the maturational viewpoint we should expect that the location and the precise output of each bubbler have been specified in advance, and as each bubbler is placed in its correct location, the lake comes gradually to have the prespecified set of local currents that will enact the required transportation routes.

In contrast, the IS framework supposes that neither the precise location nor the exact output of the bubblers need be specified in advance. Rather, as each bubbler is placed and tuned, this changes the currents on the surface, leading to adjustments in neighboring bubblers to maintain needed currents and enact new ones. Over time the local arrangement and tuning comes to be fixed, leading to a fixed set of local surface currents.

Finally, the IDS framework follows IS in supposing that the placement and tuning of the bubblers is driven by a combination of the interactions between the surface currents and the functional needs of the system. But it allows for the possibility that, although the possible outputs of each bubbler will be refined and limited over time, each bubbler might still maintain multiple possible output settings. This additional flexibility allows for there to be two different approaches to learning a new transportation route. First, bubblers can be added, and the output of existing bubblers refined, to create the right currents, as with IS. But also the bubblers can rapidly adopt one of their other modes to search for an existing configuration of bubbler settings that will create the required set of surface currents. For IDS the number and nature of the local currents are limited but not fixed; the network of bubblers will have multiple possible configurations to enact different overall transportation routes; and each individual bubbler will participate in enacting multiple configurations. Note that in this analogy it is the case that the functionally relevant effects of the bubblers—the currents and the objects they cause to move—are the result of the *interactions* among the parts of the mechanism; this is a simple illustration of a distributed interactive mechanism.

2.3 Initial Evidence for a Search Mechanism in Functional Development

The neural search process envisioned by IDS reminds me of an interesting and evocative thought experiment:

> First imagine a marvelous technology: an array of flying laser scanners that can measure the trajectories of all the hailstones in a storm. ... What would anyone do with this data? As luck would have it, there's a wonderfully geeky store in this thought experiment called the Ultimate Computer Store, which sells a great many designs of computers. In fact, every possible computer design that has fewer than some really large number of logic gates is kept in stock. (Lanier 2011, p. 41)

Lanier goes on to imagine visiting the Ultimate Computer Store with the hailstorm data and feeding those data to each computer, not as the information to be analyzed by some program on the computer, but *as the program itself*. Most of the time, the computer will do nothing at all in response to the program, or only something boring like crash or flicker the screen. But once in a while the program and the machine architecture mesh, and something really extraordinary happens.

After a while, you end up with a few million word processors, some amazing video games, and some tax preparation software—all the same program, as it runs on different computer designs. (Lanier 2011, p. 41)

In place of the Ultimate Computer Store, think the human brain. And in place of hailstorm data, substitute the sensory input associated with learning some new task. I think it is useful to imagine that learning is a process of running sensory input—which, unlike the hailstorm data, will of course change over time as behavior changes—on various neural coalitions until we find the partnership(s) that give us the behavior we want. I like this twist on the computer analogy of the brain in part because—whereas CCTM normally imagines sensory input in intentional, representational terms containing information about what is "out there" that we have to extract and appropriately transform; that is, it treats sensation as the data on which we run the programs stored in the brain (see chapter 5 for discussion)—this thought experiment instead casts sensory input as a *series of instructions* to which the regions and networks of the brain will differentially respond. The challenge of learning, then, is to find (and fine-tune) the coalition of neural partners that have the right response tendencies—the neural architecture that has the right mesh with the input—and will therefore do something useful or adaptive in light of the sensory "instructions." As I develop in much greater detail in chapter 5, I think this reflects the reality of our situation much more closely than does classic CCTM.

Note the assumption that neural reuse is arranged such that different inputs can be fed *without translation* to the same neural assemblies and still produce useful outputs. This is in fact the case for many computer programs; for instance, sorting algorithms can often just as easily sort letters as numbers; and if you feed a given algorithm pictures instead, it will do *something* with them. We have already seen specific examples of the positive functional outcomes that result from feeding, for example, tactile information to "visual" regions. Indeed, chapter 1 reviewed a great deal of general evidence that points to the same conclusion: the local neural assemblies of the brain are both metamodal and metasemantic and work with inputs of various forms and contents. Thus, it is a central tenet of the IDS framework that a search of the functional possibility underlies much of functional neurodevelopment—a process of testing multiple functional pairings and partnerships to discover what coalitions will actually produce the desired outcomes. Naturally, this raises some pressing questions that seem ready-made for an enterprising theorist of neural computation. Under what conditions might useful things be done by local assemblies working with nonstandard inputs? What kinds of implementations increase the chances of functionally beneficial outcomes given the fact of reuse?

But for now our focus is on whether and how such search could possibly be implemented in the brain. How, exactly, can the brain pass around inputs to different possible coalitions? What we appear to need is a mechanism that can test various possibilities for functional partnerships and somehow reinforce or store the promising ones without being entirely constrained by existing connections. This last point is crucial, for unless the search mechanism can somehow get beyond already established partnerships, then the potential repertoire of new behavioral competencies would be relatively constricted. Is there any evidence for such a mechanism? So far as I know, no one has yet looked specifically (although I hope that this volume will motivate work in this area), primarily because the overwhelming focus in neuroscience has been on the processes that shape local circuits—exemplified (for instance) by the debates between the maturational viewpoint and skill-learning framework. It is not until one recognizes the importance of neural reuse and of the participation of local circuits in multiple functional coalitions that it becomes clear that we need to identify the mechanisms responsible for actually establishing multiple overlapping coalitions.

But there are certainly some suggestive findings. Consider, for instance, that if there *is* some sort of search mechanism, then we would expect the learning process to be characterized by an initial increase in both the distribution and degree of brain activity as various neural partnerships were tested; that activity would then slowly subside as the proper functional partnerships were consolidated. That seems to be what Merabet et al. (2008) found when participants were learning Braille reading, and in fact, such a pattern of results is commonly reported in the literature on expertise: the neural supports for practiced skills tend to be more localized than those involved in learning novel tasks (see, e.g., Petersen et al. 1998; Petersson, Elfgren, & Ingvar 1997; Poldrack et al. 1998). This has generally been interpreted as signaling an initial increase in *effort*, followed by a decrease as the skill becomes more practiced. But within the IDS framework, this pattern looks instead like the sign of neural search.

A somewhat more specific piece of evidence comes from the literature on brain-machine interfaces for such devices as prosthetic limbs. In a typical invasive brain-machine interface (BMI), a microelectrode array is implanted into a portion of an animal's brain with some hypothesized chance of being able to incorporate the device (e.g., primary motor cortex for a prosthetic limb). Through a process of computer-assisted trial and error, the animal eventually learns to control the device, in some cases with fine enough control for such behaviors as self-feeding. Lebedev and Nicoleilis (2006) describe the neural effects of the learning process this way:

> [C]ontinuous BMI operations in primates lead to physiological changes in neuronal tuning, which include changes in preferred direction and direction tuning strength of neurons (Taylor et al. 2002; Carmena et al. 2003; Lebedev et al. 2005). In addition, broad changes in pairwise neuronal correlation can be detected after BMIs are switched to operate fully under brain-control mode (Carmena et al. 2003; Lebedev et al. 2005).
>
> Along with these physiological adaptations of neuronal firing patterns, behavioral performance improves as animals learn to operate BMIs effectively (Taylor et al. 2002; Carmena et al. 2003; Lebedev et al. 2005). Initial training to operate a BMI is characterized by an increase in neuronal firing rate variance, which cannot be simply explained by changes in limb or actuator movements (Zacksenhouse et al. 2007). As the quality of BMI control improves, initial elevation of neuronal firing variability subsides. Plastic changes in neuronal firing patterns during BMI control, leading to the physiological incorporation of the artificial actuator properties into neuronal space, could account for these changes in firing rate variance. This interpretation is in accord with the theory of optimal feedback control (Todorov & Jordan 2002; Scott 2004; Harris & Wolpert 1998). According to this theory, a motor system acts as a stochastic feedback controller that optimizes only those motor parameters that are necessary to achieve the goals of a particular task. During the brain-control mode of operation of a BMI, the goals of a motor task are achieved only by direct brain control of an artificial actuator. Thus, in terms of optimal feedback control theory, neuronal ensembles should adapt their physiological tuning properties to represent better the goal-related variables of the task performed by the BMI. (Lebedev & Nicoleilis 2006, p. 542)

Although I certainly agree that neural retuning is likely to be a crucial part of the process of learning to control a prosthetic limb with a brain implant, it is less clear that changes in tuning actually account for all of these observations. For consider, one effect of an "increase in neuronal firing rate variance" is that a given neuron *A* from the relevant population may begin by firing in synchrony with neuron *B*; but then, as their firing rates change, it will be firing instead in synchrony with neuron *C*, and so on. That is, this increase in variance actually implements a walk through neural coherence/cooperation space. Neurons will come to have many different synchronous partners over the course of this process, the end result of which is "broad changes in pairwise neuronal correlation." From the IDS perspective this walk through neural coherence space looks less like a process aimed solely at creating changes in neural tuning, and more like an initial search for, and the eventual consolidation of, a set of neural partnerships to manage the new task (at least in so far as oscillatory coherence between cells is a sign of functional cooperation). Such a process could of course be happening in parallel with neural retuning.

There are also some suggestive findings from the simulation literature. For instance, Atmanspacher and Filk (2006) describe an interesting non-monotonic relationship between the degree of randomness and the number of attractors in recurrent neural networks as they undergo training. Beginning with a fully randomized network subjected to training, they observe an initial sharp increase in the number of attractors in the system—that is, the number of different, stable patterns of activation that the network can adopt. As training proceeds, the number of attractors slowly decreases as the network achieves its optimal behavior. What is interesting about this in the current context is the nature of the trajectory that training induces through the neural possibility space. Rather than proceed via the slow shaping of a single attractor (a single stable behavior pattern) to achieve the desired end behavior, instead there is an initial flowering of rival possibilities for the time dynamics of the network, which are slowly selected out as training proceeds. Here again, from the IDS perspective this looks like a very particular kind of "search" of the possibility space: the sort of search that in biological networks could well show up as the initial increase not just in overall activity but also in the *variability* of the activity, followed by consolidation. This, of course, is what we actually observe.

2.4 Biological Mechanisms Underlying Neural Search

This brings us finally to the question of what *biological* mechanisms could support the hypothesized search for viable neural partnerships. As noted above, one might worry that any neural search would be unduly constrained by existing synaptic connections because these determine the pathways along which information can flow. But in fact neural relationships are more structurally and functionally complex than is generally appreciated. A recent review puts the matter succinctly:

> Classically, a single neuron receives multimodal information as local electrochemical signals through synaptic inputs converging onto its soma and dendrites. After integration, these confined events are translated into action potentials at the axon initial segment, which rapidly propagate to nerve terminals and trigger there quantal release of neurotransmitters. Recent electrophysiological and imaging data, however, dispute this rather naive view of neurons, highlighting unforeseen traits and dynamics of dendritic integration and communication mechanisms between nerve cells. It emerges that neurons, in addition to transferring signals through canonical

chemical synapses at the axon terminals, also interact via mixed electrochemical and electrical synapses at somatic and dendritic juxtapositions as well as through diffuse nonsynaptic volume transmission signaling (Agnati et al. 2010; Bennett & Zukin 2004; Fukuda 2007; Tamás et al. 2000). (Ovsepian & Dolly 2011, p. 19113)

Anatomically speaking, the classic synapse involves a close spatial relationship between a single axonal bouton and dendritic spine, such that the release of neurotransmitters from the bouton will have an effect largely localized to that single dendritic spine. However, there are many instances where a single bouton participates in more than one synapse as well as examples of spines and boutons without synapses. Developmentally, new spines can appear and disappear over relatively short timescales (tens of minutes), often triggered by synaptic activity. Several lines of evidence, such as the finding that dedritic spines are almost entirely replaced in neural plasticity following retinal lesions, suggest the central importance of spine remodeling in functional change. New dendritic spines preferentially form synapses with large axonal boutons already bearing a synapse. In this way neurons can come to participate in more than one functional circuit. It is not known exactly how cells regulate their participation in multiple networks, but mechanisms such as spine motility—the rapid (seconds to minutes) movement of dendritic spines—may play a role (see Holtmaat & Svoboda 2009 for a review). In addition, as reviewed in more detail in chapter 1, work with *C. elegans* and other animals suggests that neuromodulators such as the G-protein-coupled receptor *npr-1* permit the selection of a subset of functional synapses from among those anatomically available. Functionally, this allows a single anatomical network of cells to participate in diverse behaviors (Bargmann 2012).

These are just a few of the anatomical supports for functional plasticity and structural consolidation, including the formation of novel neural partnerships outside of existing connections, but these mechanisms do not describe how such partners might be initially selected, that is, activated in such a way that these structural changes are triggered. This is where volume transmission (Agnati et al. 2010), also known as nonsynaptic diffusion neurotransmission (NDN; Bach-y-Rita 1993) might play an important role.

Volume transmission refers to a class of neuromodulatory mechanisms that rely on the diffusion of molecules to cells not syaptically connected to the releasing cell. Classical wiring transmission (transmission between neurons "wired together" at the synapse) relies on the restricted synaptic space and diffusion barriers that keep the neurotransmitters relatively contained; diffusion is in this sense the enemy of reliable wiring transmission. Volume transmission, in contrast, takes *advantage* of diffusion and in particular of structural features of the brain that allow for greater diffusion in some directions than in others (called anisotropy).

Evidence for the importance of volume transmission is fairly extensive. For instance, extrasynaptic release of serotonin (5-HT) is known to modulate social behavior in leeches (Bisson, Bianconi, & Torre 2012, Hernández-Lemus 2012) and locusts (Anstey et al. 2009); the diffusion of nitric oxide (NO) modulates monoamine transporters, may act as a glutamate "extender," and is thought to be involved in long-term potentiation and long-term depression (Bach-y-Rita 1993; Kiss & Vizi 2001); and emerging evidence of the functional importance of glia in modulating intercellular communication in the brain (Fields 2010; Nedergaard, Ransom, & Goldman 2003) strongly suggests the importance of diffusion transmission because all glial chemical communication is extrasynaptic. Volume transmission is known to play a role in modulating the activity of existing wired networks (Agnati et al. 2010; Freeman 2005; see figure 2.2, for a schematic illustration) and is likely to play an important role in learning.

The effects of volume transmission depend in part on the shape of the extracellular space, which will determine the direction and extent of diffusion. Interestingly the anisotropy of diffusion—the differences between the extent of diffusion along the *x*, *y*, and *z*, spatial axes represented with the variable λ_{xyz}—is itself a dynamic property of the brain. For instance, in virgin rats the greatest anisotropy is seen relatively early in the diffusion process (around 60 seconds after initial release), whereas in lactating rats λ_{xyz} peaks much later, at around 120 seconds (Syková 2004). Moreover, brain injuries such as hypoxia, neurological disorders including Alzheimer's disease and MS, as well as normal aging are all known to decrease anisotropy in the brain.

In an especially striking set of experiments Syková and colleagues (2002) showed that the degree of reduction of anisotropy was related to loss of learning ability; superior learners (rats tested in a Morris water maze) had significantly greater hippocampal anisotropy than did inferior learners.

This appears to clearly establish the importance of directed diffusion to brain function in general and to functional plasticity in particular because in these experiments loss of behavioral plasticity appears at least partially explained by loss of differential diffusion.

What I want to suggest here is that volume transmission could also play an important role in executing the neural "search" for functional partnerships outside of anatomically consolidated (wired) networks. We know that during the recovery following a stroke, for instance, and also during the course of normal learning there can be a large increase in the number of

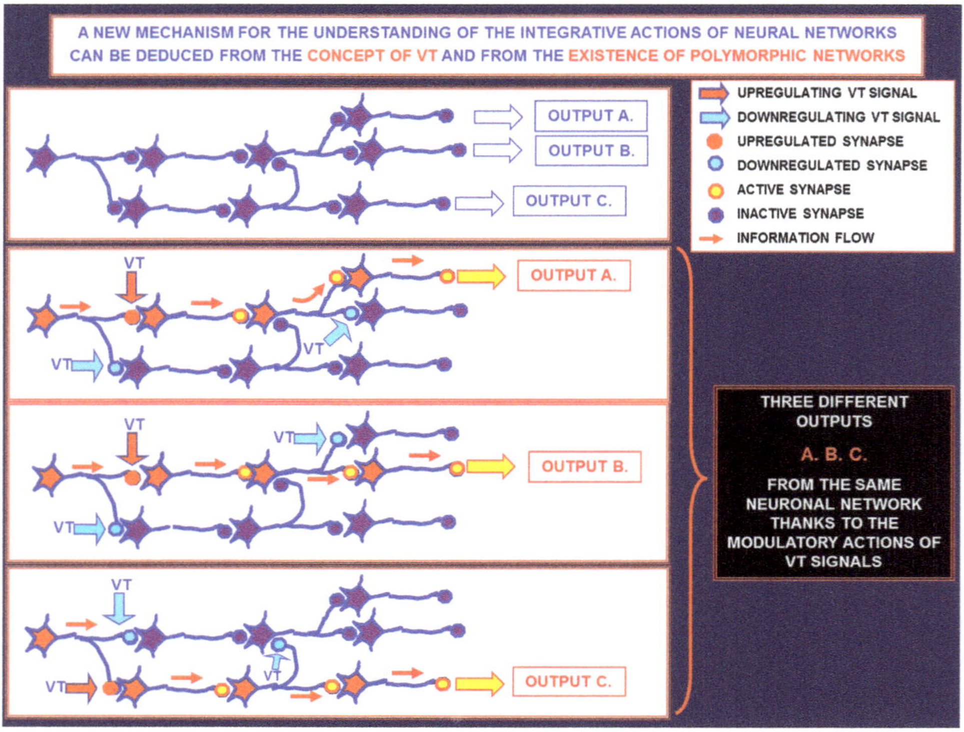

Figure 2.2
Changes in the mix of volume transmission (VT) and wiring transmission can also give rise to polymorphic networks that can give rise to different outputs according to the VT signals impinging on it. An example of a polymorphic network is illustrated for a neural network where the VT signals, by up-regulating or down-regulating synaptic contacts, can change the integrative action of the network. Thus, as shown in the schematic drawing, VT signals via modulatory actions on the synaptic contacts of the neural network can give rise to the different outputs. Figure and caption adapted, with permission, from Agnati et al. (2010).

neurotransmitter receptor sites (Bach-y-Rita 1993). The availability of additional receptors combined with the potential for extrasynaptic diffusion neurotransmission opens up the possibility that any given network or local circuit could cause activity in one or more neighboring circuits. When one considers in addition the fact that pyramidal cells, for instance, typically synapse with only about 1% of the cells within their dendritic arbor (Freeman 2005), and assuming that something like this ratio of wired cells to the total number within potential reach of diffusion mechanisms is roughly typical for the brain, the functional potential of volume transmission to activate alternate networks is clearly enormous. Any of the remaining 99% of a cell's neighbors might be potentially recruited to a new network to support a new skill. If this is right, then learning could be characterized by the formation of temporary coalitions of neural partnerships united by volume transmission, such that the most effective partnerships are eventually consolidated by anatomical mechanisms such as spine formation. Naturally, very many questions remain.

> This diversity of mechanisms and their effects adds complexity to the nervous system and raises many questions that still wait for answers. From the physiological point of view, one may ask how the neuronal firing pattern determines whether to evoke synaptic and/or extrasynaptic exocytosis in a given neuron; how activation of different synaptic inputs induces release from different release compartments and by doing so affect different targets (Velazquez-Ulloa et al., 2003). From the behavioral point of view it becomes highly interesting to explore how the timing of circuits and therefore behaviors are modulated by extrasynaptic transmission. Some psychiatric and neurological dysfunctions such as depression may be related to deficiencies in extrasynaptic transmission (Fuxe et al. 1991). ... Developmental and evolutionary biologists may also find the topic inspiring for future research [because] some of the exocytosis mechanisms involved in extrasynaptic transmission evolved before chemical synapses. For a similar reason, one may expect that during development, the extracellular levels of transmitters, modulators, and peptides may contribute to axonal pathfinding and electrogenesis before massive synaptogenesis takes place. (De-Miguel & Fuxe 2012, pp. 5–6)

In addition to these important questions, from the IDS perspective one interesting and crucial area for future research revolves around how much potential "functional coverage" each region of the brain has. That is, for any small functional unit of the brain—a single cell or a small wired network, for instance—there will be an imaginary region around that functional unit representing its potential diffusion cloud. The fibers that pass through that potential diffusion cloud give the functional unit prospective "access" to—offer the opportunity to cause activation in—some percentage of the brain C_f. A signal sent from the cloud for functional unit *A* could conceivably end up affecting events in some portion of the brain C_f and thus could conceivably establish neural partnerships with those regions. Given the functional differentiation of regions of the brain, C_f would index the potential functional diversity of *A*, the range of uses it might play a role in supporting.

The questions, then, are several. Is C_f large enough to render plausible the hypothesized role of volume transmission in setting up neural partnerships between anatomically distant parts of the brain? What is the range of values C_f for the brain? How does C_f change as a function of the size of the functional unit and of the anatomical differences between different regions of the brain? I have no answers to these questions and for now hope only that these questions appear interesting, empirically tractable, and scientifically worthwhile.

2.5 IDS Interpretation of Some Established Findings

In the course of arguing for the interactive specialization framework, Johnson (2011) discusses several pieces of evidence and describes how they fit into the IS framework. By way of conclusion I would like to revisit a few highlights of this discussion and show how these same items fit into the IDS framework. I should note at the outset that all the evidence that Johnson marshals for the increasing specialization of brain regions during development is compatible with the IDS framework, as IDS also expects refinement of the response properties of individual regions. The disagreement is only over whether to interpret that refinement as the achievement of a single specialized function (IS) or not (IDS). Thus, I focus below on just a few areas where IS and IDS offer somewhat different interpretations of the data, in hopes that this will further illuminate the IDS framework.

2.5.1 Face Perception and Developmental Prosopagnosia

Johnson (2011) reviews a number of findings related to "face blindness" (prosopagnosia): for instance, it can arise either from brain trauma or in the absence of known injury (Duchaine & Nakayama 2006); and adult prosopagnosics generally show activation of typical face-processing regions of the brain, but at least sometimes also in other nonstandard brain regions (Avidan et al. 2005; Duchaine & Nakayama 2006) that are often seen during face processing in children (Gathers et al. 2004; Passarotti et al. 2003, 2007; Scherf et al. 2007). Johnson concludes that developmental prosopagnosia is a matter of a "delayed or impaired specialization process" resulting in reduced face selectivity (Johnson 2011, p. 13). Adopting a somewhat different emphasis, the IDS framework would hypothesize that the disorder results at least in part from a delayed or divergent neural *search*. This could come in at least two forms: one in which the search process was simply arrested before it could complete but, if restarted, might still conclude successfully; and another in which the search ended up locked into a suboptimal outcome. The latter case might occur as a result of detrimental order effects in learning. Suppose, for instance, that a given task *t*, normally learned later in development, was instead learned early, with the result that it occupied a "neural niche" (the neural "environment" supporting a given function; *see* Dehaene 2005; Iriki & Sakura 2008) consisting of regions A, B, and C. But region C normally becomes part of the face network. In this case, when face recognition skills start to be developed, region C might not be functionally available, whether because of divergent differentiation caused by tuning for task *t* or simply because processing demands in C now cause interference with competing tasks including face processing. Thus, the network is locked into a suboptimal state (see Elman 1994 for discussion of a related class of cases).

It seems possible that something like this situation was observed by Kosslyn et al. (1996). In the course of investigating whether the degree of activation of particular brain regions predicted proficiency in a visual imagery task, they discovered that participants who performed the task well seemed to rely in part on Brodmann area 17 for the task, whereas those who performed poorly seemed to rely instead on a region in the parietal lobes. Although it is certainly dangerous to make general suppositions from single findings, together with the IDS framework this observation does suggest that individual differences in cognitive performance may sometimes be due to variance in the sets of neural partners that have been recruited to the task.

2.5.2 Reading and the Visual Word Form System

Reading is a recent cultural achievement for which there cannot be a set of specific evolutionary adaptations in the brain or elsewhere. Thus, learning to read is, neurally speaking, a matter of finding and shaping the right set of tools for the purpose (Dehaene 2009). Johnson cites several lines of evidence reviewed by Schlaggar and McCandliss (2007) as well as more recent studies (e.g., Bitan et al. 2009) that together show reading to be supported by "the gradual emergence of a coordinated and specialized network of regions" (Johnson 2011, p. 15). In his discussion Johnson tends to emphasize the ways in which learning to read illustrates the processes of regional refinement and specialization, whereas IDS sees learning to read as an especially good candidate for understanding the nature of the neural search process: how, precisely, *does* one find and establish the appropriate neural niche? Future research focused specifically on this question offers the potential to better understand the processes of functional brain development.

2.5.3 Executive Control and Prefrontal Cortex

Johnson reviews some very interesting evidence (Casey et al. 1997; Thomas et al. 1999) that during executive control tasks (i.e., a "go/no-go" task) children appear to use more than twice as much of their prefrontal cortex as compared to adults in the same task. As mentioned above, such findings have been generally interpreted as signaling increased effort, but the data reviewed by Johnson also include analyses showing that these findings hold up even when adults and children are matched for proficiency, suggesting that the volume of brain activity is *not necessarily* indexing task difficulty. Johnson focuses on the increasing focality of the activation in adults for the task and takes this to be a sign of the "tuning process in which one region increasingly becomes increasingly specialized for a specific computation to the exclusion of others" (Johnson 2011, p. 15). In contrast, IDS focuses on the increased volume of activation in children and takes this to be the sign of an early developmental search process; although the adults and children may well be equally practiced at this particular task, the children would certainly still be learning the ins and outs of executive control and response inhibition. Here again, there is not necessarily a deep conflict between these two perspectives, but it does illustrate the differences in emphasis between the two frameworks.

2.5.4 Resting State Connectivity and Network Development

As Johnson reviews, Fair and colleagues (2007, 2009) performed a series of coherence (functional connectivity) analyses of resting-state fMRI data in children and adults. They report that development entails both "segregation," decreased short-range connectivity indicating differentiation of the response properties of neighboring local circuits, and "integration," increased long-range connectivity between regions implementing functional networks. These findings were largely consistent with those reported in a similar study by Superkar, Musen, and Menon (2009). Although Johnson finds the observed segregation to be fully consistent with IS because it appears to be a sign of local specialization, he admits that the increased long-range connectivity is somewhat harder to explain. He considers and rejects a maturational account of the phenomenon—that the increased connectivity reflects the maturation of white-matter pathways—because the observed increases in functional connectivity occur after those fiber bundles are in place (Fair et al. 2009; Superkar et al. 2009), and he suggests instead that these observations reflect increases in functional cooperation and coactivation between these regions, perhaps consolidated by a scaled-up version of Hebbian learning. IDS agrees completely with this perspective; increased long-distance functional connections signal the establishment of functional networks of cooperating neural assemblies initiated by the process of neural search (perhaps supported by volume transmission) and consolidated by some as-yet-unknown mechanisms. In addition, IDS may have some novel light to shed on the peculiar *combination* of local differentiation combined with long-distance integration, such that neighboring neural assemblies end up participating in different functional networks: such an arrangement facilitates the functional access of each brain region to functionally nonconnected ones. If local circuits were functionally connected primarily to neighboring regions, then the access measure C_f would tend to be low for most of the brain because local integration would tend to promote the segregation of distant regions. That we see the opposite pattern—local segregation and large-scale integration—suggests that C_f may typically be fairly high, insofar as the neighbors of any given neural assembly will tend to be anatomically connected to a different set of partners. Just so long as the *neighbors* are available to activation by such mechanisms as volume transmission, this offers prospective functional access to the *partners* of those neighbors and thus to a potentially large portion of the brain. This is important because the larger C_f is, the greater will be the scope of tasks within our set of possible but unrealized potencies.

In this chapter I have laid out the interactive differentiation and search (IDS) framework for understanding the functional development of the brain in light of neural reuse. I take this framework to be a development of similar ideas from Bach-y-Rita (1993, 1995), Pascual-Leone (Pascual-Leone & Hamilton 2001), and Johnson (2001, 2011) to better fit with emerging evidence regarding the functional complexity of individual regions of the brain. I have also suggested some specific biological mechanisms that might support the mechanisms of IDS. Obviously, however, much more research is needed to establish these hypotheses.

Although I have focused in this chapter on *developmental* mechanisms, I want to pause to remind the reader of the discussion of evolvability from the end of chapter 1 and to underline the fact that evolutionary and developmental mechanisms are typically strongly linked (Anderson & Finlay in press; Kirschner and Gerhardt 2005). We are in a position now to outline a solution to the evolvability problem in a bit more detail. If the brain possesses mechanisms for functional development that include both the ability to tune local neural structure in response to task-relevant statistical properties in inputs, as well as the ability to perform a "search" for functional partnerships between structural elements at various spatial scales, then it becomes possible to see how systematic, heritable, and relatively consistent functional differentiation in the brain could occur in the absence of targeted modular or mosaic selection. Given a set of early developing and stereotyped neural projections from sensory afferents, and assuming an environment that is largely conserved over generations, local tuning mechanisms would be sufficient to produce local networks with specific and predictable functional structures and response tendencies. A set of such neural structures with different functional biases (different input–output mappings) would be enough to allow an ongoing process of neural search to identify and consolidate the sets of partnerships that reliably supported skills being acquired during development. Consistency in the early development of functional biases, and in the nature of the tasks being learned by the organism, would be sufficient in this model to produce relatively consistent large-scale functional networks without the need for direct evolutionary targeting. Indeed, in a functionally differentiated but nonmodular brain, selection pressures might work not to produce particular specializations but rather to stabilize the availability of a diverse mixture of functional properties in the entire brain (Atallah, Frank, & O'Reilly 2004)—that is, a range of cortical biases that, given sensory inputs and the interactions among regions (Johnson 2001, 2011), reliably produces the functional architecture we observe (see Anderson & Finlay 2014 for much more detailed discussion).

Thus, the developmental story offered here is also part of the evolutionary story offered in chapter 1, and vice versa. Insofar as the properties of environments are generally conserved between generations, then the developmental mechanisms described here, being sensitive to those properties, will reliably produce "heritable" functional differentiation while still being able to compensate when environments change. When this is combined with the insight that the brain has a metamodal and domain general organization, it becomes possible to see how the brain architecture I have described can be both robust to perturbations, reliably producing a diverse range of different processing operators whose cooperation can support species-typical behaviors, and yet still be evolvable without the specific genetic stabilization of domain-specific, stereotyped, modular structures.

The next chapter situates the neural reuse and IDS frameworks in the broader context of the cognitive and neural sciences and reflects in particular on the implications for the concept of local neural function. After that, I take up the question of how we might approach a neuroscientifically grounded science of the mind without abduction to computational operations for specific brain regions—without, that is, some of the metaphysical assumptions inherited from the computational theory of mind (CCTM). This will lead to a proposal for a dispositional account of the brain that—as I hope to establish in the latter parts of this book—may promote greater unification of the social, behavioral, cognitive, and neural sciences than has so far proved possible, without sacrificing the fertility of CCTM.

Interlude 2 You Are Not Your Connectome! Sorry, Understanding the Brain (or People) Will Not Be That Simple

Sebastian Seung's video "I Am My Connectome," from the famous TED Talks series, has gotten over half a million views. His book *Connectome: How the Brain's Wiring Makes Us Who We Are* (Seung, 2012) has gotten immense amounts of positive media coverage. The National Institutes of Health are spending many millions of dollars on a large, multi-institution project to *map* the human connectome. And a recent article in the *Chronicle of Higher Education,* "The Strange Neuroscience of Immortality" (Goldstein, 2012), details the impact that connectomics—the study of our brain's neural wiring diagram—is having on fringe science such as mind uploading. Because naturally, if you are your connectome, and you can exactly copy that connectome into a new medium, say an artificial computer-supported neural network, then *you* can be copied into an artificial computer-supported neural network. So long as you have a few Apple geniuses on hand to provide some good tech support, you can live forever!

All of this odd semifrenzy has grown out of the belief that connectomics is finally the big idea that will crack the neural code. The trouble is: it isn't. It is interesting, to be sure. And maybe it can excite young people to go into the neurosciences. But it will not provide the final answer we've been looking for. So please don't run out and deli-slice your brain in the vain hope of waking up in a brand-new robot body because you will be disappointed. Or, actually, you won't be; you'll be dead.

Don't get me wrong. I'm a huge fan of connectomics and believe that it is going to teach us an immense amount about the brain. I'm extremely grateful that NIH has decided to make it a funding priority, as the study of networks is a significant step forward from the more typical obsessive focus on local function. But I'm *not* a fan of brain GUTs—grand unified theories of brain function. There isn't going to be a single organizing principle for brain function, and we should stop looking for one. The brain functions as the result of the dynamic interactions among many different systems at different levels of organization, each operating according to its own rules. Consider just three extraconnectomic contributors to brain function at three different levels of organization:

2a.1 Volume Transmission

Also known as nonsynaptic diffusion neurotransmission, this refers to a class of neuromodulatory mechanisms that rely on the diffusion of molecules to cells not synaptically connected to the releasing cell. Classical wiring transmission relies on synaptic contact and diffusion barriers that keep the neurotransmitters relatively contained. But there is also a kind of chemical communication in the brain that takes *advantage* of diffusion and in particular of structural features of the brain that allow for greater diffusion in some directions than in others (anisotropy) by releasing molecules from nonsynaptic sites. When you consider that pyramidal cells, for instance, typically synapse with only about 1% of the cells within their dendritic arbor, the functional potential of volume transmission is enormous. Volume transmission is known to play a role in learning and to modulate synaptic transmission and may also play an important role in generating new long-distance partnerships in the brain. To say that volume transmission is poorly understood would be a serious understatement, but the point here is that this aspect of brain function cannot be captured by connectomics even in principle.

2a.2 Neuron-Glia Interactions

There is a growing body of evidence that glial cells—what R. Douglas Fields has called "the other brain"—are more than just the nutritional and structural background to neurons. Instead, they make important contributions to brain function. To summarize Fields (2009): we have learned that glia communicate, although not with electricity. That's why they don't have axons or dendrites or synapses. Instead, they communicate chemically, and in so doing they provide an independent, complementary network for information flow in the brain. Glia are also thought to regulate the formation of synapses, modulate learning mechanisms such as long-term potentiation, and regulate synaptic transmission because they both manage the clearance of neurotransmitters from the synaptic cleft and also release their own neuromodulatory substances. None of this crucial interaction is captured by connectomics.

2a.3 Embodiment

The brain evolved first and foremost as a control system for a particular kind of organism—you—in a particular kind of environment. Human cognition is not just human brain function but human brain-body-environment function. It is marked by the use of and interaction with the environment in myriad ways: using a pencil and paper to store intermediate results in long division or large-number multiplication; arranging a hand of cards or scrabble tiles to better see relevant patterns, matches, or potential words; rotating puzzle pieces to better discern their fit; making grocery lists, labels, signs, encyclopedias, and otherwise storing information in the world to be consulted later; and using management structures and the constraints imposed by individual social roles to accomplish complex tasks such as ship navigation or building construction. Intelligence lies less in the individual brain—and even less in that brain's internal connections—and more in the dynamic interaction of animals with the wider world and one another. Mindedness at this level of description is the activity of making the world a home that reflects the needs and abilities of its occupant. Mind aims to achieve an adaptive integration with the environment, especially including the social and cultural worlds that are so important to human cognition. Taking connectomics as a brain GUT is, in contrast, a return of an odd kind of Cartesianism, a narrow focus on the properties of an imaginary, isolated, computational processor.

So the connectome is important—crucial in fact. But so are volume transmission, neuron-glia interactions, and embodiment. And I haven't even mentioned hormones or local and large-scale oscillatory dynamics or intercellular proteins or many of the other important structures and processes that are known to make crucial contributions to brain function. So, definitely, let's study brain tissue using all the means at our disposal, but let's leave brain GUTs alone.

3 Neural Reuse in Contemporary Cognitive Science

At its core, neural reuse is a framework for understanding how the different parts of the brain are deployed and redeployed in support of cognition and behavior. The previous two chapters detailed the evidence that individual anatomical neural units—at multiple physical scales—are functionally differentiated from one another without necessarily being specialized for a single function and are used for multiple cognitive and behavioral purposes. In fact, we reached the conclusion that neural reuse may well represent a fundamental evolutionary and developmental mechanism for efficiently eliciting complex function from metabolically expensive neural tissue. This apparently simple fact about brain organization has some profound consequences for brain science.

In this chapter we think through some of those implications by way of situating the neural reuse framework in the context of contemporary approaches to mind and brain. First, in section 3.1 I take a brief look at one apparently *competing* proposal for how neural resources are typically deployed in support of the brain's function, in the form of the cognitive architecture ACT-R; then in sections 3.2 and 3.3 I discuss some other theories of function-structure relationships in the brain that centrally involve neural reuse and ask how they compare, contrast, and generally fit with the neural reuse framework. Section 3.4 brings the chapter to a close with a reflection on the meaning of local neural function in light of all the evidence reviewed to that point.

3.1 ACT-R and the Persistence of Modular Approaches to Cognition

Although I have argued at length against modular approaches to understanding the brain, it must be admitted that they remain powerful and widely accepted models for the best way to approach the study of the *mind*. Does widespread neural reuse mean that these approaches must be abandoned? Not necessarily, although I do believe that neural reuse suggests some specific methodological *reforms,* and I try to illustrate this with a discussion of the highly successful cognitive architecture known as ACT-R (adaptive control of thought-rational; the acronym originally meant atomic components of thought-revised: J. R. Anderson & Lebiere 1998).

J. R. Anderson, the originator of the ACT-R model of the mind, defines a cognitive architecture as "a specification of the structure of the brain at a level of abstraction that explains how it achieves the function of the mind" (J. R. Anderson 2007, p. 7). As currently implemented, ACT-R is explicitly modular. As of ACT-R 6.0, it consisted of eight functionally specialized, domain-specific, relatively encapsulated, independently operating, and separately modifiable components. Given this and the definition of a cognitive architecture, it might seem to directly follow that ACT-R is committed to the notion that the brain, too, consists of functionally specialized, domain-specific, relatively encapsulated, independently operating, and separately modifiable components that implement the functional modules of the ACT-R model. Certainly, recent experiments meant to associate ACT-R components with specific brain regions encourage this impression (J. R. Anderson 2007; J. R. Anderson et al. 2007). As he argues:

> As discussed above, modular organization is the solution to a set of structural and functional constraints. The mind needs to achieve certain functions, and the brain must devote local regions to achieving these functions. This implies that if these modules reflect the correct division of the functions of the mind, it should be possible to find brain regions that reflect their activity. Our lab has developed a mapping of the eight modules ... onto specific brain regions. ... (J. R. Anderson 2007, p. 74)

Given that neural reuse implies that anatomical modularity is false (see section 1.5), success in assigning ACT-R modules to specific brain regions would seem to be a problem for neural reuse, and evidence for neural reuse would appear to create problems for ACT-R. But the conclusion does not follow quite as easily as it seems to. First, ACT-R does not strictly imply anatomical modularity. ACT-R is committed to the existence of *functional* modules and to the existence of elements of the brain that implement them. If it turned out that activity in the ACT-R goal module were a better fit to the coordinated activity of some noncontiguous set of small brain regions than it was to the anterior cingulate (to which the ACT-R research group currently has the goal module mapped), then this would count as *progress* for ACT-R and not a theoretical setback. Similarly, if it turned out that some of the brain regions that help implement the goal module *also* help implement the imaginal module, this would pose no direct challenge to ACT-R theory.

Therefore, although J. R. Anderson is at pains to deny he is a functionalist—not just any possible mapping of function to structure will count as a success for ACT-R—there is a good deal of room here for alternatives to the simple 1:1 mapping that he and other ACT-R theorists are currently exploring. For its part, neural reuse predicts that the best fit for ACT-R modules, or any other high-level functional components, is much more likely to be some cooperating complex of multiple brain regions than it is a single area and that brain regions involved in implementing one ACT-R function are likely to be involved in implementing others as well. Interestingly, this is more or less what J. R. Anderson et al. (2007) found. For every task manipulation in their study, they found several brain regions that appeared to be implicated. And every one of their regions of interest was affected by more than one factor manipulated in their experiment. Thus, despite their methodological commitment to a 1:1 mapping between modules and brain regions, J. R. Anderson et al. (2007) are quite aware of the limitations of that approach:

> Some qualifications need to be made to make it clear that we are not proposing a one-to-one mapping between the eight regions and the eight functions. First, other regions also serve these functions. Many areas are involved in vision, and the fusiform gyrus has just proven to be the most useful to monitor. Similarly, many regions have been shown to be involved in retrieval, particularly the hippocampus. The prefrontal region is just the easiest to identify and seems to afford the best signal-to-noise ratio. Equally, we are not claiming these regions only serve one function. This paper has found some evidence for multiple functions. For instance, the motor regions are involved in rehearsal as well as external action. (J. R. Anderson et al. 2007, pp. 213–214)

Here, the regulative idealization promoted by decomposition and localization may have unduly limited the sorts of methodological and inferential tools that they initially brought to bear on the project. As noted already, one of the contributions neural reuse may be able to make to cognitive science is an alternate idealization that can help guide both experimental design and the interpretation of results (Anderson et al. 2010; Anderson, Kinnison, & Pessoa 2013).

Going forward there is at least one other area where we can expect theories of neural reuse and modular theories such as ACT-R to have significant, bidirectional critical contact. Right now, ACT-R is not just theoretically but also literally modular: it is implemented as a set of independent and separately modifiable software components. It does not appear, however, that separate modifiability is theoretically essential to ACT-R (although it is no doubt a programming convenience). Therefore, implementing overlaps in

ACT-R components, in light of the evidence from neuroimaging and other studies of the sort recounted here, is likely to offer scientific opportunities to both research communities (see Stewart & West 2007 for one such effort). For example, overlaps in implementation might offer a natural explanation and a convenient model for certain observed instances of cognitive interference, such as between language and motor control (Glenberg & Kaschak 2002) or between memory and audition (Baddeley & Hitch 1974), helping to refine current hypotheses regarding the causes of the interference.

The ACT-R community is already investigating similar cases in which different concurrent tasks (dialing the phone while driving) require the use of the same ACT-R module and thus induce performance losses (Salvucci 2005). Altering ACT-R so that different modules share component parts might enable it to model some cognitive phenomena that would otherwise prove more difficult or perhaps impossible in the current system, for example, the observation that children who receive pitch training starting in first grade perform better in mathematics longitudinally in third grade (Gardiner 2008). It might also form the basis for modeling the kinds of functional inheritances between such capacities as finger gnosis and magnitude representation or between motor control and memory, discussed in chapter 1. Finally, observations of interference in a modified ACT-R but *not* in human data might suggest that the ACT-R modules did not yet reflect the correct division of the mind's functions. Such conflicts between model and data could be leveraged to help ACT-R better approximate the high-level functional structure of the mind.

3.2 Classic and Contemporary Parallel Distributed Processing

The specific reforms offered above are motivated by a more general claim: the data from the neurosciences that I have been reviewing here strongly suggest that realistic models of the brain *and* mind will be achieved only when those models incorporate the neural overlaps that are evidently an important part of our functional architecture. This should not be a particularly surprising or contentious claim. In fact, from a sufficiently abstract perspective, the idea of neural reuse in cognitive functioning is nothing new. It has been a staple of debates on brain architecture at least since the advent of parallel distributed processing (PDP) models of computation (Rumelhart & McClelland 1986). For one widely cited example, consider the following from Mesulam (1990). He writes:

> A central feature of networks is the absence of a one-to-one correspondence among anatomical site, neural computation, and complex behavior. ... Figure [3.1] implies

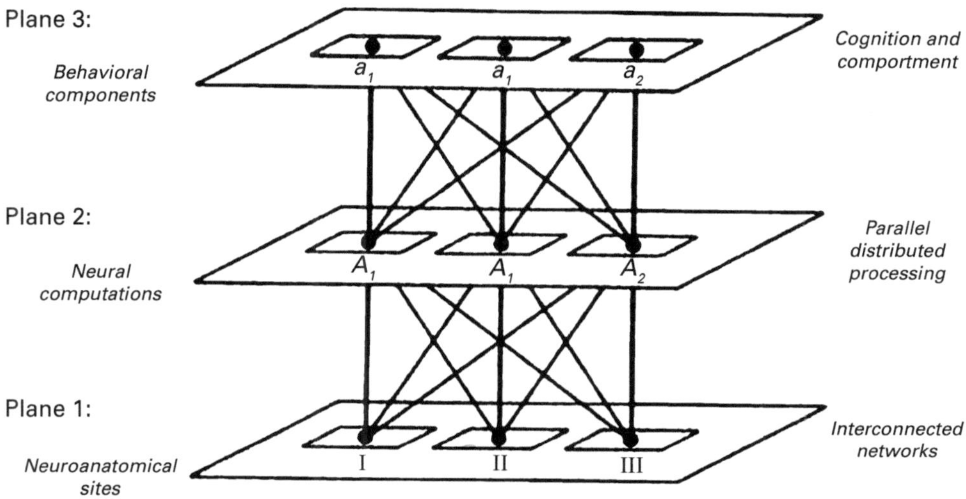

Figure 3.1
Detail of figure 3 from Mesulam (1990). Reprinted with permission.

that each behavior is represented in multiple sites and that each site subserves multiple behaviors, leading to a distributed and interactive but also coarse and degenerate (one-to-many and many-to-one) mapping of anatomical substrate onto neural computation and computation onto behavior. This distributed and degenerate mapping may provide an advantage for computing complex and rapid cognitive operations and sets the network approach sharply apart from theories that postulate a nondegenerate one-to-one relationship between behavior and anatomical site. (Mesulam 1990, pp. 601–602)

Broadly speaking, the neural reuse framework encompasses a family of network approaches to understanding the operation of the brain. They share with these an emphasis on cooperative interactions and an insistence on a nonmodular, many-to-many relationship between neural-anatomical sites and complex cognitive functions/behaviors. But there are also some important differences that set neural reuse apart from classic PDP.

First is a better appreciation of the computational work that can be done by very small groups of, or even individual, neurons (Koch & Segev 2000). Neural reuse agrees with the notion that most of the interesting cognitive work is done at higher levels of organization but also emphasizes that local assemblies have specific and identifiable functional biases. More generally, neural reuse maintains a strong distinction between that functional bias, that is, the underlying local causal features that allow anatomical assemblies to make specific contributions to overall function, and the cognitive purpose that the local network serves—the use to which it is put in any individual case.

In contrast, note that in figure 3.1 "neural computations" are located at plane 2, parallel distributed processing. This appears to indicate that computational work can only be done by fairly large numbers of neurons and that responsibility for this work is *distributed* across the network as a whole. Indeed, for some proponents of PDP architectures, this was the central point and the main challenge these architectures represented to the classic modular approaches based on standard computational theory of mind: not just distributed *processing* but also distributed *representations*. "Each entity is represented by a pattern of activity distributed over many computing elements, and each computing element is involved in representing many different entities" (Hinton, McClelland, & Rumelhart 1986, p. 77).

According to strong proponents of connectionist theory—in some cases explicitly endorsing Lashley's (1929, 1950) twin notions of equipotentiality and mass action—it made little sense to divide the network into regions and specify what each region is contributing to overall behavior; the very model of cognition rooted in the idea of the transformation of discrete representations by local processors needed significant revision (Clark 1993; Lloyd 1989; Ramsey 1997; van Gelder 1990, 1991).

Classic PDP models are indeed a powerful way to understand the flexibility of the brain, given its reliance on relatively simple, relatively similar individual elements. But the trouble for PDP models in this particular context is that there is no natural explanation for the data (for instance) on increasing scatter of recently evolved functions, nor such observations as the apparent cross-cultural invariance in the anatomical locations of brain regions supporting acquired practices such as reading (see section 3.3.3). Such observations seem to point to the existence and importance of local functional biases, but the very notion of a local functional bias sits uneasily with the guiding idealization underlying the PDP framework. Indeed, on PDP models, investigating such matters is not even a natural empirical avenue to take. This represents a significant distinction between PDP and neural reuse.

This being said, it is also true that the network theorists such as Mesulam with the strongest ties to the brain sciences have tended to simultaneously emphasize the complex functional profiles of individual regions while also stressing the importance of what I have been calling functional biases. Consider the following from a more recent work by Mesulam:

> The anatomical areas that play crucial roles in the identification of colour, movement, faces, words, objects, and spatial targets display relative rather than absolute

specializations. For example, V4 is specialized for colour perception but also participates in spatial attention, the identification of salience, and the encoding of form (Moran & Desimone 1985; Schiller 1995; Connor et al. 1996). In turn, the processing of colour information may involve not only V4 but also a part of the lateral peristriate cortex (Corbetta et al. 1990). Furthermore, neuronal ensembles selectively tuned to canonical features of faces participate, although to a lesser extent, in encoding other visual entities (Rolls 1987). It is quite likely that several ensembles, each composed of neurons optimally tuned to a different category, form an interdigitating or partially overlapping mosaic and that the predominant type of ensemble varies from one location to another. This organization has been designated "selectively distributed processing" (Mesulam 1994; Seeck et al. 1995) to set it apart from other models based on equipotentially distributed processing (Lashley 1929), parallel distributed processing (Rumelhart & McClelland 1986), and modular processing (Fodor 1983). (Mesulam 1998, p. 1022)

Clearly Mesulam is at great pains here to accommodate the observed complexity of the brain's response properties. Moreover, in this later treatment of brain networks, Mesulam emphasizes the ecological and evolutionary context within which such networks emerge. He points to some fundamental constraints facing organisms in trying to simultaneously maximize speed and flexibility of behavior: increasing the length of the synaptic chain between sensory input and behavioral output and including integrative connections between parallel streams of processing allow for greater flexibility of response, but at the expense of speed and certainty.

[P]rimate evolution represents a compromise between [two] extremes: the synaptic chain that links sensation to action has been lengthened, but not by too many synapses; and the bridge between the two has been broadened by the introduction of numerous parallel lines of communication, but with ample opportunity for integrative interaction between channels. ... These observations point to a trend towards a progressive corticalization of function and suggest that cognition, defined as the behavioral outcome of intermediary processing in cerebral cortex, becomes established as an obligatory rather than facultative correlate of information processing in the course of evolution. (Mesulam 1998, p. 1015)

In Mesulam's vision of brain organization the speed and flexibility of primate cognition are enabled by a complex combination of unimodal primary sensory processing areas stitched together by multimodal and convergence and integration zones, opening multiple avenues for shaping the behavioral control signal. Moreover, perception itself is not a passive feedforward affair but is extensively modulated by top-down feedback connections from downstream neural assemblies. As we saw in chapters 1 and 2, neural reuse shares with Mesulam's perspective both its recognition of the importance of feedback, modulation, and other forms of interaction in the brain as well as its orientation toward action. As Mesulam writes: "[a] major task of the CNS is to configure the way in which sensory information becomes linked to adaptive responses and meaningful experiences" (Mesulam 1998, p. 1014). I especially appreciate Mesulam's elegant definition of cognition as "the behavioral outcome of intermediary processing in cerebral cortex." As I suggested with the analogy of the ultimate computer story in chapter 2, conceptualizing cognition as the process of shaping inputs to cause desired behavioral outputs, rather than as primarily a matter of transforming inputs to create certain kinds of representational content, will turn out to be one of the keys to developing a new theory of the mind and science of the brain. We have occasion to meditate further on this perspective in future chapters.

One point of contrast that remains between selectively distributed processing and neural reuse concerns the patterns of functional differentiation observed in the brain. Mesulam imagines a kind of gradient in the degree of specialization from primary sensory areas to unimodal integration areas up through heteromodal integration and executive control centers of the brain.

> Nodes at upstream synaptic levels tend to contain neuronal groups specialized for encoding relatively elementary attributes of visual experience, whereas nodes at more downstream levels are organized into neuronal groups specialized for encoding more composite features. (Mesulam 1998, p. 1020)

In fact, the evidence I presented in the previous two chapters and the further evidence I present in chapter 4 suggest that although there are certainly differences in the degrees of specialization of various regions of the brain, those differences do not organize themselves according to a strict hierarchy such that degree of specialization is inversely related to distance from the sensory afferents. Rather, the feedback channels that Mesulam rightly emphasizes appear to make upstream nodes available for supporting diverse and complex tasks, and the likelihood of being so used seems to depend more on the nature of the underlying causal properties of the network than on its proximity to sensory afferent input.

Still, rather than draw a sharp contrast between selectively distributed processing and neural reuse, it is no doubt better to say that neural reuse represents a generalization and refinement of this earlier model. Neural reuse embraces the IDS framework—functional *differentiation* of neural assemblies without expecting functional *specialization*—but both reuse and selectively distributed processing envision a brain that works via the assembly of large-scale neurocognitive networks of interacting nodes. Indeed, I think it is fair to say that both reuse and selectively distributed processing emphasize the integration and interaction in contrast with the localized processing of modular architectures and emphasize the importance of cortical biases and functional differentiation in contrast with the apparent equipotentiality displayed in PDP architectures.

This twin emphasis is a feature found in other contemporary versions of network models, such as Leabra (O'Reilly 1998; O'Reilly & Munakata 2000), which tend to model the brain in terms of densely connected, locally specialized networks that are sparsely connected to one another (see figure 3.2).

In this sense Leabra appears also to be more compatible with neural reuse than classic PDP models are, as Leabra explicitly allows for regional

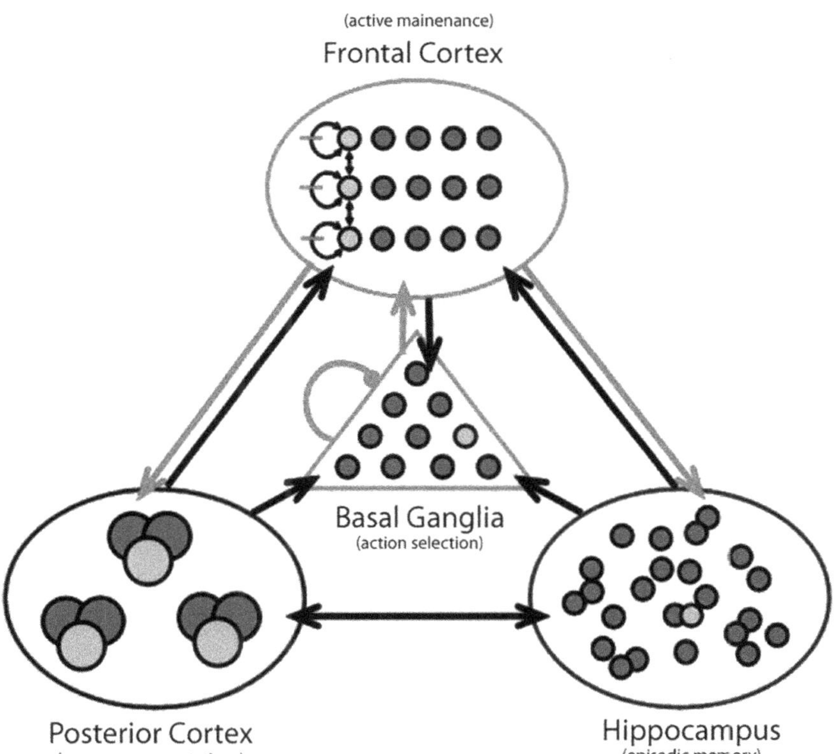

Figure 3.2
Overview of the Leabra architectural organization. Reprinted from Jilk et al. (2008) with permission.

functional biases. But insofar as this architecture reflects the influence of the selectivity assumption and represents a more componential approach to understanding the brain, there are potentially the same points of conflict with Leabra as there are with theories that emphasize these features in brain organization. Consider the following, from a recent paper describing Leabra:

> The brain is not a homogeneous organ: different brain areas clearly have some degree of specialized function. There have been many attempts to specify what these functions are, based on a variety of theoretical approaches and data. In this paper, we summarize our approach to this problem, which is based on the logic of *computational tradeoffs* in neural network models of brain areas. The core idea behind this approach is that different brain areas are specialized to satisfy fundamental tradeoffs in the way that neural systems perform different kinds of learning and memory tasks. (Atallah et al. 2004, p. 253)

There is nothing here that explicitly commits the authors to the idea that large brain regions are dedicated to specific tasks or cognitive domains—something the data presented in this volume throw into question—although that is certainly one possible reading of the passage. Moreover, O'Reilly (1998) tends to focus more on modeling *processes* over modeling parts, an approach that need not commit one to a specific story about how and where such processes are implemented in the brain—it need not be the case that individual brain regions implement the processes being modeled, for instance.

And yet, O'Reilly and his collaborators *have* assigned these processes to specific regions:

> The large-scale architectural organization of Leabra includes three major brain systems: the posterior cortex, specialized for perceptual and semantic processing using slow, integrative learning; the hippocampus, specialized for rapid encoding of novel information using fast arbitrary learning; and the frontal cortex/basal ganglia complex, specialized for active and flexible maintenance of goals and other context information, which serves to control or bias processing throughout the system. (Jilk et al. 2008, p. 204)

And, in fact, the Leabra team has gone further than this by recently integrating Leabra with the cognitive architecture ACT-R (discussed in section 3.1) to form the SAL architecture:

> When the ACT-R and Leabra research teams started to work together in 2006, they came to a startling realization: the two theories, despite their origins in virtually opposite paradigms (the symbolic and connectionist traditions, respectively) and widely different levels of abstraction, were remarkably similar in their view of the overall architecture of the brain. (Jilk et al. 2008, p. 205)

Given the 1:1 approach to function-structure mapping in the brain that the ACT-R team has taken in other work, described above, it would in my view be a step in the wrong direction should Leabra adopt *that* approach to the overall architecture of the brain. For, as with ACT-R, there does not seem to be anything *essential* to Leabra that would prevent it from explicitly incorporating neural reuse as one of its organizing principles. And the functional specializations being ascribed by Leabra to the brain regions mentioned are general enough to plausibly have many different cognitive uses, as predicted by neural reuse theories. Thus, one would hope that this statement indicates instead that the ACT-R team is recognizing in the Leabra approach a view of functional architecture of the brain that can be modeled in the ACT-R framework. Recent clarifications suggest this latter interpretation is the correct one:

> [W]hen we (Atallah et al. 2004, p. 253) wrote that "different brain areas clearly have some degree of specialized function," we did not mean cognitive functions such as face recognition. What we meant is closest to what Anderson calls "cortical biases" or, following Bergeron (2007), "working." Specifically, the posterior cortex in Leabra specializes in slow interleaved learning that tends to develop overlapping distributed representations, which in turn promote similarity-based generalization. This computational capability can be used in a myriad of cognitive functions (O'Reilly & Munakata 2000). The hippocampus and the surrounding structures in the medial temporal lobe (MTL) specialize in rapid learning of sparse conjunctive representations that minimize interference (e.g., McClelland et al. 1995). The prefrontal cortex (PFC) specializes in sustained neural firing (e.g., Miller & Cohen 2001; O'Reilly 2006) and relies on dynamic gating from the basal ganglia (BG) to satisfy the conflicting demands of rapid updating of (relevant) information, on one hand, and robust maintenance in the face of new (and distracting) information, on the other (e.g., Atallah et al. 2004; O'Reilly & Frank 2006). Importantly, most of this specialization arises from parametric variation of the same underlying substrate. The components of the Leabra architecture differ in their learning rates, the amount of lateral inhibition, and so on, but not in the nature of their processing units. Also, they are in constant, intensive interaction. Each high-level task engages all three components (O'Reilly et al. 1999; O'Reilly & Munakata 2000). (Petrov, Jilk, & O'Reilly 2010, p. 286–287)

Thus, it appears that Leabra accepts the two fundamental principles defining neural reuse: (1) functional differentiation leading to local functional biases and (2) reuse of regions in multiple cognitive contexts. Insofar as this is so, the Leabra architecture would appear to be a specific, well-developed model of learning, falling within the neural reuse framework. Yet certain vestiges of a more modular mind-set remain—note the continuing use of the term "specialization" and the labels "episodic memory," "action selection," "sensory representations," and "active maintenance" to suggest the functions of the brain regions in figure 3.2. The point of drawing attention to these apparently minor issues is *not* to criticize the Leabra team or suggest a continuing adherence to modularity in their model; I take them at their word that this is a misleading way to characterize both their intent and the facts of the model. But it *does* highlight the continuing, implicit influence of the modularist framework and the difficulty inherent in expressing alternatives to that framework while continuing to use elements of its vocabulary. This suggests the need for extra vigilance and also, wherever possible, for new concepts to replace the old—for instance, I have suggested talking about *differentiation* rather than specialization; and *functional biases* or *profiles* rather than functions. This is not a challenge specific to the PDP approach but is something that all new paradigms face as we slowly learn to see things through new eyes.

In the current case I think the root of the tension may be the way in which the Leabra team is trying to capture, in connectionist form, something like the componential structure envisioned by Herb Simon in the form of nearly decomposable subsystems. I have already suggested, in chapter 1, why I think the notion of nearly decomposable subsystems is a bad model for brain architecture. But it is worth discussing the notion of componentiality in a bit more detail because it is an apparently attractive notion that is, I think, behind some of the confusion here. In point of fact, in my initial formulation of the massive redeployment hypothesis (Anderson 2007a, 2010b), I was *also* trying to preserve the notion of low-level functional components in the form of "workings" while trying to do justice to the complex functional profiles of local regions. This is presumably why the Leabra team saw a good deal of common ground between their model and my theory, why I saw (and continue to see) vestiges of modularity in their model, and also why *they* saw vestiges of modularity in mine:

> Ironically, Anderson's critique of modularity is steeped in modularist terms such as redeployment. We are sympathetic with the general thrust of Anderson's theory and find it very compatible with the Leabra tripartite architecture (O'Reilly 1998; O'Reilly & Munakata 2000). (Petrov, Jilk, & O'Reilly 2010, p. 286)

In considering these issues I have come around to the idea that we need to abandon the notion of componentiality in the brain. But does the functional architecture of the brain justify this move? I think it does, for reasons that will become more clear as we proceed. For now, consider the case of starburst amacrine cells (SACs) in the mammalian retina (see interlude 4 for further discussion). SACs are axonless neurons with multiple dendrites arrayed radially around the cell body in a single plane of the retina. What is especially interesting about these cells is that the dendrites are *individually* differentially sensitive to motion; they release neurotransmitter only in response to motion that is congruent with the direction the dendrite points away from the cell body. SACs are important to multiple uses including motion perception and the guidance of eye movements (Masland 2005; Yoshida et al. 2001).

The natural way to approach these cells and their relationship to the larger functional networks from a componential perspective is to imagine that each dendrite is a selective low-level functional component that can be joined with other cells into networks to achieve large-scale functionality. But the biological facts appear to suggest something different. In fact, the directional selectivity of each dendrite is due in large part to the particular blend of connections these cells have to bipolar cells and other SACs such that responses in the congruent dendrites are reinforced while responses in noncongruent dendrites are inhibited (Lee & Zhou 2006). In other words the directional selectivity of the dendrite in a given situation is due not so much to intrinsic properties of that dendrite but to global properties of the network. *Global function is not built from componential local function, but rather the reverse!*

It is tempting to conclude that in this case we should simply treat the whole local network as a component that has a well-defined function and can be put together with other components to achieve motion perception or optokinesis. But I think we need to be open to the possibility that the same dynamics reproduce themselves at the higher level: whether a local network is helping with motion perception or optokinesis—and whatever functional selectivity it achieves in these different contexts—may once again be a matter of global network properties. In this case apparent local selectivity is not a stable intrinsic fact about a particular neural subsystem but a result of interactions between local functional biases and global network organization. This reverses the relationship typically assumed when one analyzes the functions of componential (component-dominant) systems, where the constituents of the overall system have well-understood and consistent input–output mappings and so can be combined into larger wholes with functional properties determined by the properties and relations of the parts. Here the functional properties of the parts in a given situation seem determined by the functional organization of the whole.

The only way to hold on to the idea of componentiality in a system like this is to imagine that the "components" are transiently (and reproducibly) assembled in real time. This points in exactly the *opposite* direction from the moral of Simon's tale of the watchmaker: there components are temporally and functionally stable subassemblies sitting on the tabletop waiting for final construction. We could *call* the transient reproducible local networks of the brain "components," but they have functional and temporal properties and interrelations that differ significantly from the kinds of components we normally imagine.

Given this, I would rather borrow terminology from a different source, Andy Clark (2010), who argues that organisms achieve behavioral complexity and flexibility by assembling heterogeneous, interacting nodes into functional networks that he calls transient extended cognitive systems (TECS; see also the "just-in-time" recruitment hypothesis: Just & Varma 2007). It appears that in the brain we have transiently assembled local neural subsystems (TALoNS). TALoNS—which are the constituents of large-scale functional networks—exhibit temporary, reproducible functional selectivity but do not have the normal functional characteristics of components. In fact, the functional-compositional relationships are in some cases very nearly reversed—in a disanalogy with linguistic componentiality, here the meanings of the parts are (largely) determined by the structure of the whole (see interlude 3 for discussion of a similar case). Or, to stretch a different analogy, TALoNS—like the knife blade discussed in the introduction to this book—do not have fundamental functions but rather sets of causal properties useful in various circumstances. Which of these properties is most relevant and contributes most to overall functional outcome is determined by the totality of those circumstances and the interactions they engender. Here again, this is not a criticism of classic or contemporary PDP models—and, indeed, these approaches may be the best equipped to help us understand these dynamics—but rather suggests that we need to bring to bear a fresh set of concepts and tools when interpreting and explaining such models. I discuss PDP models again in part II of this volume when more of the theoretical framework of reuse is in place and when we can thus better appreciate the continuing relevance of PDP models and also the various ways in which they may need to be reconsidered in light of the evidence presented here.

Where there *is* a serious challenge specific to both classic and contemporary PDP is in dealing with the discoveries reviewed in chapters 1 and 2 of nonsynaptic communication between neurons and the rapid modulation of network configurations by extraneural mechanisms including glial intervention and genetic expression. These features of brain function challenge the adequacy of *any* approach that relies on modeling brain networks solely in terms of neuron-neuron interactions and those solely in terms of modulations of synaptic strength. Brain networks appear to be polymorphic (Getting & Denkin 1985), and none of the leading models do justice to this fact. This is a challenge for the neural reuse framework as well, insofar as it represents a set of findings that need to be accounted for and assimilated into the theoretical framework being developed. It is my contention that these findings both strengthen the neural reuse framework's claim to be offering a fundamental characterization of the functional architecture of the brain and also point the cognitive neurosciences in a rather different methodological direction from the one they are currently pursuing.

3.3 Neural Reuse for Learning and Development

The sort of reuse featured in PDP approaches to the brain is primarily a consequence or side effect of the architecture itself: widespread reuse of functional elements simply *is* how networks operate. In contrast, the development of my own framework was motivated first and foremost by functional, developmental, and evolutionary considerations. As noted in chapter 1, I was led to postulate the possibility of widespread neural reuse as part of the answer to the question of how animal brains could come to support new cognitive functions. In this I am far from alone. In fact, over the past several years at least three other frameworks have been proposed that postulate neural reuse as part of the solution to this particular evolutionary and developmental puzzle. Two of these theories build on the core notion of the sensorimotor grounding of conceptual content to show how it could implicate many more aspects of human cognitive life: the "neural exploitation" hypothesis (Gallese 2008; Gallese & Lakoff 2005); the "shared circuits model" (Hurley 2005, 2008). The third, Dehaene's "neuronal recycling" theory focuses instead on the process of neural reuse in the learning of complex skills such as reading and mathematics (Dehaene 2005, 2009; Dehaene & Cohen 2007). Others have clearly been thinking along similar lines, including Morten Christiansen and Nick Chater (2008), Luiz Pessoa (2008), Gary Marcus (2004, 2008), Steven Scher (2004), William Bechtel (2003), Dan Lloyd (2000), and Stephen Kosslyn (1999; Kosslyn & Koenig 1995). These models have some interesting similarities and equally interesting differences, but taken together they point toward new research-guiding idealization of brain organization, even if, as I suggest, they have not yet gone far enough. In the following sections I briefly discuss each of these models and what they might collectively mean for cognitive science.

3.3.1 Neural Exploitation Hypothesis

The neural exploitation hypothesis is a direct outgrowth of conceptual metaphor theory and embodied cognition (reviewed in chapter 1) and largely sits at the intersection of these two frameworks. The main claim of the framework is that "a key aspect of human cognition is ... the adaptation of sensory-motor brain mechanisms to serve new roles in reason and language, while retaining their original function as well" (Gallese & Lakoff 2005, p. 456). This claim is the conclusion of an argument about the requirements of understanding that runs roughly as follows:

1. *Understanding requires imagination.* In the example most extensively developed by Gallese and Lakoff (2005), understanding a sentence such as "He grasped the cup" requires the capacity to imagine its constituent *parameters*, which include the agent, the object, the action, its manner, and so on.
2. *Imagination is simulation.* Here, the neural exploitation hypothesis dovetails with concept empiricism in arguing that calling to mind individuals, objects, actions, and the like involves reactivating the traces left by perceiving, doing, or otherwise experiencing instances of the thing in question.
3. *Simulation is therefore neural reuse.* Simulation involves reuse of the same *functional clusters* of cooperating neural circuits used in the original experience(s).

As much of the evidence for these claims has been laid out already in earlier chapters, it is not recounted here. The reader will of course notice that the theory as stated is limited to the adaptation of sensorimotor circuits, and we have already seen that reuse in the brain is much more broad-based than this. However, because the neural exploitation model has been expanded to include not just the case of concept understanding but also that of human social understanding (Gallese 2008), and because it incorporates a detailed computational model for how the reuse of networks might actually support semantic inheritance (based on work by Feldman & Narayanan 2004), it actually has broader applicability than is evidenced in the two main statements of the view (Gallese 2008; Gallese & Lakoff 2005).

The core of the computational model is a set of *schemas*, which are essentially collections of features in two layers: *descriptions* of objects and events and *instructions* regarding them. These two layers are systematically related to one another and to the sensorimotor system, such that event schemas can be used both to recognize events and to guide their execution, and object schemas can be used both to recognize objects and also to guide actions with respect to them. The schemas are also connected to the conceptual system, such that the contents of our concepts are built from the same features that form the schemas. The general idea is that the features' connections to the sensorimotor system give semantic substance to the concepts as well as a natural model for understanding as the activation of neurally instantiated features and schemas. Like Gallese and Lakoff (2005), Feldman and Narayanan (2004) focus primarily on cases of understanding that can be directly ("He grabbed the cup") or metaphorically ("He grabbed the opportunity") mapped to basic perception-action domains.

But there is no reason in principle that the model need be limited in that way. As the authors note, by adding layers of abstraction, one can move from concrete action execution plans to abstract recipes such as mathematical algorithms. Given this flexibility it seems that action schemas need not be limited to providing guidance for the manipulation of independent objects (whether concrete or abstract) but could presumably also become control systems for the manipulation of neural assemblies. That is, the same action schema that might normally be used to control rhythmic speech production could be reused to guide silent memory rehearsal, and more abstract schemas might form the basis of control systems for other applications.

Of course, this emendation would constitute a significant departure from the model as originally formulated. In particular, it would turn a system in which neural reuse was driven by *grounding*—the inheritance of semantic content from one level to another—into one in which reuse was driven by the need to create control systems for functionally relevant outcomes. Although it is far from clear that this switch *precludes* the possibility that grounding plays a role in driving neural reuse, it certainly moves it from center stage, which may have undesirable theoretical consequences for the theory as a whole and for the way it interfaces with related ideas in linguistics, philosophy, and psychology. On the other hand, without some emendation that significantly broadens the kinds of cases that it can cover, the neural exploitation hypothesis risks being inadequate to the full range of available empirical evidence.

3.3.2 The Shared Circuits Model

The shared circuits model (Hurley 2005, 2008), like the neural exploitation hypothesis, is an attempt to show how higher cognitive function can emerge from the reuse of more primitive action-oriented processors. The model is organized around five control layers of similar structure, which are differentiated by the increasing abstraction of inputs and outputs (table 3.1). Each layer consists of an adaptive feedback loop that takes state information as input and generates control information as output. The first, lowest layer is a simple perception-action feedback loop that monitors progress toward action goals (reaching a target) and adjusts motor output in light of perceptually generated state information. It is, in this sense, a model of the

Table 3.1
Outline of Hurley's five layers

Layer	Takes as input	Implements
1	Perception	Basic adaptive feedback control of action
2	Perception and efference copy of motor output	A forward model of the perceptual effects of action
3	Observations of actions or action-evoking objects	Mirroring for priming, emulation, and imitation
4	Observations of actions or action-evoking objects	Simulated mirroring for action understanding
5	Simulations of actions or action-evoking objects	Counterfactual simulation of actions for strategic deliberation

simplest sort of thermostat; and the idea is that behavioral control systems might consist, at the most basic level, of multiple such control systems or circuits. Layer 2 takes input from the external world but also from layer 1 and implements an adaptive feedback loop monitoring the original layer. That is, layer 2 is in essence a forward model of layer 1. As is well known, incorporating such models into adaptive control systems tightens overall control by allowing for the prediction of state information so appropriate action can be taken without waiting for the (typically slower) external feedback signal. The more hysteresis in the system—the longer it takes control interventions to produce expected results—the more improvement forward models can offer.

Circuit *sharing* really begins with layer 3, in which the same control circuits described by layers 1 and 2 take as input observations the actions (or situations) of other agents. Hurley's suggestion is that the mirror system (Decety & Grèzes 1999; Gallese et al. 1996; Rizzolati et al. 1996) should be modeled this way, as the activation of basic control circuits by representations of the state, action, and situations of other agents. Layer 3 also implements output inhibition, so agents do not automatically act as if they were in another agent's situation whenever they observe another agent doing something. Layer 4 incorporates monitoring of the output inhibition, supporting a self-other distinction; and layer 5 allows the whole system to be decoupled from actual inputs and outputs to allow for counterfactual reasoning about possible goals and states and about what actions might follow from those assumptions. The idea is that the same circuits normally used to guide action in light of actual observations can *also* be fed *hypothetical* observations to see what actions might result; this can be the basis of predictive models. By the time we achieve the top layer, then, we have the outline for a model both of deliberation about possible actions and also of multiagent planning, which could serve as the basis for high-level social awareness and intelligence.

Like the neural exploitation hypothesis, one of the main explanatory targets of the shared circuits model is the possibility of mind reading and intelligent social interaction. And like the neural exploitation hypothesis, it is built entirely on the foundation of sensorimotor assemblies. However, unlike the neural exploitation hypothesis, the shared circuits model does not revolve around the inheritance of semantic content from one level to another but rather around the inheritance of function. The core capacities of the higher layers are based on exploiting the functional properties of the lower layers; all the layers are essentially control loops containing predictive models because they are reusing the basic implementation of the lowest levels. This is an advantage in that it is easier to see how the shared circuits model could be used to explain some of the specific instances of function-driven inheritance canvassed in chapter 1; for, although Hurley models layer 1 on low-level sensorimotor circuits, there seems no reason in principle that the general approach could not allow for other kinds of basic assemblies on which other functional layers could be built. It is also a potential weakness in that it is less easy to see how it could be used to account for the central findings of concept empiricism or conceptual metaphor theory. Can the sort of functional inheritance allowed by this model *also* allow for semantic inheritance? The inheritance of a basic feedback structure does not seem to lend itself to any obvious examples of this sort. This is not a criticism of the model as it stands—it was meant only to account for our understanding of instrumental actions; but it suggests that there is no very simple way to generalize the model to a wider set of cases. On the other hand, there seems no *fundamental* conflict between inheriting a function and thereby inheriting semantic content or domain structure.

As we will see beginning in chapters 4 and 5, I think we should understand the brain as fundamentally for controlling action and interaction with the environment. More specifically, I believe the brain evolved to manage various organism-environment relationships. Insofar as the two models above imagine that the basic control systems of the brain can be reused to support more abstract or environmentally decoupled cognition, I am in complete agreement with this approach. However, insofar as they assume that the mechanisms always generate *representations*—representations, for instance, of the beliefs in another mind—then I must part company. I would rather say, more generally, that the mirror system helps us to interact with one another (Gallagher & Hutto 2008) and to see how other organisms might interact with their environment. Sometimes this will be facilitated by the generation of particular representations, but in other cases interaction might be facilitated rather by the retuning of perception. Consider, in this regard, an alternate proposal for how to describe the function of layer 3.

I mentioned at the outset that the hierarchy of levels was characterized by an increasing abstraction of input and output, but it might be better to say that layers are increasingly sensitive to higher-order properties or invariants. In layer 3, for instance, input will be both impoverished and higher order as compared with lower layers. It will be impoverished because it will be missing a great deal of the richness of embodied experience—tactile experience, proprioceptive feedback, and efference copy are all absent when one is observing as opposed to acting. One is left with the *visual* perception of an action, and an action viewed from the first-person perspective looks different from the same action viewed from the third-person perspective. This last fact points to one reason that the information must be *higher order*: because the visual experience of another agent's action will differ in most, if not all, of its low-level particulars, the system must be sensitive not to these but to high-level invariants of the action that are common to the two situations. Moreover, by hypothesis, layer 3 responds not just to actions, but to situations in which actions are possible—not just to another agent *reaching* for a banana but to the banana being *within the reach* of another agent. All this suggests to me that the best way to describe layer 3 is as being sensitive to environmental affordances—relationships between an organism's abilities and things in the environment (Chemero 2009). We perceive what the world affords us: we can reach for the door, jump that gap, walk down the hallway, and so on. What the mirror system is allowing us to do, in the framework I am developing, is to reuse our ability to see what the world affords to *us* to also see what the world affords to *others*, and this can be the basis for knowing (*seeing*) what they are likely to do.

3.3.3 The Neuronal Recycling Hypothesis

The neuronal recycling hypothesis (Dehaene 2005; Dehaene & Cohen 2007) originates from a set of considerations rather different from those motivating the neural exploitation hypothesis and the shared circuits model. Those are neutral on the question of how and over what timescales the brain organization they propose came about, whereas Dehaene is interested specifically in those cognitive capacities—such as reading and mathematics—that have emerged too recently for evolution to have generated cortical assemblies specialized for these purposes. Such cultural practices must be learned, and the brain structures that support them must therefore be assigned and/or shaped during development.

There are two major ways to explain how recent cultural acquisitions, which emerge and are maintained in a population only by learning and not via genetic unfolding, can be supported by neural structures, as of course they must partly be. One way is to take our capacity to acquire such practices as reading and arithmetic as evidence for domain-general learning mechanisms (Barkow et al. 1992) and fairly unconstrained neural plasticity (Quartz & Sejnowski 1997). The other way is to suggest that cultural acquisitions must find their neural niche (Dehaene 2005)—a network of neural structures that already have (most of) the structure necessary to support the novel set of cognitive and physical procedures that characterize the practice. The neuronal recycling hypothesis is of the latter sort.

Note the interesting implication that the space of possible cultural acquisitions is partly constrained by our cortical biases. The phrase "neural niche" is clearly meant to echo the idea of an ecological niche and suggests both that acquired cognitive abilities "belong" in specific neural locations (i.e., can only survive where the neural climate is appropriate) and that the neural ecology may partly determine the characteristics that these cultural acquisitions possess by limiting what is even possible to learn (and therefore which "cognitive animals" survive). Assuming the set of cortical biases is relatively consistent in a species, and at least partially determined by selection pressures, we should expect to find evidence of at least three things: first, the neural manifestations of acquired abilities should be relatively consistent across individuals and even cultures; second, these practices should have some common cross-cultural characteristics; and third, the same sorts of cortical biases, as well as some of the same possibilities for learning, should be present in close evolutionary relatives, including nonhuman primates.

As evidence for the first expectation, Dehaene and Cohen (2007) note that the visual word form area, functionally defined as a region specifically involved in the recognition and processing of written words, appears in the same location in the brain across participants, whether the participants in question are using the same language and writing system or using different ones. Similarly, the intraparietal sulcus has been implicated in numeric tasks regardless of the culture or number representation system used by the participants. As evidence for the second expectation, they point to work by Changizi and colleagues (Changizi & Shimojo 2005; Changizi et al. 2006) that writing systems are characterized by two cross-cultural invariants: an average of three strokes per written letter and a consistent frequency distribution for the types of contour intersections among the parts of those letters (T, Y, Z, etc.). Finally, the third expectation has been supported by some interesting and groundbreaking work by Atsushi Iriki and colleagues (Iriki 2005; Iriki & Sakura 2008) who uncovered evidence for real-time neural niche construction in primate brains (specifically *Macaca fuscata*) as the result of learning to use simple tools. The location of the observed neuromorphological changes following tool training is roughly homologous to the regions associated with tool use in the human brain (Culham & Valyear 2006). Thus, the theory suggests a novel pathway by which *Homo sapiens* may have achieved its current high-level cognitive capacities.

The neuronal recycling hypothesis outlines a universal developmental *process* that, although illustrated with specific examples, is meant to describe the way any acquired ability would come to have a neural instantiation. In this sense it is broader in conception than the neural exploitation and shared circuits theories described above (although, as noted, their scope might well be increased with a few modifications). How much neural plasticity would be required in any given case will vary with the specifics of the acquisition, but one strength of the neuronal recycling hypothesis is that it makes clear some of the limits and costs that would be involved. The greater the distance between the function(s) required by a given practice, and the existing cortical biases, the harder the learning process will be, and the more likely it will be that the learning process will disrupt whatever *other* functions the affected brain regions support.

On the other hand, the more the requirements of the acquisition match what is already possible, the less novel and potentially less valuable the cultural practice is likely to be—unless, that is, it is possible to combine existing capacities in new ways, to use old wheels, springs, and pulleys to form new machines. It is interesting to note in this regard that although the neural exploitation and shared circuits theories outlined earlier tend to envision neural assemblies being put to fairly similar new uses—for example, forward models in motor control being used to support forward models in social interaction—the neuronal recycling hypothesis suggests that neural assemblies might be put to uses quite other than the ones for which they were initially developed. As should be obvious given the marshaling of evidence in chapters 1 and 2, this possibility is also central to the neural reuse framework.

Of course, the question arises: Exactly *how* is such reuse, especially reuse that leads to very different functional outcomes, in fact possible? Although I hope the IDS model offered in chapter 2 points toward an answer to that question, it must be admitted that the question remains largely open. Going forward, supporters of neural reuse need to provide at least three things: specific models of how information could flow between redeployed assemblies; particular examples in higher animals of how different configurations of the same parts can result in different cognitive and behavioral outcomes; and a more complete discussion of how (and when and whether) multiple uses of the same local region can be coordinated. It is to the credit of both Hurley and Gallese that they each offer a (more or less concrete) proposal in this regard (see Gallese 2008; Gallese et al. 1996; Hurley 2005, 2008). That neither seems wholly adequate to the task should not be surprising or overemphasized; the neurosciences are replete with what must be considered, at best, partial models of the implementation of function by neural structure. That there are details to work out is to be expected. What the neural reuse framework primarily offers is a unique guide to discovery—a sense of what to look for in understanding brain function and how to put the pieces together into a coherent whole. For instance, if neural assemblies are put to many different uses, then the focus on explaining cognitive outcomes must shift from determining fixed local function and the exclusive exploration of single-voxel effects to uncovering the complex and context-dependent web of relations between local networks. The framework focuses attention in particular on exploring the functional relationships between regions and on understanding how specialized behaviors can arise from the cooperation of relatively unspecialized apparatus.

3.4 Whither the Concept of Local Function?

Considering all of the evidence presented so far, the functional architecture of the brain appears to have the following characteristics:

1. Local neural units are used and reused in multiple cognitive/behavioral circumstances, across traditional task categories. Large-scale function is achieved by local genetic and global network dynamics that produce sets of interacting transiently assembled local neural subsystems (TALoNS) for various purposes.
2. The synaptically connected element of the brain network (the wiring transmission, or WT, network) has multiple states or configurations because each synapse can be modulated by extrinsic factors including genetic modulation and glia-neuron interactions. Absent evidence to the contrary, I am working under the assumption that there is no one-to-one relationship between network states and tasks. That is, I believe that some of the observed functional diversity of brain regions is due to modulation of the

local network that changes its causal-functional properties; and some of the observed diversity is due to the fact that the causal-functional properties a network has in each of its configurations are in fact useful in more than one cognitive/behavioral circumstance.

3. In addition, there is a nonsynaptic volume transmission (VT) system overlaid on top of the synaptic network that is perhaps better modeled as a field than as a network. The VT field is poorly understood but clearly operates according to a different set of rules from the WT network. It is presumably more deeply stochastic, exhibits a different temporal dynamics, and is modulated by a different set of factors including the amount and anisotropy of the intercellular space. It also interacts with the WT network in important ways, not just modulating elements of the WT network but also allowing for large-scale integration and coherent performance (Agnati & Fuxe 2000), and it potentially facilitates the discovery during learning of alternate network configurations via a search process different and distinct from the incremental change in synaptic strength characteristic of quasi-Hebbian learning and also captured, for instance, in typical backpropagation algorithms (e.g., Rummelhart, Hinton, & Williams 1986a). That is, rather than being restricted to the slow alteration of overall WT network weights to acquire a new capability, VT allows for the testing of new configurations, the successful instances of which could then be incorporated into the WT network. Moreover, this learning process often simultaneously establishes multiple WT networks (Merabet et al. 2008), leading to an "overconnected" system that exhibits both flexibility and degeneracy (Sporns 2011; Bargmann 2012).

Thus, there is clearly a deep sense in which local neural assemblies are polymorphic and "multifunctional." But how should we pin down the meaning of "multifunctional" in concrete terms? One promising option is the notion of programmable neural networks (Donnarumma, Prevete, & Trautteur 2012), itself closely related to the idea of polymorphic networks (Getting & Denkin 1985). Donnarumma et al. envision a network such that the weights of the connections are actually stored as multiple alternative matrices that can be dynamically swapped, thus changing the behavior of the network over very short time scales. Although the mechanism they implement for this switching in the programmable artificial neural networks they have built has no plausible biological analogue, I do think the idea is worth pursuing. One could model a given local network of the brain as a set of neural nodes plus multiple matrices specifying the connection weights they take on under different circumstances. Under this scheme it would be possible to map a specific set of uses to each of the weight-state configurations of the network, and in some cases it might even be possible to establish workings for each of the configurations that account for the various uses of the network. Programmable neural networks thus take up the challenge of multifunctionality in natural brain networks directly.

Among the strengths of this approach is that it would allow for a systematic investigation of cognitive interference, the mechanisms underlying which are relatively poorly understood. Cognitive interference occurs when the simultaneous performance of multiple tasks produces performance deficits in some or all of the tasks. Some tasks—visual and auditory memory, for instance—are compatible, but many are not, presumably because they use overlapping resources in incompatible ways. From the current perspective there are several questions raised by the incompatibility of task pairs. For instance, are the observed performance effects a matter of the costs of neural configuration switching? And if so, what are the mechanisms that allow for sufficient retention of state across configurations to permit even degraded performance on two tasks? Or, if the performance issues are instead a result of trying to use a network—locked in one configuration in virtue of being cognitively occupied—for an alternate use, then why do we generally see bidirectional performance effects (that is, decrements in performance in both of the interfering tasks)? Are there cases of unidirectional (or nearly unidirectional) cognitive interference? If so, that might be a sign of two tasks that generally use the same local network in different configurations, but the network was unable to switch configurations to accommodate the second task, leading to a performance decrement primarily in one of the two interfering tasks. Alternately, it may turn out that it is not generally possible to get cognitive interference across configurations; that is, if a network is locked in one configuration by a given task, then a second task that requires that same network to be in a different configuration will simply be impossible to perform until the first task has been completed. This latter outcome would mean that all observed cases of cognitive interference are those where the two tasks use the network in the same configuration, and the interference would be due to conflict between the sort of activity required by each task. Any of these outcomes would be scientifically very interesting, and programmable neural networks offer a paradigm within which to pursue these and other related questions.

There are, however, some weaknesses to the approach. One already mentioned is that the particular mechanisms proposed by Donnarumma et al. are biologically implausible, and moreover, as a model of brain function it fails to incorporate the VT network. Thus, significant work would have to be put into making the model more biologically realistic before it would be of maximal usefulness in cognitive neuroscience. Another weakness is that, at least on the biological side, we await technological advances sufficient to begin the process of enumerating the possible connection states of a given network; we have to be able not only to see the synapses but also to know when they are functionally effective. That is, although the theoretical work can proceed apace, the practice of mapping functions to actual biological networks will require that we can identify the various states of those networks to which we would like to ascribe function. This we are not yet able to do, at least in the brains of humans.

The final weakness I mention is not specific to this proposal but is shared by nearly all approaches to function-structure mapping in the brain: although mapping sets of functions to local networks represents an advance over the one-to-one structure-function relationships typically sought in the cognitive neurosciences, this model still assumes that the goal in the cognitive neurosciences is discrete mappings between well-defined cognitive operations and well-defined biological structures (Bechtel 2012; Bechtel & Richardson 2010). I have no particular argument with the practice of mechanistic analysis and decomposition, especially when it is sensitive to the importance of high-level dynamic interactions (Bechtel & Abrahamson 2010; Bechtel & Richardson 2010). When it works it can provide elegant models that offer genuine insight into both the normal operations and the specific malfunctions of cognitive systems, as has been beautifully demonstrated in the case of human spatial memory, for instance (Craver 2007). And quite clearly, much of the work I am doing in this book can and should be characterized as offering mechanistic explanations of just the sort Bechtel, Craver, and others describe. But given the picture painted above—of temporally dynamic networks with multiple physical connection states and physical and functional boundaries made fuzzy by the diffusion of neurotransmitters to synaptically unconnected neurons (which some researchers suspect actually comprise the *majority* of interneural communication, see Bach-y-Rita 2004)—I myself have become pessimistic that it will *generally* be possible to identify the specific neural mechanisms (the "working parts"; Craver 2007) to which we can assign cognitive operations. Call this the "identification problem." Moreover, the CCTM paradigm also assumes that it will be generally possible to break cognitive processing into discrete operations that interact in a linearly tractable, additive, componential fashion. Given the fluidity of behavior, the importance to cognition of interactions with elements of the external world (including other beings), and our remarkable sensitivity to (and robustness in the face of) environmental perturbations—just to name a very few considerations—this increasingly seems to me to be a very ungrounded assumption; call this the "decomposition problem" (Anderson et al. 2012). And even if cognitive processes *are* generally decomposable in the necessary way, it strikes me as incredibly unlikely that we have already identified the right set of candidate operations—weight perception, angle estimation, face recognition (Kanwisher et al. 1997), mind reading (Young, Dodell-Feder, & Saxe 2010), agency detection (Blakemore et al. 2003), and so on.

For all of these reasons, in the next chapter I turn to developing an empirical alternative to the analyze, decompose, and localize approach to function-structure mapping that permits us to do good science while doing better justice to the complexity of the functional profiles in the brain. In particular I argue that we should supplement *componential analysis* with a *dispositional vector* account of brain activity.

Interlude 3 The Dynamic Brain: What Your Brain Is Doing When It's Not Doing Anything

Over the past few years there has been growing interest in something called "resting state functional MRI," a technique for seeing what your brain is doing when you aren't doing much of anything at all. It turns out that brains at rest are pretty restless, consuming far more energy than they do when *doing*. More interesting, "resting" activity is not random but highly coherent, consistent, and predictable. The discovery of the brain's characteristic resting behavior led some years ago to the postulation of a "default network" for the brain—a set of regions that consistently cooperate to do ... well, what, exactly, we don't know. But surely it must be something interesting. Your brain would hardly waste all of that energy dancing to the beat of its inner drummer if it didn't serve some function, right?

Our ignorance regarding the function of all that fluctuation isn't for lack of trying. The discovery of the brain's default network has led to hundreds of studies relating the default network to the brain's anatomical structure as well as to mood disorders such as depression, developmental problems such as autism, and degenerative diseases such as Alzheimer's. It has even been suggested that resting-state activity holds the key (cue deep voice and echo effect) *to understanding consciousness itself* (e.g., Raichle 2010). Now, when neuroscientists start brandishing the "c" word, there are two predictable reactions: increased public interest and attention and increased scientific scrutiny and criticism. Both have happened here, generating a cadre of enthusiastic adherents and an equally committed group of critics who question whether we should continue wasting *our* energy figuring out why the brain appears to be wasting *its* energy. Or, as one prominent neuroscientist put it to me recently, "It's just such a fad. I kind of hate it."

Hyperbolic allusions to cracking the mystery of consciousness excepted, I don't think anybody should hate it. But to see why we should if not *love* then at least *care about* the brain's intrinsic activity requires us to think about brain function in a new and unfamiliar way.

The brain is, in essence, a collection of oscillators: billions of biological yo-yos going endlessly up and down. The electrical charge in individual neurons rises and falls, as do the local concentrations of neurotransmitters such as dopamine and serotonin, the amount of oxygen in the local blood supply, and the brain's overall electrical field (creating the brain waves you can see with an EEG). Each of these oscillations has a different beat for different circumstances—the fast "beta" waves seen in the alert working brain and the much slower "delta" waves of dreamless slumber. These rhythms interact in various ways, from the shared swaying of a slow dance to the counterbalanced ups and downs of two kids on a seesaw, all the way to the complex, syncopated interplay of a jazz band. And like all pendulums, each of the brain's various oscillators also has its preferred swing, a way it will move if left to its own devices. The combination of all these individual, periodic, preferred oscillations *is* the brain's resting state.

Which brings us to the first reason to care about intrinsic activity: what we typically think of as brain function—seeing, thinking, deciding, acting—is actually a *disturbance*, an alteration of the brain's natural harmonies. When we think about brain function, then, it is not enough to ask "why this activity?" we must also ask "why this *change*?" Why did this (perceptual, behavioral, electrical, chemical) input cause *this* exact deviation? If we want to understand the brain's particular sensitivities, we need to know not just *that* some part of it reacts to some stimulus or task but *how much* it reacts, and what else reacts (and interacts) with it. Resting-state fMRI is one method for ascertaining the baseline from which such changes must be measured.

Similarly, just as in music no note has meaning in isolation, so too the brain's local activity can only be fully understood when set against its background. Consider in this regard hippocampal place cells—those neurons famous for firing whenever an animal is at some particular spot in the world. The apparent one-to-one mapping between hippocampal cells and environmental locations has led their function to be understood in terms of simple location representation, as if each cell is designed for saying: "you are here." But these cells have been in the news lately because of the discovery that they are also sequentially activated *prior* to novel spatial exploration, which indicates that their function is not so simple and underlines the point above that to understand the brain's activity *now,* one must also consider what its activity *was* and *will be* (Dragoi & Tonegawa 2011).

Moreover—and more to the current point—these cells fire not just when an animal is at a given location but just before and just after too.

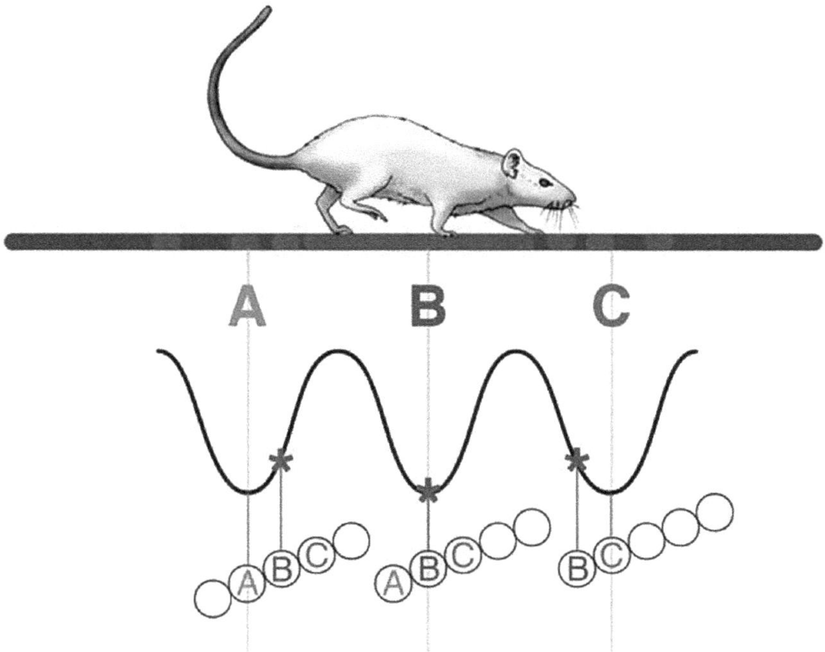

Figure 3a.1
Place cell B fires at different times relative to the background theta-wave as the rat moves from locations A through C. Reprinted from Buckner (2010) with permission.

Interestingly, the differences among concurrent firing (the "you are here" signal), prospective firing (signaling in advance of being at a location), and retrospective firing (signaling after the animal has left a location) are marked not by any difference in the neuron's activity itself but rather by its relationship to the background theta-band (~6–10 Hz) oscillation of the whole hippocampus. In its retrospective role the cell fires earlier, and in its prospective role later, in the theta cycle than it does when the animal is actually at the location in question (figure 3a.1) (Buckner 2010).

In other words, what that cell's activity *means*—what it is actually signaling—depends on how that activity relates to the ongoing background oscillations.

Thus, understanding the resting state is important because it draws our attention to the ever-present background against which function needs to be measured and meaning assessed. In the dynamic brain, local activity is always a change from what was happening before and occurs in relation to all the other things happening now.

It is hard to think about the brain this way, especially given the continuing influence of the computer metaphor for the brain and of the focus on localization (vision here, language over there, motor control in this bit) that we inherited from an earlier age of neuroscientific investigation. In a computer what matters is the nature of the local processing: what's happening in this chip at this time. Background processes are irrelevant to the process of interest, which can therefore be safely isolated and studied as such. But the brain is not that sort of organ: the background is not irrelevant, and relationships between oscillations do a lot of the functional work.

This leaves us with a lot of thinking still to be done. What does it mean for brain function to be defined not just by the intrinsic characteristics of the supporting neural activity, but also by the deviation that it represents from some default? Can the change from one dynamic equilibrium to another be itself a functional event? How is it possible for an organ to do work not with oscillations themselves but with relationships between them? We are a long way from answering these questions, but we've made some promising beginnings. One theoretical development of particular interest is called liquid-state computing, an attempt to understand how information processing can be carried out by variously coupled oscillators that respond to inputs the way a pond does to a stone. We certainly need more such pioneering efforts. For now, however, the important point is this: measuring, characterizing, and reflecting on the resting dynamics of the brain are important early steps toward understanding the *functional* dynamics of the brain in terms much more appropriate to the biological reality than electrons whizzing around in isolated silicon processors.

4 Do Brain Regions Have Personalities of Their Own? Toward a Dispositional Neuroscience

Over the past few chapters I have tried to paint a picture of the overall functional architecture of the adult human brain that (1) permits a kind of comprehensive abstract image of its functional structure, (2) illuminates its evolutionary and developmental origins, and (3) does justice to the significant functional complexity of its individual parts. I have also suggested that many elements of this picture sit uneasily not just with some leading theories of neural architecture such as modularity but also with the prevailing interpretive framework adopted in the cognitive neurosciences, which centers on the analysis, decomposition, and localization of component cognitive operations. This latter suggestion has so far been made using a fairly gentle tone, but here I am more blunt: *we cannot continue to do human cognitive neuroscience that way.*

The primary reason we cannot is that we currently lack the technology necessary to identify individual networks and their various functional states in the living brain. Without the ability to do this, any specific functional attribution to a particular structure of the brain will be scientifically unreliable because the set of observations of the functional activity/relevance of that structure will contain a heterogeneous mixture of data collected while the structure was in an unknown number of different connection states. With no way to sort the observations by states, attributions of underlying fundamental function to TALoNS will be no more reliable than specifying the elemental properties of earth or air or seawater, for it will not be possible to know which observations a given functional operation needs to explain. It should go without saying that we must also curtail the common practice of reverse inference—of assuming that the observation of activity in a given region of the brain signifies the same thing each time, so that, for instance, activity in V1 that indicates that a specific visual operation is being engaged, activity in M1 means a motor operation is being engaged, and activity in the amygdala necessarily signifies emotional processing.

Unless and until we are able to identify the individual connection states of the brain's networks—to identify the actual working parts, the TALoNS of the mechanism (a prospect that will continue to be complicated by the volume transmission system)—this interpretive pathway is closed to us.

Naturally, I do not expect the preceding paragraph to close the world's cognitive neuroscience labs. But I do hope that this book can help the scientists in those labs to reconsider how to interpret their data and in some cases how to design their experiments. For that wish to have any chance of fulfillment, of course, I need to provide some alternative empirical and interpretive pathways. This chapter offers two. The first possibility is one that I think will occur to many in the field because investigating functional connectivity is becoming increasingly popular and applying it here would appear to offer a way to retain much of our current interpretive practice with a relatively minor methodological shift (see, e.g., Klein 2012). I argue, however, that this promise is in fact illusory, and although functional connectivity is and should remain an important avenue for investigating brain function, it is not likely to solve the fundamental interpretive problems presented by the functional complexity of individual brain regions and networks. Thus, I outline another, more novel approach to investigating the brain that I hope can allow us to do justice to the observed functional complexity while still offering epistemic purchase on underlying causes.

To do this, I highlight emerging work in which the aim is to measure the responses of the brain in a multidimensional manner and try to discern the structure that underlies these responses. Just as persons have characteristic dispositions to respond in particular ways in a variety of circumstances—a set of dispositions we understand and explain (in part) in terms of their individual personalities—so I suggest by analogy that we might usefully focus the cognitive neurosciences on uncovering the "personalities" of brain regions. I argue that this empirical avenue promises to shed light on the functional dispositions of individual regions, their underlying causal powers, and their propensities to cooperate with sets of other regions, thus simultaneously doing justice to both functional differentiation and functional integration in the brain. This approach is elaborated starting in section 4.2.

4.1 Network State Identification via Functional Connectivity Analysis

Functional connectivity analysis refers to a family of techniques for investigating relationships indicative of functional cooperation or integration in the brain. In general the idea is to take a set of observations of activity in the brain—whether derived from fMRI, EEG, or some other technique—and search for deviations from statistical independence in the activities of individual neural regions. Typical methods involve looking for correlated activity, but one might also look for other indications of functional relationships such as phase or spectral coherence (Sporns 2011; Sporns, Tononi, & Edelman 2000). And in fact there are a set of closely related techniques known as "effective connectivity analysis" that apply such empirical tools as Granger causality or transfer entropy to estimate causal relationships between brain regions (Seth 2005; Seth & Edelman 2007; Sporns 2011).

Perhaps the best known use of these techniques is in the investigation of the "resting state" or "default mode" of the brain (Grecius et al. 2003; Honey et al. 2009; Raichle et al. 2001)—the set of interregionally coherent oscillations that emerge when one is not engaged in any particular cognitive or behavioral task (see interlude 3). But estimating functional connectivity has emerged as an important technique for investigating interregional cooperation in particular cognitive tasks (Bitan et al. 2009; Cantero & Atienza 2005; Gazzaley, Rissman, & D'Esposito 2004; Varela et al. 2001), and even as a tool for understanding psychological and neural disorders such as autism (Barttfeld et al. 2011; Just et al. 2004; Minshew & Keller 2010), Alzheimer's disease (Stam et al. 2007), and schizophrenia (Skudlarski et al. 2010).

Given the increasing popularity of this technique and its undeniable importance in investigating functional integration in the human brain, one might be tempted to approach the conundrum of specifying network states by taking them to be reflected in different functional connectivity states. Such an approach might proceed as follows: for each region of interest (ROI) in the brain, one would catalog the various sets of functional partnerships that the ROI engages. So, for the sake of illustration, assume that our ROI cooperates with regions *A*, *B*, and *C* during some set of tasks $\{t_1\}$; with *C*, *D*, and *E* during task set $\{t_2\}$, and with *A*, *F*, and *G* during task set $\{t_3\}$. If the set of partnerships reflects the underlying network state of the ROI, then it would seem that one could in this case assign three different cognitive operations or workings to the ROI, reflecting its three different network states. Each operation would explain some aspect of each of the task sets—that is, the operations would be hypotheses about what the ROI does for or provides to its partners so that together they can support the tasks in each set. (I am here making no assumptions about the degree of similarity among the tasks in these sets; naturally generating hypotheses could be more or less difficult depending on their degree of similarity; see Penner-Wilger & Anderson 2008 for a discussion.)

The trouble with this approach is in the assumption that local network states are necessarily reflected in patterns of functional connectivity. This *could* be true but need not be. Consider two kinds of changes that could take place in a given ROI: changes in its peripheral input-output (I/O) relations and changes in its central processing organization. In our current example suppose that the differences in functional partnerships observed when switching from tasks in $\{t_1\}$ to those in $\{t_2\}$ reflect only changes in I/O relationships; in this case the ROI would be providing the same service to (would be one of the TALoNS for) a different set of partners. In contrast, suppose switching from tasks in $\{t_1\}$ to those in $\{t_3\}$ requires both I/O changes and changes to central processing connections—the activation or deactivation of some set of synapses. In this case the ROI would do something different for these different partnerships. Finally, we can imagine a task set $\{t_4\}$ in which the ROI *also* engages partners *A*, *B*, and *C* but with a different set of central processing connections from those used to support $\{t_1\}$. Here the ROI would be doing something different for these partners—and without the ability to directly track the underlying network state, we would not be in a position to notice this. In fact, following the investigative logic outlined above, $\{t_4\}$ would end up being an invisible subset of $\{t_1\}$, rendering suspect any hypothesis regarding the TALoNS' underlying operation (for the set $\{t_1 + t_4\}$ would in fact involve the contribution of two different TALoNS). Moreover, in the scenario as described, the correct set of tasks for attributing a component operation to the ROI would be $\{t_1 + t_2 - t_4\}$. So in fact any attribution made on the basis of $\{t_1\}$ would be suspect for multiple reasons.

As I suggested in section 3.4, it may be that a systematic investigation into cognitive interference and its underlying neural causes will put us on a path where we *could* specify the sets of tasks supported by a given network in a given underlying state. Similarly, models of the network that show how it could contribute to some group of tasks but not to some others while in the same underlying configuration could provide evidence leading to the generation of task sets that more clearly correspond to underlying network states. Or there may come along some technological solution to determining which synapses in a network are and are not functionally effective at a given time. Under these conditions we could apply cross-domain modeling (Penner-Wilger & Anderson 2008) to generate hypotheses for the local operation performed or service offered by the network in that state. But so long as we lack the ability to detect this kind of heterogeneity in task sets—instances when the tasks in question are in fact supported by networks in different underlying states—I think a healthy skepticism is the proper attitude to take toward hypotheses concerning the operations implemented by individual TALoNS. Moreover, the discussion above assumes that *if* we could identify individual network states, *then* each implements an operation that contributes to overall function in a componential way. Although this scenario is likely to obtain in many cases, given the importance of interactivity and context to brain function, it seems unlikely to obtain in all cases. Thus, we appear to need alternative frameworks for the study of brain function, and I argue below that we in fact have some.

4.2 Multidimensional Functional Representations for Neuroscience

Although it is not possible, given the current state of the science, to specify the fundamental operations of local networks, we nevertheless often have immense amounts of information about their functional properties. Thus, my suggestion for how to proceed is simple and straightforward: *we should enumerate and try to understand these functional properties*. I have recently been working with Josh Kinnison, Luiz Pessoa, and Lucina Uddin to do just that (Anderson et al. 2013; Uddin et al. 2014). I think it is worth describing this work in some detail, as it offers one model for how to do cognitive neuroscience without the analysis, decomposition, and localization of component cognitive operations.

The approach relies on the existence of large collections of data—in the cases I describe, neuroimaging data (Fox & Lancaster 2002; Yarkoni et al. 2011)—but the techniques are general and can be applied to information about brain function derived from any technique. In chapter 1 I described the work we did to quantify the diversity of tasks that a given ROI supports. We have recently extended that work to generate *functional fingerprints* (Passingham, Stephan, & Kotter 2002) for our ROIs, such as those shown in figure 4.1

These plots were constructed by determining how often each region of the brain was active in around 2,000 experimental tasks from the 20 different BrainMap task domains shown in the figure. These domains were chosen because they represent a diversity of task conditions and have fairly good representation in the data set. But there is nothing particularly special about these domains; they simply represent a reasonable place to start. As is discussed starting in section 4.3, below, future work can and should explore various other options for quantifying the multidimensional activity budgets for individual regions of the brain. In any case it is quite clear, even given these preliminary analyses that different regions have different activity budgets—that is, different functional fingerprints that index their

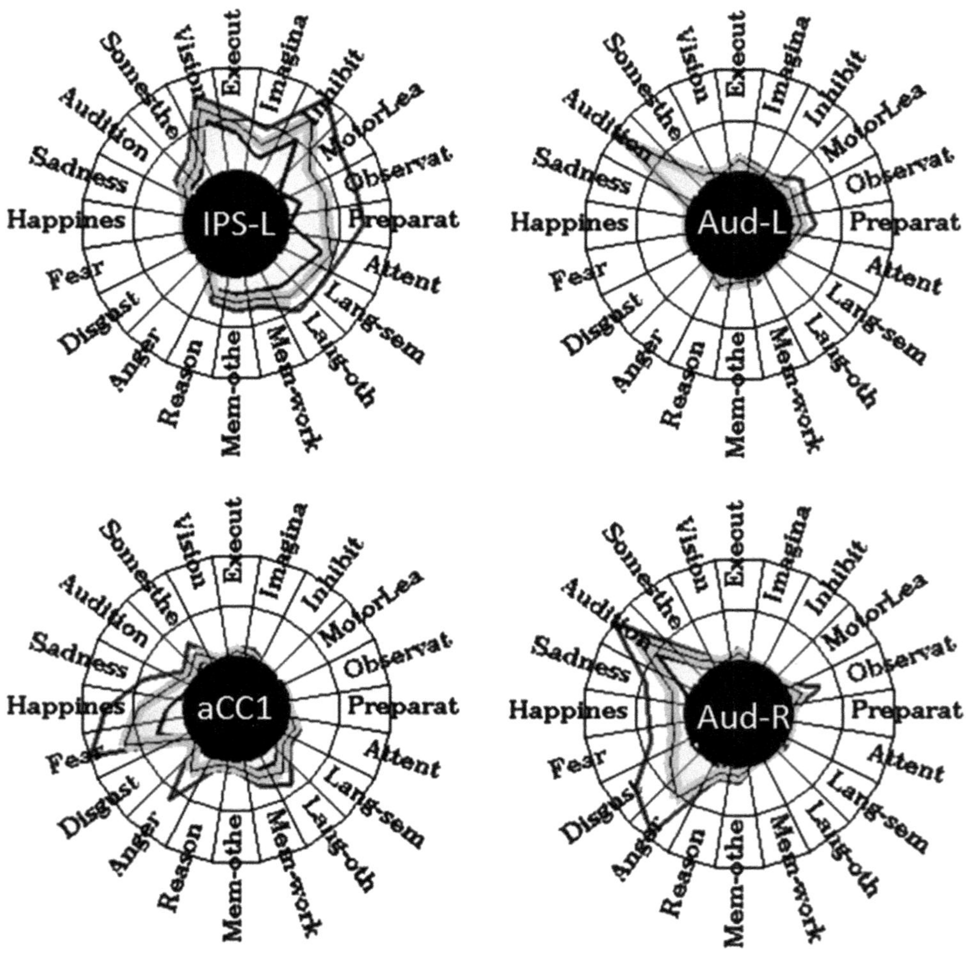

Figure 4.1
Functional fingerprint plots for four regions of the brain: left inferior parietal sulcus, left anterior cingulate, and left and right auditory cortices. The plots show the relative degree of activity of the regions under the 20 task conditions listed (middle line; inner and outer lines represent confidence intervals).

underlying functional biases. Some regions such as left auditory cortex are relative specialists, whereas others such as the left inferior parietal sulcus and left anterior cingulate show a much broader set of responses, although each structure achieves its relatively high functional diversity with a different mix of task responsiveness.

Looking at the data in this way allows us to quantify aspects of the brain that remain hidden when more traditional methods are used. In one study we performed a systematic, voxelwise exploration of the functional fingerprints in the brain using a spherical "searchlight" with a radius of 10 mm. That is, for each location in the brain, we drew an imaginary sphere with a 10-mm radius centered at that location and then generated a functional fingerprint using all of the activations that fell within the sphere. Doing this allowed us to quantify the relative similarity of the fingerprints across the whole brain. We discovered, for instance, that the functional fingerprints tend to change slowly from location to location, such that neighboring regions of the brain have relatively similar functional fingerprints, a finding that is certainly consistent with the idea of a regional functional differentiation. Somewhat more interestingly, we discovered that, in general, voxels with similar functional fingerprints were more likely to coactivate with one another—that is, to be active during the same experiment.

This is a fairly striking finding, as the data set used to calculate coactivation likelihood was the completely independent Neurosynth database (Yarkoni et al. 2011) containing at the time of the study more than 4,300 imaging studies across a wide variety of task domains and conditions. We also excluded any voxels within 20 mm of the seed voxel from the analysis. So we know that the result was not an artifact of overlap between the studies used to generate the functional fingerprint and those used to calculate coactivation frequency, nor of the generally slow change in functional fingerprints across the brain and the fact that neighboring regions will naturally tend to coactivate. Instead, it appears to be a fundamental aspect of the functional architecture of the brain. It also underlines the potential scientific utility of quantifying the multidimensional functional fingerprint of brain regions. What we showed was that what is at best a partial measure of the functional biases of a region—using only a couple of thousand studies in a selected number of task categories—is still sufficient to predict which other regions are likely to be active under the same circumstances (coactive in a single experimental condition). With increasing similarity of functional fingerprint, you also get increasing likelihood of responding under the same conditions—and this strongly suggests that the functional

Table 4.1
A listing of the networks investigated and where they were originally described

Network	Abbreviation	Function/label
Frontoparietal (Toro, Fox, & Paus 2008)	FrontPar_T	Task-positive
Cinguloparietal (Toro et al. 2008)	CingPar_T	Task-negative
Dorsal attention* (Yeo et al. 2011)	DorsAtt	Goal-directed attention
Ventral Attention* (Yeo et al. 2011)	VentAtt	Stimulus-driven attention
Control* (Yeo et al. 2011)	Control	Control
Default* (Yeo et al. 2011)	Default	Default
Reference (Van Dijk et al. 2010)	Ref	Not applicable; defined to be an experimental control
Frontoparietal (Dosenbach et al. 2007)	FrontPar_D	Rapid adaptive control
Cingulo-opercular (Dosenbach et al. 2007)	CingOper_D	Stable set control

Asterisk indicates that the network was defined using resting-state data.

fingerprints are successfully capturing, at least in part, the underlying functional biases of the regions.

To further investigate the relationship between similarity of functional fingerprint and functional cooperation, we explored these properties using several brain networks that have been described in the literature (listed in table 4.1). Because these networks consist of functionally connected regions, we wondered if they would also tend to be "functionally assortative," that is, composed of regions with similar functional fingerprints. Interestingly, we discovered that networks are often, but not always, functionally assortative. We investigated this in two ways.

For each constituent of each network we first asked whether it was more similar to other members of its home network or more similar to the members of any of the other networks. Interestingly, although most pairs of networks showed significant distinctiveness in the functional fingerprints of their constituents, in many cases the networks were essentially indistinguishable. For instance, a clear difference is detected between FrontPar_T and CingPar_T, task-positive and task-negative networks, respectively. On the other hand, FrontPar_D and CingOper_D, executive control networks defined based on behavioral task results and resting-state data were essentially indistinguishable (see Anderson et al. 2013 for full description and discussion).

To quantify this functional assortativity further, for each of the networks listed in table 4.1, we formally quantified the assortativity of the networks relative to a set of null models. Functional fingerprints can be represented in various ways, but one way to think about them is as a multidimensional vector: a string of numbers that specifies a location in a multidimensional space. To determine assortativity—functional similarity between members of the networks—we computed the average euclidean distance between the functional fingerprint vectors *within* each network and compared this to the average distances observed in the null model distribution. Here a functionally assortative network would be one in which the average distance was *smaller* than mean distances in the null model distribution; and a disassortative network would be one in which average distances were greater. Assortativity was quantified in terms of percentiles—that is, where it fell in the distribution of permutations of the model. The results are shown in figure 4.2. Interestingly, most of the networks were positively assortative, and some robustly so (above the 90th percentile). In contrast, the default network was not assortative, and one version of the task-negative network, CingPar_T, actually was somewhat *disassortative*. (The control network was defined in a nonstandard way [Yeo et al. 2011] so it is difficult to comment on its assortativity value.) This may suggest that the default mode network is composed of more heterogeneous regions than the task networks investigated. Alternately, it could be that the relative lack of assortativity in the

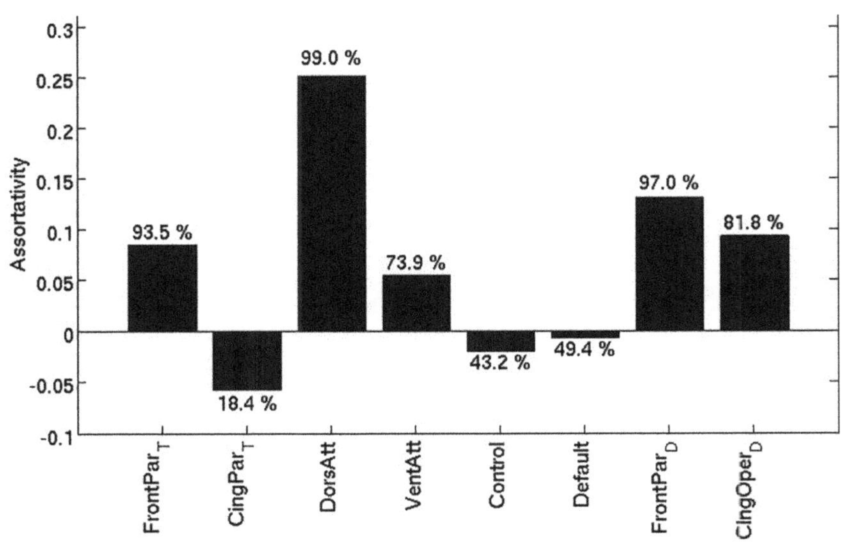

Figure 4.2
Network assortativity results. Percentile scores indicate the degree of assortativity relative to a null distribution.

default network indicates that it should be further divided into more functionally homogeneous groupings. Apropos of this, it is important to note that Yeo et al. (2011) did warn that some of the networks they described could be further fractionated. If positive functional assortativity turns out to be a typical property of brain networks, our analysis suggests that the default network could probably be further divided. In any case this possibility deserves further exploration.

In fact, we are not the only ones to suggest that the default network might actually consist of multiple components. For instance, Andrews-Hanna and colleagues (2010) have proposed breaking the default network into two dissociated subsystems: one in dorsal-medial prefrontal cortex and another in medial temporal lobe, along with a midline "core" involving anterior-medial prefrontal cortex and posterior cingulate that is common to (integrates) both subsystems. To investigate, we generated network functional fingerprints for the three proposed elements of the default network by taking the union of the functional fingerprints of the network constituents (here using a reduced set of domains for ease of visual comparison). The results are shown in figure 4.3.

As one can see, the fingerprints of the two subsystems are complementary, and the core does indeed appear to reflect elements of both subsystems, which is consistent with a role in integrating across the functionally dissimilar subsystems. This finding appears to suggest that functional assortativity is indeed a useful tool in quantifying the characteristics of brain networks and also once again underlines the scientific utility of specifying functional fingerprints. Without functional fingerprints, it is not possible to even *notice* functional assortativity, much less use it for scientific investigations. Here the fingerprints were able to provide evidence to help illuminate a debate about the unity or disunity of a particular network in the brain.

We have also explored the nature of these brain networks further by building functional fingerprints for networks as a whole that reflect the union of the functional fingerprints of their constituents (figure 4.4). Building the network fingerprints for all the networks investigated, we noticed something else quite interesting that will open up new avenues of future exploration: even apparently functionally selective networks such as the attention networks are not composed of functionally selective components! This further underlines the claims made in earlier chapters that individual regions of the brain appear to offer their services within multiple functional partnerships. It also suggests that we have a great deal to explore with respect to understanding functional specificity in networks: why should

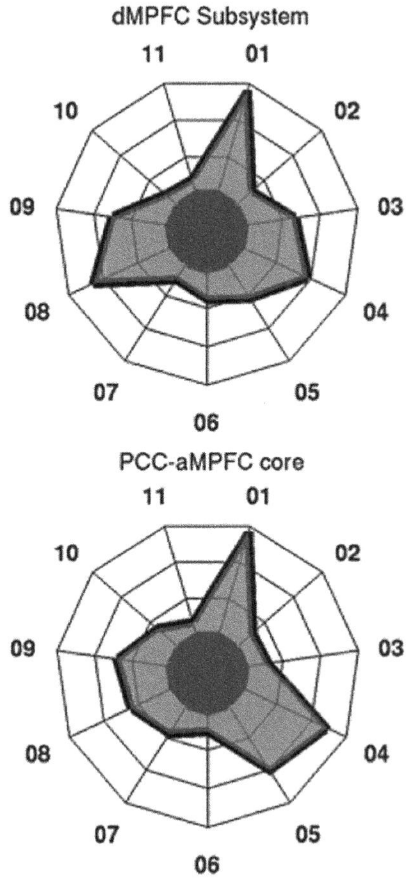

Figure 4.3
Network functional fingerprints of the proposed subdivision of the default network.

that particular mix of underlying functional biases lead to an attention network, while *this* mix leads to an adaptive control network? These are crucial questions, and I believe that working with functional fingerprints can help answer them.

Here it is worth recalling the case of starburst amacrine cells (SACs) in the mammalian retina (see section 3.2 and interlude 4). If one were to measure the responses of these cells over time, they would appear to be sensitive to motion in *all* directions and in this sense multifunctional. And yet the cells have individually direction-selective *parts* and play a role in different high-level uses, precisely because of their relationships with other cells. Thus, they appear to illustrate the principle of achieving temporary

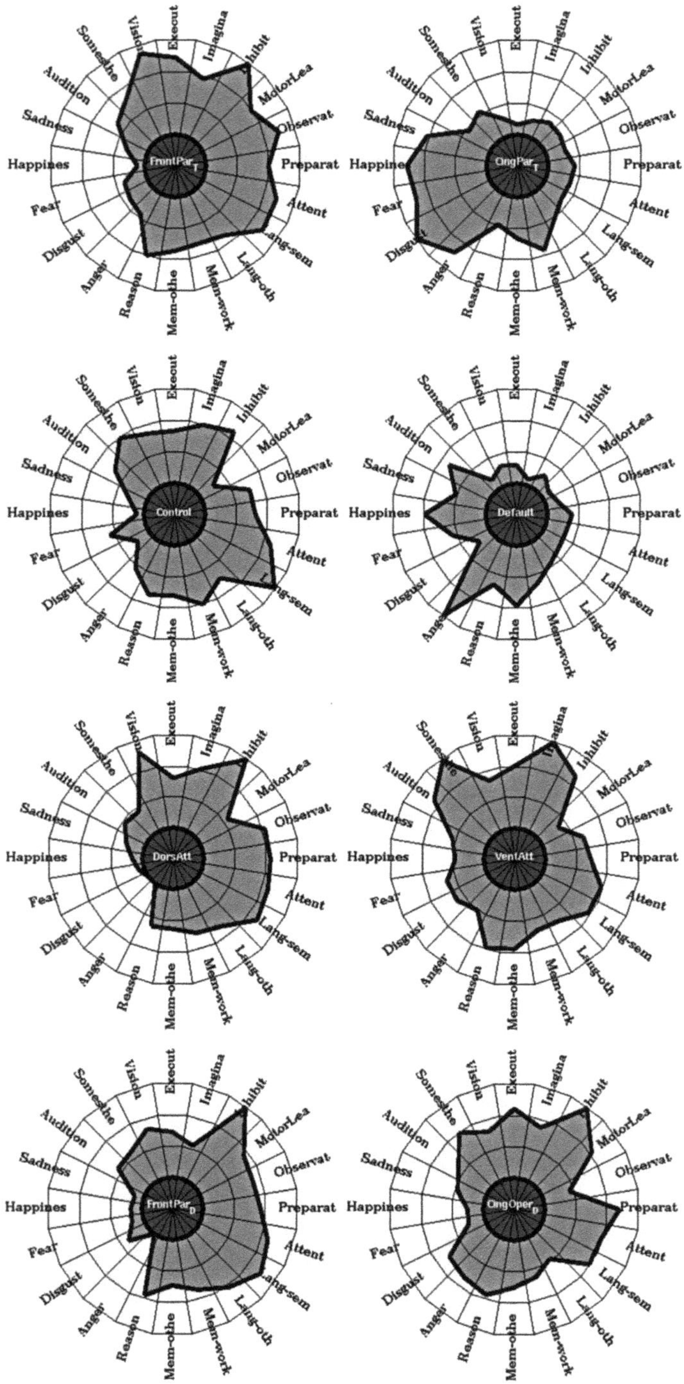

Figure 4.4

Network functional fingerprints. Not just regions but also networks generally show unique functional fingerprints.

local specificity and specific high-level function via network interactions. The interactive mechanisms important in that case, including selective reinforcement and inhibition of response tendencies, seem likely to play an important role in determining the functional properties of large-scale networks as well and to partially determine the functional contributions of their constituents while acting as part of that particular network. This further underscores the crucial importance of interactive dynamics in the brain (Anderson et al. 2012), which must therefore be approached with tools different from those used to understand primarily componential systems. We pick up this thread again in chapter 6 with a discussion of biased pattern competition as a fundamental mechanism for achieving high-level function via network interactions.

The techniques I have been describing can of course also be used to thoroughly investigate the functional properties of individual ROIs. In one series of investigations Lucina Uddin, Josh Kinnison, Luiz Pessoa, and I (2014) turned our attention to the insula. We showed, among other things, that the tripartite insula subdivision suggested by Deen, Pitskel, and Pelphrey (2010) based on anatomical connectivity data does indeed break the insula into functionally distinct elements, but with an important caveat.

Our analysis showed that all six insula subdivisions cooperate with a very functionally diverse set of neural partners. Notably, the functional partners of the insula are active at least some of the time in all 32 of the cognitive domains we investigated, leading to very high values for functional diversity. Nevertheless, examining the functional fingerprints in detail reveals that the partners of each individual subdivision have a distinctive set of functional biases or preferences. And yet, although the functional fingerprint of each region possesses unique functional elements, there are also cases of shared biases, especially bilaterally. For instance, both left and right dorsal anterior insula show functional peaks in mathematics and working memory, a feature not seen in any of the other subdivisions. In addition, both left and right ventral anterior insula show peaks in olfaction, sadness, and anger; and both left and right posterior insula show peaks in music, somesthesis, and action execution. Moreover, all six subdivisions are active across a set of tasks spanning all of the domains investigated. Thus, although the results certainly support that contention that these subdivisions of the insular cortex are functionally differentiated from one another, the notion that dorsal anterior insula is more "cognitive," ventral anterior insula more "emotional," and posterior insula more "somatomotor" (Chang et al. 2012; Deen et al. 2010; Kelly et al. 2012; Kurth et al. 2010; Taylor, Seminowicz, & Davis 2009) is in fact an oversimplification of the reality. Using functional fingerprints as an investigative and representational tool permits a much more nuanced account of the functions of these structures than is possible using simple generalizations of functional attributions (Klein 2012; Price & Friston 2005).

In addition to looking at the network diversity of the regions functionally connected to these insula subdivisions, we also estimated information flow through the overall brain network by measuring the "strength" of insula. Strength measures the overall degree to which a structure is functionally connected to other structures in the brain. Using this measure, we discovered that the left and right dorsal anterior insula are among the most important structures in the brain (figure 4.5). This finding, when combined with information about the functional fingerprint of these structures, offers evidence for the hypothesis that anterior insula is part of a "salience network" comprised of this region along with dorsal anterior cingulate (Seeley et al. 2007) that works to detect salient events in the environment and to initiate appropriate control signals (Menon & Uddin 2010). The connectivity profile, functional fingerprint, and node strength together seem

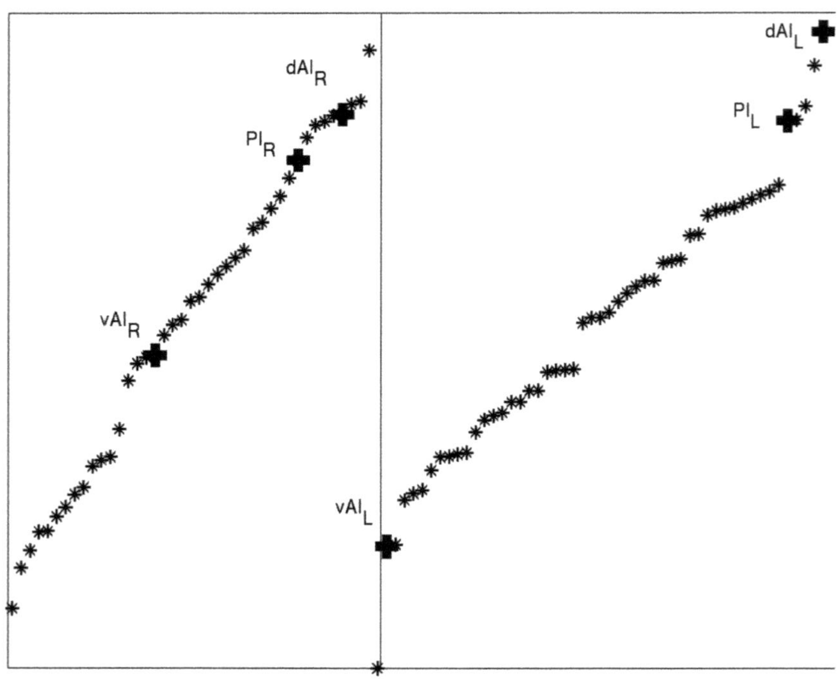

Figure 4.5
Relative node strength of insula and 160 other ROIs in the brain.

to support recently proposed theories positing a unique role for the AI and salience network in orchestrating access to attention and control systems (Menon & Uddin 2010). The results reinforce the emerging picture of insular cortex as standing at the crossroads of interoceptive, affective, attentional, and cognitive control systems of the human brain (Cole & Schneider 2007; Deen et al. 2010; Dosenbach et al. 2007; Taylor et al. 2009). As such, it is perhaps not surprising that this region should contain a critical node for information flow.

It is important to note that my collaborators and I are not the only group trying to work with functional fingerprints or their equivalent. For instance, both Cauda et al. (2012) and Chang et al. (2012) use some version of functional fingerprints in their own investigations of the insula; Cauda et al. use a five-dimensional vector consisting of action, perception, cognition, interoception, and emotion, whereas Chang et al. use a 15-dimensional vector based on a selection of "topics" derived from an analysis of the full texts of the neuroimaging articles in the Neurosynth database (Yarkoni et al. 2011). The topics are represented by such keywords as pain, conflict, inhibition, and anxiety but in fact consist of clusters of words related to these keywords and to each other; their functional fingerprints thus represent correlations between regional activity and the appearance of these topics in the articles reporting the activations. Torta and Cauda (2011) generate six-dimensional behavioral profiles for cingulate cortex across six task domains (attention, pain, emotions, language, memory, and action execution) and five experimental paradigm classes (semantic discrimination, tactile/motor discrimination, film viewing, go/no-go tasks, and visual pursuit); Lancaster et al. (2012) perform a "behavior analysis" for six regions of the brain, estimating the strength of association between these regions and some 52 different task domains; and Hanson and Schmidt (2011) use machine learning techniques (multivariate classifiers) to estimate the relative responsiveness of voxels in the fusiform face area when participants were shown faces, animals, cars, and novel objects. In each case information from the functional fingerprints was crucial to the scientific conclusions, which further showcases the utility and versatility of this technique. Given the state of our knowledge of the brain, and the limitations of more traditional techniques, we need much more such work. In fact, almost all of the work that I recommend in this book is already being done *somewhere* by *someone*. What is lacking is *concentrated, coordinated effort* unified by an appropriate guiding framework. That framework, of course, is one of the things I am hoping to be able to provide; with luck it will lead to the concentrated coordinated effort of many different research groups that is needed.

4.3 From Behavioral Description to the Specification of Underlying Functional Dispositions

One of the things you will have noticed about the work described above is that, although every lab and study employed some form of multidimensional representation of regional brain activity, each used a different number and different sorts of dimensions. Although this may give the appearance of disorder and confusion, it is actually a positive sign and indeed necessary for future progress. To see why, we have to step back a pace and consider at a fundamental level what the practice of making scientific measurements entails and produces.

An experiment is a deliberate intervention into the causal structure of a partially or relatively isolated system. We intentionally manipulate—vary the value of—some physical condition and record an observation of the value of another. In this respect scientific practice is simply a more carefully controlled extension of everyday practice; we are constantly intervening in the world and seeing what happens, and of course conditioning and learning depend crucially on our native sensitivity to contingencies in the environment.

Now, in the natural world as well as in the lab, the signal that our intervention produces—the series of measurements we record—is generally *mixed*, that is, dependent on numerous causal factors that we would ideally like to disentangle (see Stone 2004 for a very nice discussion of the basic epistemic situation). Here is a simple example: we measure weight, but physics teaches us that weight is in fact the product of two more fundamental causal factors, mass and the acceleration produced by gravity. This realization was pretty important to the history of physics and gave us better purchase on the underlying structure of physical reality. Similarly, the varying price of a stock over time is a mixed signal driven by multiple economic factors including the money supply, corporate profits, and perceived innovation, whereas the price of 100 stocks is a set of mixed signals, all being driven by the same set of fundamental factors but to different degrees. The price of a tech stock might be relatively less sensitive to earnings than the price of an energy stock, for instance. That is to say, the importance or weight of each of the factors is different for each stock and thus has more or less impact on the price.

Returning to the brain, then, we need to ask: When we measure the activity of 1 or 100 (or 1,000) different regions of the brain, what is the underlying nature of that set of mixed signals? What (and of what kind) are the factors that contribute to the signal we record? This question can be taken in two ways: every node of an EEG or microelectrode array, for instance, records mixed electrical signals from multiple physical sources—multiple neurons or neural assemblies. The field of neuroscience has been very technically adept in confronting this problem and has devised numerous spike-sorting (e.g., Lewicki 1998) and source localization (e.g., McKeown et al. 1998) algorithms to disentangle the signals attributable to the different physical sources of the mixture. What we have arguably been less attentive to, however, is the fact that these *physically* unmixed signals may nevertheless represent *psychological* mixtures. If this is so—and the evidence presented over the last chapters certainly seems to point in this direction—then the scientific question we need to be asking is not: What are the cognitive operations implemented in individual neural regions? Rather, it is: What are the psychological factors that account for the differential activity of the brain? I call these neuroscientifically relevant psychological (NRP) factors. That we have rarely thought to ask such a question is testament to the subtle but significant effect of the widespread assumption that the observed functional *differentiation* in the brain is the result of underlying functional *specialization*. But it need not be so.

What I am getting at is this: in the same way that analysis of the people's responses in a variety of circumstances (or to a variety of questions) can reveal a common set of factors defining individual personalities, so too can the analysis of the multidimensional functional fingerprints of brain regions and networks reveal a set of primitive psychological factors, or "ingredients," that capture the underlying functional contributions of regions and networks to overall behavior (Barrett & Satpute 2013; Gold et al. 2011; Lindquist & Barrett 2012; Lindquist et al. 2012; Poldrack 2010; Poldrack, Halchenko, & Hanson 2009). This intriguing empirical possibility appears to respect three organizational features of the brain emphasized in the work reviewed thus far: (1) the functional diversity of individual regions of the brain, (2) the functional differentiation of individual regions of the brain, and (3) the frequent functional overlap between the constituents of different brain networks. According to this approach, psychological states such as anger and fear, as well as processes such as attention and cognitive control, would involve different mixtures of many of the same domain-general ingredients.

It is expected that more than one brain region will likely be involved in implementing or contributing each of those psychological ingredients to a given state or process, and each brain region will have a role in generating or contributing more than one ingredient. That is, brain regions and networks will differ not necessarily in terms of their primitive component operations but rather in their loadings on this set of primitive NRP factors (see also Pessoa 2012, 2013 for further discussion).

This is why the current diversity of approaches to the multidimensional measurement of brain activity is a good and necessary development. For the more information we have about the response tendencies of a neural assembly—that it responds to olfaction and emotion and during go/no-go tasks and visual pursuit—the better empirical position we will be in to discern the underlying NRP factors and their relative weightings that account for the observed signals. I am proposing this approach as an explicit alternative to the currently dominant practice of trying to identify the individual component operations or computational algorithms implemented by these regions. You will recall from the introduction that I think that we have already jumped the gun in cognitive neuroscience by adopting functional principles pulled from computational theory before doing very much thorough descriptive analysis. So I believe that at this stage to presume to offer a set of functional principles would just recapitulate this mistake.

Naturally, a call for research aimed at identifying the NRP factors that account for brain activity must rest on the assumption that we have not already identified them. In this regard it is worth considering the challenge to current practice offered in a recent paper by Russell Poldrack (2010). In it, he makes a direct comparison between phrenology and the current search for the mapping between brain function and mental functioning.

> Imagine that fMRI had been invented in the 1860s rather than the 1990s. Instead of being based on modern cognitive psychology, neuroimaging would instead be based on the faculty psychology of Thomas Reid and Dugald Steward, which provided the mental "faculties" that Gall and the phrenologists attempted to map onto the brain. Researchers would have presumably jumped from phrenology to fMRI and performed experiments manipulating the engagement of particular mental faculties or examining the individual differences in the strength of the faculties. They almost certainly would have found regions in which activity was correlated with the strength of each faculty across subjects. In support of this assertion, table [4.2] provides a demonstration of some modern neuroimaging data that the intrepid post-phrenologist might have appealed to in order to demonstrate the neural reality of his proposed faculties. (Poldrack 2010, p. 753)

Poldrack proceeds to argue that, *because* in this imagined age of phrenological imaging the faculties that are being localized are obviously composites, across many studies it would soon become clear that individual neural regions are in fact engaged by the manipulation of multiple faculties. Moreover, because in our actual age this is exactly what we see with imaging

Table 4.2
A mapping of some of Gall's phrenological categories onto the brain

Faculty	Modern neuroimaging equivalent	Associated regions	References
Impulse to propogation	Viewing of romantic lover vs. other individuals	Basal ganglia	Aron et al. (2005)
Friendly attachment or fidelity	Viewing friend versus a stranger	Right temporoparietal cortex	Sugiura et al. (2005)
Poetic talent	Generation of creative versus uncreative narrative	Right medial frontal cortex	Howard-Jones et al. (2005)
Ambition and vanity	Activation for judgment about self versus others	Medial prefrontal cortex	Ochsner et al. (2005)
Sense of locality	Scenes versus nonscenes	Parahippocampal cortex	Epstein & Kanwisher (1998)

Adapted from Poldrack (2010).

guided by the component operations specified by cognitive psychology, this strongly suggests that those operations are composites too.

> Neuroimaging studies rely upon a theory about the structure of the mind that specifies the component operations that comprise mental functions, which I will refer to as a *cognitive ontology* (Bilder et al. 2009; Price & Friston 2005). ... [C]orrectness of the ontology would be reflected in selective association between structures and functions. That is, if a specific structure or network is activated in association with only one putative cognitive process, then one could argue that the reality of this process has been established. A review of the literature suggests that selective association between mental processes and brain structures is currently impossible to find. (Poldrack 2010, pp. 753–754)

Note that the argument only follows given the additional assumption that brain regions are selective for cognitive operations. Without this assumption, overlaps in the neural implementations of these operations are not by themselves evidence for the composite nature of the operations. But although if that premise is false the argument is not logically sound, Poldrack is certainly correct to point out the *risk* involved in borrowing wholesale the categories from a discipline that has typically maintained its autonomy from the neurosciences (Block 1997; Fodor 1997). It would be something of a miracle if the functional categories developed for an autonomous science of the mind were simply the same as those that turn out to do the most explanatory work in a science of the brain. And yet if we are to be naturalists about the mind, we do need a way to make the fundamental explanatory concepts from the two sciences cohere.

Poldrack's solution to this problem is to give the brain its voice in establishing the cognitive ontology via an iterative process of formalizing the cognitive ontology with the participation of relevant experts in the field (Miller et al. 2010; Poldrack et al. 2011; http://www.cognitiveatlas.org). Using specialized empirical techniques to mine collections of functional brain imaging studies (http://www.openfmri.org) for associations between specific regional activations and elements of the developing ontology, Poldrack hopes to put the cognitive neurosciences on firmer functional footing. According to this approach, where selectivity—one-to-one mappings between brain regions and cognitive operations—can be found, confidence in those elements of the ontology will be thereby reinforced; where selectivity is lacking, the relevant operations must be reconsidered. Poldrack, Halchenko, and Hanson (2009) describe a study intended to demonstrate the feasibility of this approach as follows:

> In this study, we first examined how well classifiers can predict which set of eight cognitive tasks a person is engaged in, on the basis of pattern of activation from other individuals, and we found that such predictions can be highly accurate. We next examined the dimensional representation of brain activity underlying this classification accuracy and found that the differences among these tasks can be described in terms of a small set of underlying dimensions. Finally, we examined how these distributed neural dimensions map onto the component cognitive processes that are engaged by the same eight diverse tasks by mapping each task onto an ontology of mental processes. The results demonstrate how neuroimaging can in principle be used to map brain activity onto cognitive processes rather than onto tasks. (Poldrack et al. 2009, p. 1364)

It is worth examining these steps in detail as they are interesting in their own right but also, I think, point the field in a rather different direction from the one Poldrack advocates. The first step is part of the growing trend of using machine learning technologies for "brain reading" (O'Toole et al. 2007), that is, for predicting properties of the experimental environment—the level of the independent variable or the nature of a visual stimulus, for instance—from patterns of activity in the fMRI results of the experiment. In a representative study Haynes and colleagues asked participants to decide either to add or subtract two numbers that had been presented to them (Haynes & Rees 2006). On the basis of fMRI data the experimenters were able to determine with up to 70% accuracy whether the participants would sum the presented numbers or whether they would subtract one number from the other (see Haynes et al. 2007 for a review).

In the current case Poldrack et al. (2009) used fMRI data from eight experiments previously run in Poldrack's lab. The eight experiments tested risk taking, classification, rhyme judgments, working memory, gambling decisions, semantic judgments, reading pseudowords aloud, and response inhibition. Brain activity at a given time can be represented as a list of numbers indicating the degree of activity in each location (voxel) in the brain. That is, it can be described as a high-dimensional vector or as a specific location in a high-dimensional space—in this case ~214,000 dimensions, one for the activity level of each voxel. One can think of the brain state of each participant while performing each task as a labeled dot (p, t)—participant, task—in this multidimensional space, with the combined activity of all the participants in each task forming a "cloud" of points in that space. Data in this form were fed to a multiclass linear support vector machine, which is a machine learning classifier that can learn to draw hyperplanes through the multidimensional space that divide the space in such a way that—in the ideal case—each region in the space contains all and only locations labeled as occurring during a particular task, that is, contains one and only one of the clouds. In general, one trains the classifier on labeled data and then feeds it unlabeled data, which the classifier will label based on the region of space in which the new data fall. The accuracy of the linear classification of unlabeled instances is an indication both of how linearly separable the data actually are and of how consistently brains during the relevant task can be described by a dot inside the relevant region. Linearly separable data are those that tend to cluster by class label in regions of space without overlapping the regions occupied by (the cloud formed by) data with different class labels. Accuracy is thus also an indirect indicator of how different the brain states in these conditions appear to be—how isolated they are in the multidimensional brain activity space. In this study they achieved accuracies as high as 90% and were in all cases labeling with accuracies well above chance, indicating that the data were fairly separable and the brain states fairly consistent and fairly different in each task.

Because it is somewhat difficult to visualize regions in a 214,000-dimensional space, much less interpret the psychological or neural meaning of the data clouds, the authors sought to extract some more intelligibility from the results by applying dimensional reduction techniques. As discussed further in chapter 5, the point of dimensional reduction is to permit one to focus on the important and to ignore the unimportant variance in a signal—that is, to reduce the data description to a set of variables that maximally captures or expresses the differences between different conditions. This often allows for a level of interpretation and explanation of the nature of the phenomena of interest that is generally impossible with very-high-dimensional data. Here the authors used a neural network classifier with a six-node hidden layer. Because the neural network also performed classification very accurately, this indicated that the major differences between the tasks—at least according to the brain—could be captured by six major variables, that is, by variance along six important dimensions. Such dimensions—at least had they been based on a more comprehensive analysis—would be candidates for some of the NRP factors that I am arguing we should seek.

Figure 4.6 visualizes this result by showing the dimensional loadings onto the task categories, that is, the value along that dimension that characterizes the task, as different wedge sizes. Risk taking can be characterized as exemplifying a lot of dimension 5 and a bit of 1, 4, and 6; in contrast semantic judgment is not dimension-5-like at all but is instead described as having medium amounts of 2 and 3 with a bit of 1, 4, and 6. As I have mentioned already and shall elaborate on in greater detail below, perhaps the best way to understand this is by analogy with the common dimensions of personality: openness, conscientiousness, extraversion, agreeableness, and neuroticism. Just as your friend may be open, agreeable, and extraverted but not particularly conscientious or secure, working memory is 1-y, 3-y, and 5-y but not very 2-y or 6-y.

Naturally, we want to get beyond characterizing tasks as being 1-y and 5-y and move on to something with a bit more psychological content. To do this, the authors coded each task according to whether or not it involved one of a number of cognitive operations—elements in their ontology—and estimated the strength of association of these elements with the six dimensions of variance suggested by the brain data. They also took the further step of estimating the dimensional loadings onto brain regions. The results are shown in figure 4.7. As can be seen there, dimension 1 relates most strongly to speech, hearing, working memory, and control (inhibition), and to a few small regions of the brain, whereas dimension two appears to be more associated with language per se, and many widely distributed brain regions can be characterized as sharing in or exhibiting this psychological dimension. Certainly the process of understanding what these dimensions mean will not be easy, but just as clearly this sort of analysis is very, very promising.

A related set of techniques involving pattern analysis and multidimensional scaling has also been used to systematically explore the underlying

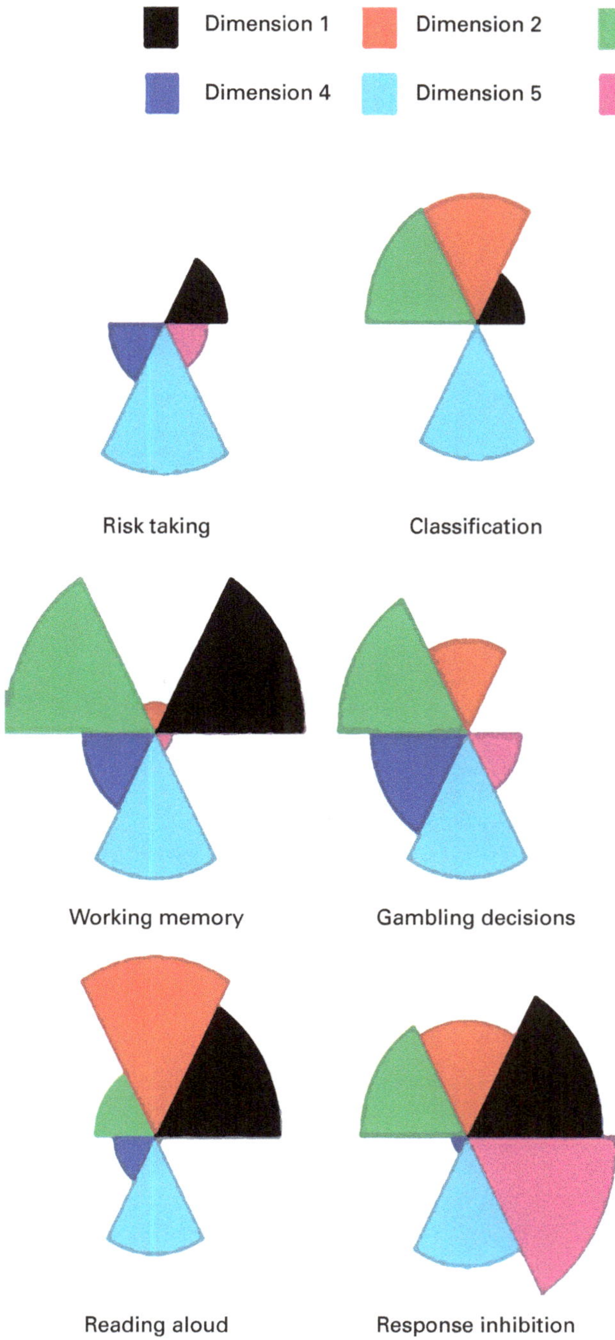

Figure 4.6
Loadings of each of the six predictive factors onto the original task categories. Reprinted from Poldrack et al. (2009) with permission.

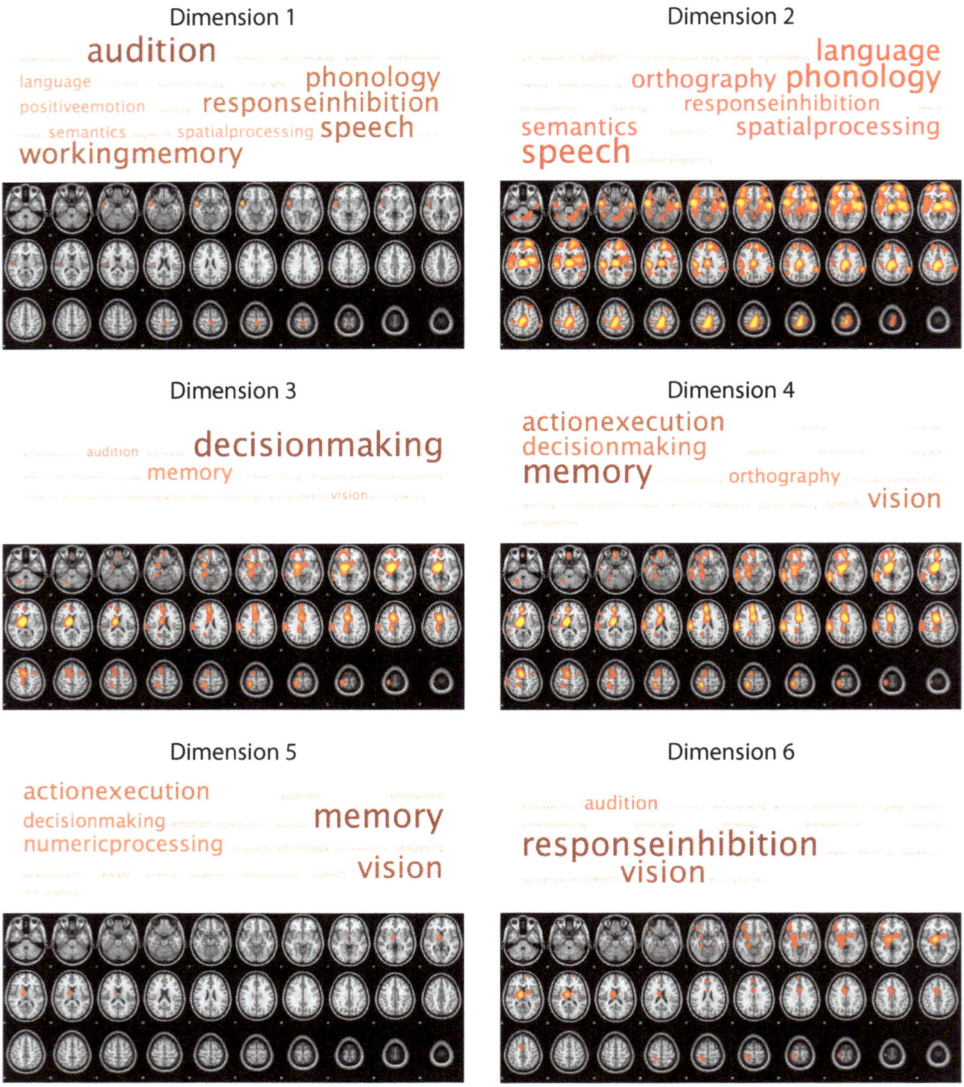

Figure 4.7

Associations of mental concepts with dimensions and dimensions with brain regions. The size of the word represents the strength of the association, and the brightness of the color (cool-to-hot scale) the value of the dimension. Reprinted from Poldrack et al. (2009) with permission.

structure in the brain's responses and to relate that structure to relevant aspects of behavior. Kriegeskorte, Mur, and Bandettini (2008a) demonstrated that it was possible to analyze the distributed patterns of neural responses to various stimuli and use the similarity in those *patterns* to generate a brain-derived measure of the similarity relations among the stimuli. Given data on the pairwise similarity of multiple stimuli (or multiple tasks), one can use multidimensional scaling (Edelman et al. 1998; Shepard 1980) to discern the underlying structure of the similarity space, uncovering the dimensions along which the brain takes the stimuli and tasks to vary. Not surprisingly, the shape of the space and the interpretation of the underlying dimensions vary depending on which region of the brain is analyzed. For instance Kriegeskorte et al. (2008b) showed participants simple images of trees, animals, tools, faces, and the like and found that although in early visual cortex the stimuli seemed to be arrayed along primarily visual dimensions such as color and shape, in inferior temporal cortex the differences in responses were better explained by more abstract dimensions such as the distinction between animate and inanimate objects. Crucially, these techniques can be used not just to generate but to test hypotheses for the underlying dimensions that best characterize differential brain activity (Kriegeskorte et al. 2008a; Jacoby & Armstrong 2014).

Thus, although in section 4.2 I showcased the power of simple multidimensional functional representations for understanding and analyzing neural assemblies, in point of fact I believe that for these multidimensional representations to achieve their full scientific promise we need also to follow the lead of these authors and use the techniques of dimensional reduction, independent and principal component analysis, factor analysis, multidimensional scaling—the full suite of the tools we have for finding structure in multidimensional data—to systematically explore and interpret the underlying dimensional structure of the brain's activities. This, I believe, will lead us not just to a better understanding of the brain but to alternate possibilities for the underlying psychological factors that best explain brain and behavior. If we want to understand the fundamental elements of the psychological mixtures that we measure in the lab and in this way get an empirical handle on the underlying functional dispositions of brain regions, this is the way the science must turn.

4.4 From Interpretable Dimensions to Neural Personalities

Although we generally think of imaging experiments as ways of learning something about the brain, the work showcased above demonstrates that one can *also* use these data to let the brain tell us something about the experiments—to reveal the underlying attributes of the task situation to which the brain differentially responds. Of course, one *might* think that we do this routinely in the cognitive neurosciences. For instance, researchers might see a region of the brain implicated in theory-of-mind tasks activated in what they believe to be an empathy task and conclude that empathic responses involve the attribution of mental states to others (Schulte-Ruther et al. 2007). This is one way of using the brain to inform us about the nature of the empathy task. But one of the points of this volume is that because *this* brand of reverse inference requires that the region in question is involved in supporting *only* theory-of-mind operations, it is undermined by the apparent functional flexibility of the brain (Poldrack 2006). In contrast, the power and promise of the technique pioneered by the authors featured in the last section are that it allows the brain to influence the categories we use to understand it while both offering a path to scientific intelligibility and doing justice to the underlying functional complexity of individual regions of the brain. As I have been arguing throughout this book, and as is clearly shown in the results reported in these studies, individual regions of the brain are active across multiple tasks and load on multiple psychological dimensions (the fact that many brain regions in figure 4.7 load on none of the dimensions is largely a consequence of the sparse set of voxels that they used out of methodological necessity to generate the dimensional reduction, although of course no one would expect a brain region to load on every NRP factor, nor the reverse; see Poldrack et al. 2009 for the details). As the authors note:

> [Figure 4.7] revealed that the brain functions across these diverse tasks were organized on a small set (three to six) of unknown functional features, which suggests that all the tasks recruited similar brain networks, to different degrees. (Poldrack et al. 2009, p. 1369)

Because of such findings, it is here that must I part company with the vision expressed by Poldrack (2010). Given the fact that the techniques employed here appear to allow us to better understand both these tasks and the brain, without a single one-to-one mapping between cognitive operations and individual brain regions anywhere in sight, I fail to see the scientific motivation for pushing onward with the project of specifying component operations that selectively engage individual neural regions, *especially* given the fundamental biological facts that appear—at least in the current state of the art—to significantly complicate the interpretive problem. Attributing a single cognitive operation to a local, polymorphic neural assembly of necessity distorts its functional role, but expressing function in terms of its differential loads on a set of underlying NRP factors in fact *captures* that functional profile across its different network states. We saw above that it is possible to do good science when expressing multidimensional function using simple and coarse task categories; I am confident that the quality will only improve with the use of factors that better capture the underlying variance (Gold et al. 2011).

Now, one might think that the fact that Poldrack et al. (2009) discovered voxels in the brain that were diagnostic of the different tasks being performed by the participants—that is, they were in fact able to predict task type from brain state—means that to identify selective brain regions we just need to look at the brain using machine learning classifiers. That is, perhaps the moral of the story so far is that although looking for single-voxel effects using the general linear model—the sort of data we canvassed in chapter 1—did not uncover much selectivity in the brain, using other techniques promises to do so. It is certainly true that diagnosticity is indicative of a special informational association between a voxel state and a task. But selectivity generally also implies a special *causal* association: one wants to be able to say that regions diagnostic for a given mental operation are the ones most likely to *implement* that mental operation. There are, however, reasons to doubt this inference. It is worth examining this possibility in some detail, as it will help us to understand the importance and the limitations of applying machine learning techniques to brain-imaging data and also, in the end, will reinforce the case for functional diversity as a fundamental feature of most brain regions.

In the current case, as noted above, the authors used a support vector machine (SVM) for their analysis. In general an SVM analysis of brain-imaging data works as follows: researchers divide their data into a training set and a test set. In the training set will be some number of examples of the signal data gathered under each of the task conditions of interest. The signal is transformed into an input vector—a string of numbers representing the signal in each voxel in a given condition. Elements of the input vector might correspond to the activation values of individual voxels or average values for different ROIs, the degree of deviation from baseline, or more complex constructions. For the purposes of this discussion it is safe to think of them as the activation values of individual voxels in each condition (although the authors used a statistical measure—the *z*-score—that quantified the significance of the deviation of each voxel from an average value, that detail is not important here).

SVM uses the information present in the input vector to build a model that will allow it, given an unseen example from the test set, to decide under which of the conditions it was gathered. The model consists of a set of features each assigned a weight, multiplied by the value of the feature. For instance, a binary linear classifier over an input space of dimension n looks like this:

$$\text{prediction} = \text{sign}(b + \sum_{i=1}^{n} w_i * x_i)$$

where the ith weight is w_i and the ith component of the input vector (the list of numbers that describe the patterns to be classified) is x_i and the bias value is b. The weight vector—the string of weights multiplied by the input vector—is thus the *predictive model* that the SVM generates for the data. If the sum above is positive, the instance is classified one way; if it is negative, the instance is classified the other way.

The SVM algorithm has two goals: to find the hyperplane with the widest margins between the data clouds generated by the input vectors and to minimize the sum of the absolute values of the weights. The weights determine the orientation of the hyperplane, and minimizing the weight vector gives the hyperplane with the widest margins. The most diagnostic voxels are those with highest weights—those whose removal would make the biggest difference to the prediction because their removal would tend to influence the orientation of the hyperplane and thus change the decision surface. It is for this reason that they have the special informational association to the task in question. But I am hard pressed to conclude that the voxels whose state is most important to determining the orientation of a hyperplane used in a classification decision are the voxels whose state is most important to supporting a task (or implementing an operation) in the brain.

Moreover, it is obvious that a voxel can be diagnostic of more than one task—in the simplest case it can have more than baseline activity during one task and less than baseline during another. One might argue that in this case the increase of activity is the important factor; diagnostic voxels with increased activity during a task are those most likely to be involved in implementing that task. But here, too, the inference is problematic, driven I believe by the least appropriate aspects of the computer metaphor of the brain. In the kind of computer with which we are most familiar, activity is an indication of computation, and inactivity is a sign of rest. But the brain is not like that. In the brain, processing is indicated by deviations from endogenous dynamics and can be a local increase or decrease. In point of fact we do not have a particularly good theory of the relationship between brain dynamics and cognitive processing. Liquid-state machines (Branicky 1995; Maass & Markram 2004; Maass, Natschläger, & Markram 2002) offer one model for understanding how a collection of oscillators could implement computational processes, but the idea is in its infancy, and I believe we also need alternate models of processing. In any case it seems likely that simply equating increased activation with responsibility for processing is inappropriate for the brain.

Finally, given that there are always a greater number of correlated voxels than diagnostic ones, it seems especially risky to claim that only the diagnostic ones are part of the implementation of the processes of interest. The reason this claim is tempting to many scientists is that we have been trained to look for selectivity. Correlation clearly is not providing it, and diagnosticity appears to produce stricter associations between structure and function. Although it is arguably the case that a psychological difference that makes no detectible neural difference should be considered suspect (Lenartowicz et al. 2010; note the claim is that such constructs would not be candidates for NRP factors, *not* that the psychological depends *only* on the neural; see also chapters 5 and 6), the techniques showcased above are sensitive enough to reveal neural differences—such as diagnostic patterns of distributed neural activity—that are sufficient to establish differential responsiveness without thereby demonstrating selectivity in any region or network. Thus, a central point of this book is not just that we do not *get* selectivity in the brain but that *we don't need it. We can stop looking for it.*

It is worth pausing to emphasize the nature of the bet I am placing here. My expectation is that there will prove to be no parsing of the brain that gives a one-to-one mapping between regions and NRP factors or primitives, but there may be a parsing of *psychological experiments* that reveals which psychological primitives are in fact being manipulated. In the general case more than one brain region will be involved in implementing or enacting each of those psychological primitives, and each brain region will have a role in more than one primitive. That is: brain regions and networks will differ in their loadings on a set of primitive NRP factors. In this framework *selectivity* would be the extreme case where a region or network had a loading of 1 on factor X, and 0 on everything else; and what might be called *solitarity* would be the case where only one brain region loads on a given primitive factor. Such functional arrangements are certainly possible, but my bet is they will be relatively rare.

All this having been said, the following point is crucial to appreciate: the analytic techniques and interpretive framework I am advocating will reveal cases of solitarity and selectivity and their frequencies. That is, if we assume there are selective components in the brain and interpret our experiments based on this assumption, then we will risk seeing selective components in the brain whether they are there or not because we will interpret variety of activation by assigning single mental operations that ostensibly account for the variety. If, on the other hand, we give up on this methodological assumption and allow for the possibility that regions and networks differ *not* in terms of implementing some specific stable mental operation but in terms of their loadings on a set of fundamental NRP factors, then we will see selectivity where there is selectivity, solitarity where there is solitarity, and irreducible functional complexity where there appears to be irreducible functional complexity. This empirical approach will reveal to us where in the brain the best candidates for componential understanding are, along with their activity profiles, which will help us specify the nature of those components, and where in the brain componential explanation appears less likely to be fruitful. In short, I believe that the framework I am advocating offers the best way forward for the field, *even for those scientists who do not ultimately wish to jettison the idea of functional components.*

Thus, I want to propose that—at least for the time being—we give up on the global framing notion of component operations in a selective, computational brain and build our neuroscience instead around the idea of NRP factors differentially expressed in a dispositional brain (Lindquist & Barrett 2012). We should make the transition in our science from the qualitative analytic reduction that we currently practice to the quantitative descriptive compression that we *should* practice, as exemplified by Poldrack et al. (2009; see also Baucom et al. 2012; Douglas et al. 2011; Gold et al. 2011; Haxby et al. 2011). In short, we should base our science around the specification of *neural "personalities."* The developments I have been describing suggest that in fact the field is already unconsciously recapitulating the history of the study of personality (albeit with more sophisticated analytical methods); I believe it would benefit enormously from a *conscious and thoughtful* consideration of that framework, for it appears to have the potential to unify and clarify some of the best work being done to advance the field.

This is not the place to offer a detailed history lesson, but highlighting some elements of the historical development of personality studies will I think prove instructive.

4.4.1 The Lexical Hypothesis

Among the foundational assumptions underlying much research in the nature of personality is what has come to be called the lexical hypothesis:

> This assumption was first articulated by Klages (1932) and then elaborated by Allport (1937), Cattell (1943), Norman (1963), and Goldberg (1982): *Those individual differences that are most salient and socially relevant in people's lives will eventually become*

encoded into the language; the more important such a difference, the more likely it is to become expressed in a single word. (John, Angleitner, & Ostendorf 1988, p. 174, emphasis in original)

Having identified the large but finite number of aspects of personality encoded in the language, it came to seem possible to analyze the patterns with which such descriptors appeared—to discover which terms tended to apply to the same persons and which ones rarely did, in short, to determine the way these terms clustered in the environment—and to use those patterns to make inferences about underlying causes.

> The most potent method of attacking the tangle is to work out correlation coefficients between the inconveniently multitudinous variables abounding ... and seek some smaller number of "behind the scenes" or underlying variables, known as factors. (Cattell 1946, p. 272)

This expresses in brief the sort of path I am advocating for the cognitive neurosciences. Obviously, one can not expect the lexical hypothesis to hold when applied to identifying those psychological traits most important as descriptors and/or explanations of brain activity. But a corollary might hold true or be true enough for our purposes: *in describing their work, cognitive and behavioral neuroscientists tend to use vocabulary indicative of the most salient and scientifically relevant aspects of their experimental designs and of the conditions their participants experienced.* We can call this the neuroscientific lexicon hypothesis. Insofar as it holds true, we can similarly look for patterns in the occurrence of these descriptors and for consistent associations between these descriptors and the activity of the brain and use these to identify some smaller number of fundamental causal factors. Although there are general objections that can be raised to this approach (I discuss a few below), the bottom line is that we need a place to start, and this offers what appears to be a promising one.

In fact, such a direction is already being pursued in a few labs. In addition to the work described in sections 4.2 and 4.3, Nielsen, Hansen, and Balslev (2004) describe a method to discover associations between words in scientific abstracts and particular structures in the brain. Yarkoni et al. (2011) describe methods for associating words in the full text of neuroimaging studies with single-voxel activations such that both forward (given a descriptor, what brain activity is likely) and reverse (given brain activity, what descriptors are likely) inference are supported. Voytek and Voytek (2012) describe methods for identifying significant relationships between neuroscientific concepts in the articles stored in the enormous PubMed database, such that it is possible to identify likely associations among brain function, structure, and disease. My hope is that such work will become more widespread, more formalized, and more coordinated and that the framework being articulated here will encourage scientists to move beyond simple term-structure associations and embrace the project of identifying both the set of underlying factors and the differential factor loadings for each ROI in the brain.

This being said, there have been at least three important (and largely legitimate) criticisms leveled against the notion that important scientific knowledge can be extracted from patterns of language use, and it is important to consider these and work to mitigate their impact: (1) the forces that govern language development are poorly understood and may not have produced many terms well adapted to the very particular project of understanding brain function; (2) vocabularies and other linguistic conventions will differ between labs and language communities; (3) natural language terms are ambiguous, poorly defined, and often polysemous—the same word can be used by different people in different ways and by the same person differently in different situations (see John, Angleitner, & Ostendorf 1988 for a discussion of these criticisms in the specific case of personality psychology). To this I add a fourth criticism specific to the neuroscientific lexicon hypothesis: (4) the language scientists use will itself be partly determined by the theoretical framework within which they are working, which will impact both their judgments about salience (what to report) and their choice of description (how to report it).

I believe that all of these criticisms should be taken seriously. Projects like the Cognitive Atlas (Miller et al. 2010; Poldrack et al. 2011) are an important part of addressing the first set of concerns by helping the field develop a shared technical vocabulary—although, as I have indicated above, I think it would be better to structure the project around the goal of identifying the set of underlying NRP factors that best explain our observations of differentiated brain activity rather than to try to identify the particular cognitive operations implemented in individual brain regions. This would involve specifying an ontology of psychological terms that characterize the psychologically relevant features in the task environment and the kinds of changes—cognitive and behavioral—they typically elicit. In point of fact it appears that the terms in the Cognitive Atlas (http://www.cognitiveatlas.org) are largely appropriate for the sort of analysis I am advocating, although in my view they will need to be supplemented with more interactive, relational terms. Beginning in chapter 5, I have some more to say about the nature of the psychological vocabulary we are likely to need for our ultimate scientific purposes.

Interestingly, I think problem 2 is likely to help mitigate problem 4, and insofar as projects such as the Cognitive Atlas remain open to all, the diversity of viewpoints and vocabularies can in fact be adequately captured. If so, then the very process of analyzing cross-term correlations and structure-function associations for underlying factors can itself be an impetus for theoretical progress, as proposals for factors that do justice to the full range of associations are likely to require not just engagement with alternate theoretical frameworks but also attempts to reconcile differences in perspective.

4.4.2 Traits, States, and Factors

Research in personality distinguishes between a state and a trait: traits are said to relate to or explain (or in some cases to constitute) one's enduring, consistent patterns of behaviors, whereas states may be more transient conditions, perhaps induced by elements of the current situation. For instance, one may be feeling anxious because of an upcoming test (state anxiety) despite not being an anxious person in general (trait anxiety). The distinction is of course intelligible, but in reality the difference may be less clear cut, and the two can interact in important ways. For instance, although I am generally introverted, some situations (an important work-related social event, for instance) may require behavior uncharacteristic of that trait—a kind of state extroversion. Conversely, having just been forced into an extroverted state for a period of time, I may react by becoming *more* introverted than usual and even exhibit behavioral changes along other dimensions (become less agreeable than usual, for instance). And obviously, traits are only *relatively* enduring and can be impacted by extended exposure to particular kinds of situations (extended state anxiety from a stressful environment can lead to trait anxiety), mental disorders, and normal aging and development.

What does all this have to do with the brain? It is important to remember that although what we are hoping to measure are the enduring response tendencies that both distinguish ROIs from one another and offer biopsychological explanations of behavior, they are neither immutable nor causally isolated from more transient variables. States and situations can interact to make a region more or less likely to exhibit its characteristic responses. There can be neural sensitization or desensitization due to recent experience, and elements in a situation that might call for conflicting behavioral responses might result in suppression or paralysis.

This also reminds us that we need to be clear about what we take psychological factors in this context to *be* and how, therefore, factors relate to states and situations. My proposal is that such NRP factors should be understood as a region's disposition to help determine the outcome of an organism's response to a situation and to influence the character of the organism's interaction with its environment, that is, to help manage some aspect of the organism-environment relationship. Thus, characterizing these NRP factors, and eventually coming to quantify their differential loadings on the regions of the brain, will be in part a matter of identifying the relevant relationships that the brain is interested in managing and designing experiments that manipulate the values of these relationships. I take up this topic in chapter 5, where I also offer some preliminary suggestions for what some of these factors might be.

Put in the language of the dynamic brain, NRP factors represent the openness to deviation from endogenous dynamics under various circumstances, which deviation of course has its downstream causal consequences. Such dispositions can be modulated by transient states, and in the brain the relationship is further complicated by the fact that an organism's states will also have their neural expression. In the end, I do *not* think that we will uncover anything like a one-to-one correspondence between such organism states and brain states; the state of an organism will be determined by a complex combination of brain, body, and environment (including the social environment). But clearly one important causal link permitting the modulation of NRP factors by organism states will be the neural expressions of those states.

4.4.3 Persons and Situations

This brings us to another of the important debates from the history of personality psychology, which centered around the question of whether the primary source of explanations for behavior were persons or situations. The debate was sparked in part by Walter Mischel (1968), who developed the observation that both situations and traits influence behavior into a critique of trait explanations of behavior. Largely on the basis of a meta-analysis of many studies relating personality traits to behavior, Mischel argued that the predictive value of traits by themselves was very low. In predicting individual instances of behaviors, situations proved more influential. Put differently, because individual behavior is not invariant across situations, then insofar as traits *are* invariant, they of necessity fail to capture the causal forces that determine behavioral outcomes. Thus, scientific focus should remain on situations and environments (thereby remaining closer to a behaviorist perspective). Note that this sketch is an intentional oversimplification of Mischel's argument, meant to capture the way his book was initially perceived in the scientific community, a perception that Mischel himself was at pains to dispel:

A decade ago I published a book that was widely taken as a broadside attack on personality. Many also saw it as an attempt to replace dispositions and indeed people with situations and environments as our units of study. These effects of *Personality and Assessment* (Mischel, 1968), these widespread perceptions of it as a situationist's manifesto aimed at undoing the role of dispositions, were far from my intentions. (Mischel 1979, p. 740)

In any case, a long debate ensued, and although the details need not concern us, the outcomes should. Here are a couple of the key lessons of the debate.

1. Although traits are often not very good at predicting individual instances of behavior, they can nevertheless be useful for predicting aggregate behavior. Consider the following:

The major value of traits lies, however, not in their usefulness in predicting specific behaviors, but in their value as predictors of aggregated behavior, that is, of behavior in the long haul averaged over many situations, occasions, and responses. Relatedly, although traits are often not discernible in individual acts, this does not make them unimportant with respect to individual lives. Just as one item by itself cannot provide a strong measure of a trait but the accumulation of many can, one unexceptional act by itself is not apt to significantly affect the life of a person or of his or her acquaintances, but the accumulated consequences of many can. (Epstein & O'Brien 1985, p. 532)

For the NRP factors this suggests that we should not expect these to relate strongly to individual behaviors—nor to strongly predict individual instances of brain responses—but rather to be related to long-term tendencies in response and behavioral outcome visible primarily in aggregate. This is both a methodological and a theoretical point. Methodologically, it means we need to rely more on *collections* of measurements across multiple situations and less on individual experiments, as is indeed becoming more common in the field (Eikhoff et al. 2009; Laird et al. 2009; Wager, Lindquist, & Kaplan 2007; Yarkoni et al. 2011). Theoretically it suggests that we need to acknowledge from the beginning that to understand behavior-function-structure relationships we cannot focus exclusively on situations, states, or factors but must be constantly aware of the relationships among these.

2. The best explanations arise from a recognition and specification of the interactions of multiple contributors.

Ultimately, Mischel's attack on the generality of behavior, along with two other developments, provoked a paradigm crisis in personality research. One of the other developments consisted of a series of studies by Endler, Hunt, and their associates (e.g., Endler & Hunt, 1966, 1968, 1969; Endler, Hunt, & Rosenstein, 1962) that demonstrated that the amount of variance accounted for by situations and person-situation interactions was greater than that accounted for by persons. This ultimately led to the view that interactionism provides a reasonable resolution of the person-situation debate, as behavior can never be determined by person or situation variables alone but is always a result of the interaction between them. (Epstein & O'Brien 1985, p. 515)

Interestingly, Epstein and O'Brien reject this interactionist perspective in favor of their own solution to the dilemma (outlined in moral 1, above) in part because although the interaction explained a *plurality* of the variance in behavior, the amount explained was still not particularly high. Thus, we have the call to essentially give up on explaining variance in individual instances of behavior in favor of explaining aggregate stability. I myself do not see these two positions as being mutually exclusive options. In this regard it is worth considering an interesting analysis of the debate offered by Buss (1979).

Buss suggests that in one common conceptualization of the debate, the pure personalist position supposes that behavior is determined entirely by the person: $B = f(P)$. In contrast, the pure situationist perspective supposes that behavior can be determined entirely by situations: $B = f(S)$. Finally, one obvious way to characterize the interactionist perspective is as suggesting that in fact $B = f(P, S)$. When the matter is framed this way, with a single dependent variable, behavior, codetermined by two independent variables, the natural way to approach the study of behavior is with factorial designs analyzed with ANOVA. In the linear case the variance in the dependent variable is conceptualized as having three additive components, one for the impact of each independent variable and one for their interaction. In this way one can capture the individual contributions of each factor to the outcome as well as the impact of the state of each independent variable on the contribution of the other. And indeed, when one explores this possibility, it is often the case that the interaction accounts for most of the variance. Buss cites reviews by Argyle and Little (1972); Bowers (1973); Ekehammar (1974); and Endler and Magnusson (1976). And yet this notion of interaction is perhaps more limited and specific than is sometimes appreciated:

However, as noted by Overton and Reese (1973), the ANOVA model is a linear model, and by the term "interaction" within the ANOVA model, one means specifying the nonreciprocal relationship between environmental and person variables which best accounts for, predicts, or "explains" the behavioral variability. (Buss 1979, p. 197)

In contrast, there is another view of interaction that conceptualizes behavior as simultaneously a dependent and independent variable, with reciprocal causal interactions between the factors influencing the behavior and the behaviors themselves. So not $B = f(P, S)$, but rather $P \leftrightarrow S$. Here the idea is that the person's behavior changes the situation even as the situation affects the person's behavior (as mediated by traits and states, a complication not shown in this reduction). A person's traits may lead him or her to seek out (or avoid) particular situations, which in turn elicit patterns of behavior, which change the situation, and so on.

What does this mean for the brain? First, it reminds us that when we discuss "interactions" among the variables of interest we should be clear which underlying causal conception we have in mind. It may be that different conceptions are more appropriate for different experimental interventions and analyses; or it may be that the field will eventually settle on a conception that is agreed to adequately capture the empirical situation. I do not try to settle that debate here, although, as will be clear from the remainder of this book, I am endorsing the latter conception of interaction as being the most appropriate. Recent relevant work on gene × environment interactions in psychological disorders (e.g., Caspi & Moffit 2006) suggests that the field could yet benefit from more careful and comprehensive consideration of the kinds of interactions that might be investigated.

Second, and more specifically, it means we need to be experimentally and theoretically open to the discovery of interactions (of whatever sort!) between system-level variables such as regional brain activation, person-level variables such as emotional state, theoretical constructs such as NRP factors, and environmental variables such as situations (and their relevant aspects). In the picture that I am trying to sketch here, differential brain activity is partly determined by NRP factors (and their specific regional loadings) in specific situations as modulated by organism states. This differential activity in the brain impacts ongoing behavior in such a way as to alter situations, states, and—at least over developmental time—regional loadings on NRP factors. To put this into the language of the IDS framework, the development of regional functional biases involves the tuning of the NRP factor loadings, driven by a combination of biological mechanisms including Hebbian learning and the neural search for and establishment of different local and long-distance functional partnerships. The functional partnerships of neural regions both partly determine and are partly determined by functional biases. Understanding the brain will centrally involve untangling all of these relationships.

In the next chapter we review some of the current work that is relevant to these issues, which will lead to the identification of a third important sense of "interaction": the apparent fact that many of the aspects of situations to which we are sensitive are best described as organism-situation *relationships,* indicating the available opportunities for environmental interactions. I can give one simple example here, which is developed in more detail later. Much of the investigation of the human perceptual system has focused on our estimation of basic physical quantities such as illumination, temperature, and weight. In the case of the last, a long-standing puzzle in the field is the fact that, given two objects of equal weight but different sizes, the larger one will be perceived as lighter; this is known as the size-weight illusion. How should we account for this apparent misperception of the true physical state of the world? Recent work suggests that what is being perceived in this case is not weight but "throwability" (Zhu & Bingham 2011). According to their analysis, when one conceives of the relevant property in this interactive, relational way, human perception is in fact accurate; there is no perceptual illusion here. Rather, the puzzle was caused by a misappropriation of the categories of physics for use in psychology; we evolved to be primarily (fundamentally) sensitive to aspects of the world different from those that are fundamental to classical physics. As should be clear from this chapter, I think a wholesale reconsideration of our psychological categories may be in order.

4.5 The Kind of Intelligibility Being Offered Here

I think it is worthwhile to end this chapter with a few words about the *kind* of science that is being proposed, the sorts of explanations it can offer, and the nature of the understanding that we can hope to gain in this manner. So far I have been arguing that the current dominant framework for organizing the cognitive and behavioral neurosciences, centrally involving the analysis and decomposition of psychological tasks into component mental operations, which are then localized in individual regions of the brain, not only fails to do justice to the functional and biological complexity of the brain but is guided by a borrowed conception of the *kinds* of things mental operations can be that likely systematically distorts the science in important ways. In particular—an argument that is developed in greater detail beginning with the next chapter—the notion of a *computational operator* (or symbol processor) is likely to be a mismatch for a brain system with endogenous dynamics in an organism characterized by (and evolved for) continuous interaction with the environment.

In its stead I am suggesting that we adopt a framework according to which individual regions of the brain exhibit not functional specialization (the implementation of a single mental operation) but rather relative functional differentiation—the development of regional functional biases.

I propose that we capture these biases in terms of different loadings on a set of common, neuroscientifically relevant psychological (NRP) factors that together capture as much as possible of the observed variance in brain activation in light of situational and behavioral outcomes. In interpreting these factors we need to be open to relational, interactive properties of situations; and in explaining brain-behavior regularities we need to be mindful of different kinds of potential interactions among the causal variables. Moreover, just as with persons, the affiliations that a region has—the partnerships it forms—can tell us a great deal about the nature of the region's functional preferences and can be an important factor in altering those preferences. Although we have yet to define anything like an adequate or complete set of such NRP factors, I have shown that even preliminary proposals for multidimensional representations of regional brain activity have scientific utility and facilitate the discovery of important fundamental functional properties of the brain. This suggests that such functional fingerprints can serve as an empirical quantitative *and* qualitative index of underlying causal powers.

Thus, although the account offered here specifies only *dispositional* explanations for psychological and neuroscientific observations, this does not mean that such explanations are merely *descriptive*. I have suggested that NRP factors should be understood as a region's disposition to help shape an organism's behavior in a situation, to help determine the character of the organism's interaction with its environment, or to manage some aspect of the organism-environment relationship. NRP factors and their specific loadings, then, index or express the underlying causal properties of a region and specify possible causal relationships between that region and an organism's behavior (Vanderbeeken & Weber 2002). As such they can offer genuine causal explanations and support both prediction and retrodiction (as indeed we saw in sections 4.2 and 4.3, above).

What a framework like this does *not* offer are categorical explanations of observations in terms of underlying causal mechanisms. As I have indicated already, I have no argument with the practice of giving categorical explanations of this sort; specifying underlying mechanisms can be an extremely illuminating scientific endeavor. It may be that we will eventually be able to provide mechanistic explanations in the brain that also account for the observations and underlie the specified dispositions. But specifying dispositions is generally an indispensible step toward the goal of specifying underlying mechanisms. And it should be noted that mechanistic explanations need not *replace* dispositional ones in this case; for the dispositional explanations will often be more economical (Vanderbeeken

& Weber 2002). Moreover, in this case at least, I believe the dispositional properties will remain an essential part of the bridge between any successfully specified brain mechanisms and the understanding of resulting behavior, for they will illuminate the psychological conditions under which the mechanisms do their work. Finally, as I elaborate in part III of this book, I believe that understanding brain activity using dispositional NRP factors offers the opportunity to better integrate neuroscience with the other behavioral sciences—including political sciences, economics, sociology, and sociolinguistics—if only because findings in these disciplines tend to be organized around dispositionally relevant properties such as practices and preferences. This is a promissory note with a distant redemption date, and yet such consilience would represent a scientific achievement of the utmost impact and importance.

Interlude 4 The Eyes Have It: Unraveling the Brain by Tugging on a Retinal Thread

I recently attended a fantastic lecture by physicist-turned-neuroscientist Sebastian Seung of MIT, held as part of the Bio-X seminar series at Stanford. The lecture showcased recent research on the structure and function of the retina and in particular on the mechanisms behind the cells that allow us to perceive motion, which, I think you will agree, is a pretty important ability to have. Now you might wonder why I am writing about eyes in a book about the brain, but as my ophthalmologist wife likes to remind me, the retina is part of the brain—the part of the body where the brain has pushed itself outside of the skull to get a look at the world—and so she actually examines far more brains than I do! The work demonstrates the potential of the new science of connectomics to move neuroscience forward. Plus, the pictures are beautiful (you can see more at the eyewire.org wiki, which is where I got most of the visuals presented here).

The story starts with a retinal cell called the starburst amacrine cell (SAC). This is a pretty interesting neuron in its own right. First, it has no axon, just the starburst of dendrites—those splayed threads you see here in figure 4a.1. SACs secrete two different kinds of neurotransmitter molecules—acetylcholine (ACh) and GABA—that is, both a primarily excitatory and a primarily inhibitory neurotransmitter, which is pretty unusual. Moreover, each dendrite is relatively independent of the others, such that each sends a signal when it detects motion that matches the direction of the dendrite itself (figure 4a.2; Masland 2005).

So this one cell is able to detect motion in any direction, and it sends a signal from just the dendrite that indicates that direction. Interestingly, one of the main mechanisms that permits this sort of single-cell multifunctionality is the pattern of connections to and between these SACs, which interacts with cell morphology in just the right way. Loosely speaking, the inputs to each individual dendrite are arranged so that light-detecting bipolar cells connect serially along the dendrites, an arrangement in which

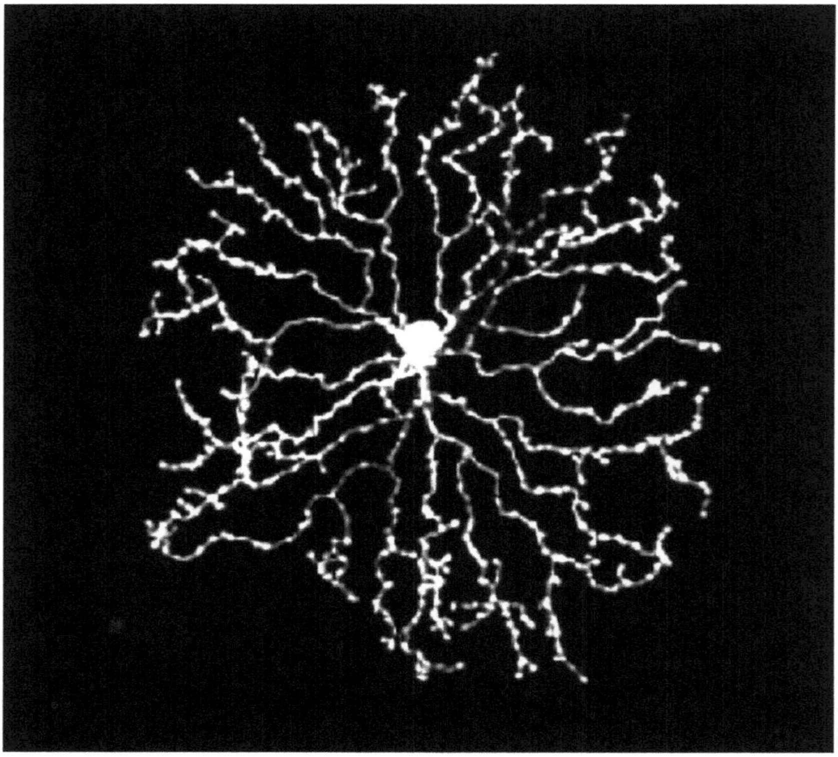

Figure 4a.1
Starburst amacrine cell. Reprinted from wiki.eyewire.org, under the creative commons attribute share alike 3.0 license.

a neighboring bipolar cell makes a neighboring synapse on a dendrite, making it more likely that motion in that direction will cause signals that reinforce along that dendrite. In addition, neighboring SACs selectively reinforce and inhibit one another so that dendrites going in the same direction reinforce each other while dendrites going in different directions inhibit one another. The resulting directional selectivity of the dendritic signal is so useful that these cells play a role in multiple higher-level functions, including motion perception and also the control of eye movements.

These details are interesting in their own right, I think, but also underline some important principles of structure-function relationships in the brain: (1) Even individual cells can be multifunctional in various ways, doing different things at different times and having multiple higher-level uses. (2) The source of neural function is not just a matter of the intrinsic,

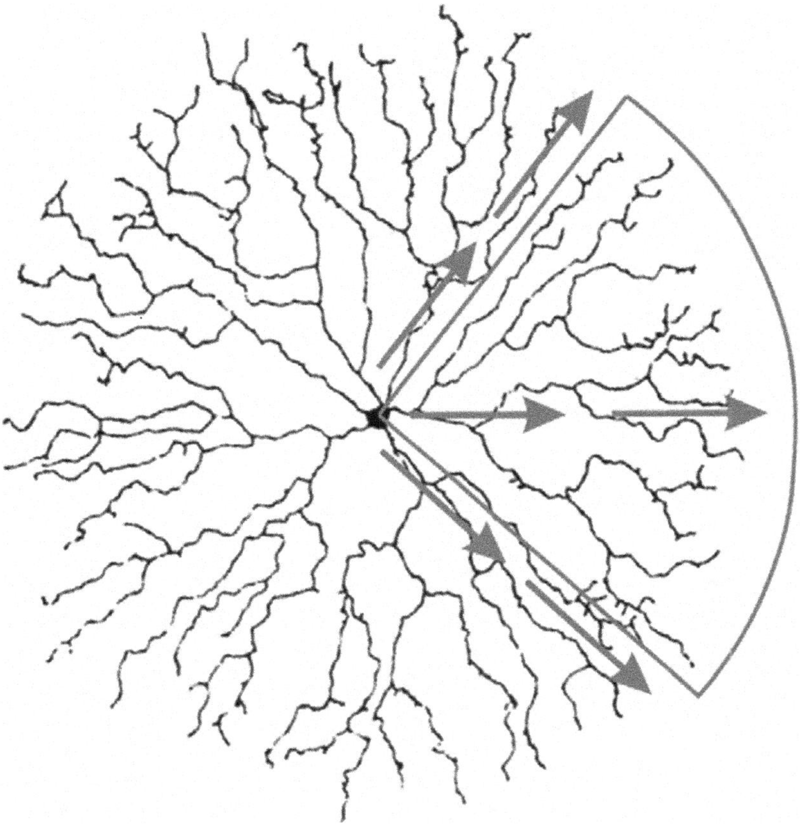

Figure 4a.2
Directional sensitivity in the starburst amacrine cell. Figure reprinted from Masland (2005) with permission.

local properties of the cell but of global properties of the network. Note that this example appears to show that one can get *subcellular* functional selectivity as the result of global network properties. This is a reversal of our usual, componential, bottom-up way of thinking about brain function. (3) Physical structure matters. If these dendrites were not arrayed spatially in this way, the cell could not have the functional properties it does.

This last point is further underlined by an intriguing hypothesis for the mechanism of directional selectivity of the so-called JAM-B cells. The gorgeous figure 4a.3 shows a number of these cells in mouse retina, with the long axons traveling along the back of the eye to the optic nerve in the center and the dendritic arbors all pointing in the same direction.

Figure 4a.3
JAM-B cells in mouse retina. Reprinted with permission of S. Seung.

JAM-B cells respond only to upward motion (remember, the image on the retina is upside-down), and the question is how? Well, the asymmetry of the dendritic arbors gives these cells a distinctive depth profile, as shown in figure 4a.4. Seung's hypothesis is that an important part of the mechanism for directional selectivity in these cells involves inhibitory synapses from SACs—they will prevent the cell from responding when the direction of motion is not up. Now, in fact, SACs make very few connections to JAM-B cells, and so one would think that their influence would be minimal. But here is the crucial part of the hypothesis: SACs exist in a lower layer of the retina, and so the connections they *do* make to JAM-B cells will tend to be close to the cell body, in the region indicated by the large white arrow in the figure.

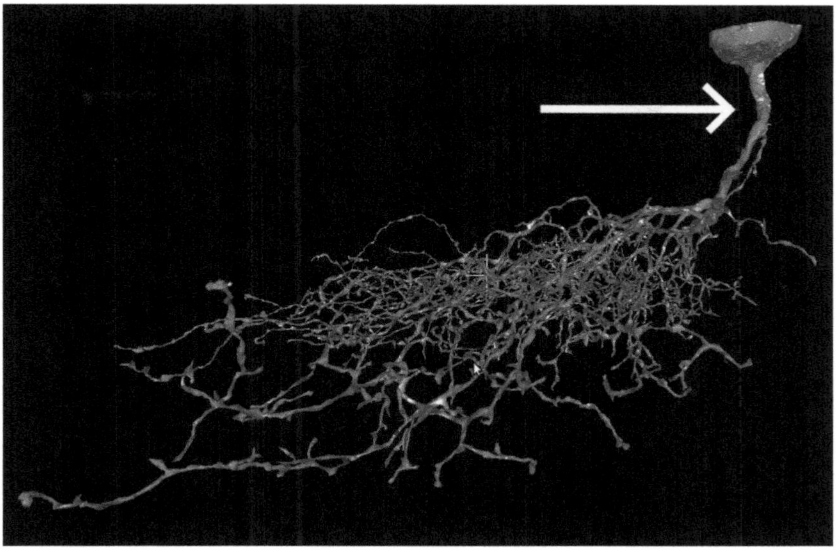

Figure 4a.4

Possible location of the interface between SACs and JAM-B cells is indicated by the large white arrow. Modified from blog.eyewire.org, reprinted under the creative commons attribute share alike 3.0 license.

Thus, the few connections SACs do make to JAM-B cells are perfectly positioned to inhibit signals traveling to the cell body. Put differently, an important determinant of function in this case is the *relative spatial position* of the neurons in the circuit. As important as it is to know *that* cells are connected, it can also be crucial to know *where* they are connected. The details of our embodiment matter in sometimes surprising ways. This, in fact, is the theme of the next chapter.

Part II Bodies

5 Brains and Their Bodies

Over the course of this chapter I lay out a picture of the place of organisms in, and their epistemic relationship to, their environments that is explicitly inspired by ecological and evolutionary considerations. This is followed in chapter 6 by an account of the brain mechanisms that mediate and modulate the sensorimotor coupling in support of environmentally situated adaptive behavior. The reason for this approach is simple: if we are ever going to understand brains—what they do and how—it is crucial to understand what it is they *need* to do for the organisms that have them. And I claim that traditional cognitive science has gone astray in answering this question, leading in part to the picture of brain organization and function that I have been criticizing here.

Because in making this case I need to traverse some *highly* contested ground, offering ample opportunity for misunderstanding, I want to begin by saying a little bit about where I come down in the end. The brain is a dynamic information-processing system that responds to and transforms structured signals from the environment in the service of generating adaptive behavior. Sensory inputs induce patterns of activity in the brain that depend on both the nature of the input and the dynamic state of the brain. Those patterns interact in various ways—as I argue in chapter 6, biased pattern competition is one of the primary mechanisms of action selection—and change over time in a way determined by the connection state of the brain, which is itself dynamic at various timescales and affected by the evolving patterns (see Spivey 2007; Sporns 2011 for extensive discussion).

Now these patterns are commonly called "representations." There is of course some sense to this, especially considering that they frequently exhibit sufficiently robust covariance with aspects of the environment and an organism's response to it that it is often possible to decode these patterns to guess what the organism is seeing (Kay et al. 2008; O'Toole et al. 2005) or what they are about to do (Haynes et al. 2007; see Haxby 2012; Norman et al. 2006 for reviews). Although the limitations of these techniques should be kept in mind—accuracies are far from perfect even in the restricted-choice contexts typically imposed in the experiments—it is impressive that it is possible at all. One might, of course, question whether what can be used as a (predictive) representation by fMRI researchers is being used as a representation *by the brain*. But my main objection to this way of speaking is that it comes freighted with the baggage of reconstructive perception and the symbol systems hypothesis (discussed extensively below) that the field needs to move definitely beyond. In its stead I offer not a representational but a *performative* theory of mind and brain in which we understand these patterns primarily as the evolving control states of a complex dynamic system. According to the view I develop, *some* of these control states will be usefully understood as representations. Rick Grush's emulator theory of representation (Grush 2004), Mark Bickhard's interactivist framework (Bickhard 2009), and my own guidance theory of representation (Anderson & Rosenberg 2008) are all attempts to capture and specify a brand of content-carrying intentional states that are also primarily control states, partly on the assumption that these are just the kinds of representations one is most likely to find in animals and other action-oriented systems. But in my view *many* if not *most* of our neural states will *not* be usefully understood as representations, and here the concept of *affordances* (Chemero 2003, 2009) will have important work to do for us. What I have been calling the computational, componential theory of mind (CCTM) takes representations to be central and foundational to cognition and action, but I take them to be peripheral and emergent. Representations emerge in the cognitive systems that have them as part of strategies to fine-tune control processes (Grush 2004) or to extend their reach, as with the use of stored traces of perceptual experience for such purposes as simulating, predicting, and planning (Clark 1997, 1998a, 1999). They are, in this sense, side effects of the elaboration of our basic sensorimotor systems and their application to ever more complex and abstract cognitive domains. As I argue in chapters 6 and 7, even our acquisition of abstract cognitive skills such as mathematics and natural language is best understood in terms of the deployment, in a special kind of symbolic environment, of our basic sensorimotor strategy of iterative interaction with the world.

Thus, although it will seem at some times to some readers that I am offering a dogmatic antirepresentationalism, I hope that those readers will find the patience to let the whole presentation play out. My real purpose in offering these critiques is to present the organism-environment relationship in (what may be to some) a new light and to show that there is in fact a diversity of strategies and structures available to the organism to solve the problem of adaptive behavior; the exclusive focus on representations too often blinds us to this important fact.

5.1 Reconstructive Perception

Traditional cognitive science is captured by a particular picture of our fundamental epistemic situation according to which sense organs are conduits for inputs called "sensations" on the basis of which the individual organism generates a representation of the causes of that input and produces motoric outputs to the world. The traditional scientific study of the brain begins with those inputs and focuses on the nature of the processes that transform them first into our representation of the world and then into our responses to it. As Quine succinctly puts the matter: "the stimulation of sensory receptors is all the evidence anyone has to go on, ultimately, in arriving at his picture of the world" (Quine 1969, p. 75).

One of the first "discoveries" of a science that begins with the stimulation of our sensory receptors is the paucity and unreliability of the data that they apparently provide. Steven Pinker characterizes the consensus view well:

> When [organisms] apprehend the world by sight, they have to use the splash of light reflected off its objects, projected as a two-dimensional kaleidoscope of throbbing, heaving streaks on each retina. The brain somehow analyzes the moving collages and arrives at an impressively accurate sense of the objects out there that give rise to them. (Pinker 1997, p. 212)

Viewed from a sufficiently narrow perspective, this can seem to be an accurate description. Light from our rich three-dimensional world enters the eye and is immediately reduced to a two-dimensional retinal image consisting of just local light intensity values and sparse spectral information. This reduced projection is compatible with—could have been produced by—a literal infinity of visual worlds (figure 5.1). Figuring out which of these objects actually *did* cause our perception is the problem of inverse optics, and the problem is intractable in the general case.

Thus, to generate the richness of our perceptual experience and accurately guide behavior using this kind of information, it appears that the brain that stands between inputs and outputs must employ sophisticated, heuristic reconstructive techniques, adding information as needed in the form of assumptions (or perceptual biases) that make solving the inverse optics problem tractable: light generally comes from above; illumination

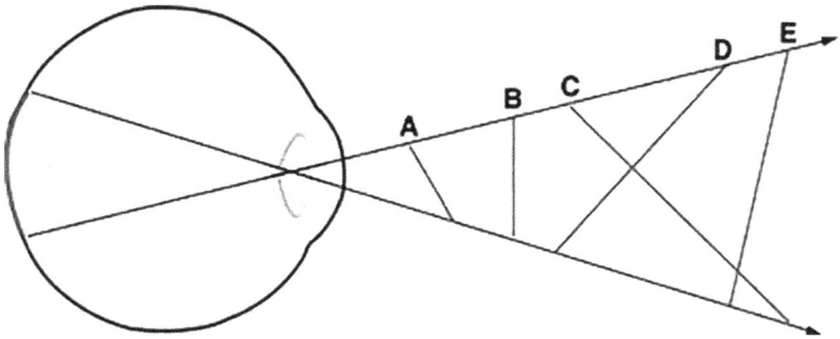

Figure 5.1
An infinite number of objects at various distances and orientations can produce the identical projection onto the retina. From D. Anderson & Stufflebeam (2012), reprinted with permission.

varies smoothly, whereas reflectance admits of discontinuities; and so on (Marr 1982; see also Edelman 2008; Warren 2005 for excellent extended discussions). Put differently, traditional cognitive science understands perception by analogy with scientific inference: from incomplete and fragmentary data, one generates hypotheses (or models) for the true nature of the world, which models are then tested against and modified in light of further incoming sensory stimulation. Fundamentally, then, perception is reconstructive and inferential, and the machinery that supports it operates via the algorithmic transformation of symbolic representations. As Richard Gregory notes:

> The key notion of cognitive psychology since the collapse of behaviourism is that we build brain descriptions of the world of objects, which give perception and intelligent behaviour. Perceptions are not regarded as internal pictures or sounds, but rather as language-like descriptions coded, we suppose, by brain structures of what may be out there. We carry in our heads predictive hypotheses of the external world of objects and of ourselves. (Gregory 1998, p. 1693; see also Gregory 2002)

In this view, it is only when all this perceptual construction of the world gets done that real cognition begins. We form beliefs based on our representation of the world, reason logically about these beliefs, integrate them with our desires, make decisions, and form intentions. Cognition is postperceptual—even in some sense aperceptual—representation-rich and deeply decoupled from the environment. In fact, on this view it has to be this way, for reconstructed representations are what the system fundamentally has to work with; the world, once sieved through our senses, simply provides insufficient information about itself. Our understanding of the nature of the epistemic problem that must be solved, then, drives us to hypothesize a particular kind of complex solution, captured paradigmatically in CCTM. I quote a characterization of one version of that hypothesis—the *physical symbol systems* hypothesis (PSSH)—at length:

> A physical symbol system is built from a set of elements, called symbols, which may be formed into symbol structures by means of a set of relations. A symbol system has a memory capable of storing and retaining symbols and symbol structures and has a set of information processes that form symbol structures as a function of sensory stimuli, which produce symbol structures that cause motor actions and modify symbol structures in memory in a variety of ways.
>
> A physical symbol system interacts with its external environment in two ways: (1) it receives sensory stimuli from the environment that it converts into symbol structures in memory; and (2) it acts upon the environment in ways determined by symbol structures (motor symbols) that it produces....
>
> Symbols are patterns. ... The way in which symbols are represented in the brain is not known; presumably, they are patterns of neuronal arrangement of some kind. ...
>
> We call patterns symbols when they can designate or denote. ... Symbols may designate other symbols, but they may also designate patterns of sensory stimuli, and they may designate motor actions. Thus, the receipt of certain patterns of sensory stimulation may cause the creation in memory of the symbol (say "CAT") that designates a cat (not the word "cat," but the animal). Of course, this does not guarantee that there is really a cat out there: That depends on the veridicality of the processes that encode the stimulus into the symbol designating a cat. Similarly, a motor symbol may designate the act of "petting" (with some parameters to assure that the cat will be the object of the petting). (Vera & Simon 1993, pp. 8–9)

Here perception is the process of transforming sensory stimulation into symbolic descriptors that represent (or "denote") things in the world; perception is input to cognition, which consists of modification operations carried out on symbol structures; and behavior consists of the motor actions that are the output of cognition. The acceptance of this basic picture is what accounts for the talk of neural "representations" that suffuses work in cognitive neuroscience—representations for faces (Kanwisher et al. 1997), of places (Epstein & Kanwisher 1998), of intentions (Haynes et al. 2007), of actions (Gallese et al. 1996), of other minds (Young et al. 2010), and more. Acceptance of this framework accounts for the fact that one of the fundamental jobs of cognitive neuroscience has been to find out what is represented where and how each representation is transformed into or impacts the others. Acceptance of this framework accounts for the interest in specifying the innate "knowledge" or stored assumptions underlying general perceptual reconstruction (Marr 1982), grammar (Chomsky 1957,

1988), folk psychology (Carruthers 1992), or the nature of objects (Spelke 1982), for insofar as sensation is natively inadequate to the task of specifying the world, then it must be supplemented with knowledge—innate or at least easily acquired—and it becomes a central part of the job of psychology to investigate the nature and extent of this knowledge. Acceptance of this framework even accounts indirectly for the componential assumption that is built into most theories of the functional structure of the brain, for insofar as representations are specifiable and language-like—having a syntax as well as semantics—and insofar as perception can be cleanly separated from cognition and cognition from action, then certainly the most natural assumption with which to begin a study of the brain is that one is dealing with a componential physical system that implements the symbolic processes characterized by Vera and Simon, above.

The point of this litany is not to take specific issue with any of this work but rather to show just how deeply the framework has rooted itself in the cognitive (neuro)sciences, how naturally it all flows from the initial analysis of our fundamental epistemic position, and how many steps back we need to take in reassessing the matter. For if it is correct—as I argued in the first part of this book—that the brain is very unlikely to be a componential system, then it appears we also need to reexamine the claim that it is (fundamentally) a symbol system, else the notion of neural componentiality will continue to reassert itself—as I believe has happened, for instance, in the merger between contemporary PDP models and explicitly symbolic cognitive models (Jilk et al. 2008; see section 3.2) after years in which one of the main distinguishing features of PDP models was their rejection of componentiality (McClelland & Rumelhart 1986; Rumelhart & McClelland 1986). And if we are to reject the symbol system model, then it appears we must question the analysis of our epistemic circumstance that leads to it.

5.2 Seeing and Looking

There are also broader motivations to seek alternatives to the traditional approach. For as John Dewey suggests, this framework is in fact a recapitulation in modern form of a set of Cartesian principles that have been roundly and rightly criticized in many different contexts—epistemic, metaphysical, scientific, social, and political (Warren 2005; see also interlude 5):

> The older dualism between sensation and idea is repeated in the current dualism of peripheral and central structures and functions; the older dualism of body and soul finds a distinct echo in the current dualism of stimulus and response. Instead of interpreting the character of sensation, idea and action from their place and function

in the sensory-motor circuit, we still incline to interpret the latter from our preconceived and preformulated ideas of rigid distinctions between sensations, thoughts and acts. The sensory stimulus is one thing, the central activity, standing for the idea [is another], and the motor discharge, standing for the act proper, is a third. (Dewey 1896, pp. 357–358)

So what is wrong with this picture? The key to understanding that is to make the subtle but important shift from treating visual perception as fundamentally a matter of the reception of sensations of light to appreciating the true nature of *looking*. Ironically, Pinker was on to something with his reference to the heaving, throbbing epistemic misbehavior of retinal stimulation: for indeed the retinal image is hugely unstable because looking is an *activity*.

Let us take, for our example, the familiar child-candle instance (James 1950, Vol. I, p. 5). The ordinary interpretation would say the sensation of light is a stimulus to the grasping as a response, the burn resulting is a stimulus to withdrawing the hand as response, and so on. There is, of course, no doubt that is a rough practical way of representing the process. But when we ask for its psychological adequacy, the case is quite different. Upon analysis, we find that we begin not with a sensory stimulus but with a sensorimotor coordination, the optical-ocular, and that in a certain sense it is the movement which is primary, and the sensation which is secondary, the movement of body, head, and eye muscles determining the quality of what is experienced. In other words, the real beginning is with the act of seeing; it is looking, and not a sensation of light. (Dewey 1896, pp. 358–359)

Although in the traditional framework it is natural to begin with sensory stimulation and focus on its transformation in the brain, this approach ignores the fact that any such scientifically isolated perception-action sequence was in fact preceded by an action that was itself preceded—and not just preceded but accompanied—by a different perception, and so on through the whole history of the organism, creating a framework of sensorimotor coordination apart from which the isolated sequence cannot be understood. The failure to appreciate this deep but simple point has marred the science of vision in particular, and damaged cognitive neuroscience along with it, because it has led to a fundamental misunderstanding of the nature of the tasks that brains must execute and thus, I fear, to a fundamental misunderstanding of the function of the brain.

In part the fault lies with the Modern obsession with vision as the epistemic sense *par excellence* (Levin 1993; Rorty 1980). Perhaps because of the ubiquity of visual art, it can seem natural to suppose that the eyes deliver static pictures to the brain. But when one considers olfaction (or, rather, chemical sensing more generally), it is clear that this sense is very nearly useless without the ability to move. This is both because the detection of chemicals without the ability to move in response to them is without purpose, but more importantly, because all of the useful information about chemicals lies in their *distribution* in the environment, and the detection of gradients requires the ability to change location and position. Put differently, chemical detection is not chemical perception until it is in fact chemotaxis. Not surprisingly, recent models suggest that the primary evolved function of the olfactory bulb was chemotaxic navigation (Jacobs 2012). Because chemical cues tend to be sparse, evanescent, and subject to various kinds of disruption (e.g., from turbulence), reliable reception requires constant sampling, integration with navigational systems, and sensitivity to relevant potential disruptions. Thus, for instance, are "olfactory systems ... notably integrated with mechanosensory systems to measure turbulence, such as vibrissae (mammals), antennae (insects), antennules (crustaceans), and lateral lines (fish) (Dehnhardt & Mauck 2008; Thewissen & Nummela 2008)" (Jacobs 2012, p. 10694). Moreover, because chemotaxis was an early evolving capacity, one would expect that later evolving sensory systems would leverage the same mechanisms—such as the use of multimodal integration and of physical movement for epistemic purposes—that worked so successfully in olfaction (indeed, that are required for successful olfaction).

This brief reference to olfaction is meant simply to emphasize the fact that our perceptual apparatus has evolved and develops as part of a moving body embedded in the world. The visual input to the brain is not like the snapshot of a camera but involves multiple, mutually informing information flows from the eyes, the ears, the limbs, and the rest of the acting, sensing body. Dewey has put us on the right track, but the preeminent theorist of active perception is J. J. Gibson (1966, 1979).

> The active observer gets invariant perceptions despite varying sensations. He perceives a constant object by vision despite changing sensations of light; he perceives objects by feel despite changing sensations of pressure; he perceives the same source of sound despite changing sensations of loudness in his ears. The hypothesis is that constant perception depends on the ability of the individual to detect the invariants and that he ordinarily pays no attention whatever to the flux of changing sensations. (Gibson 1966, p. 3)

The data of visual perception are not the momentary impacts of light on the retina but the relationships between *changes* across multiple sensory modes as one's posture and position in the environment changes. Louise Barrett, in her excellent recent book, puts the matter succinctly:

For example, when you look at a rectangular table, you usually don't actually see it as a perfect rectangle because you can do that only if you look at it directly from above. Rather, you see a set of different and constantly varying trapezoidal forms with different angles and proportions as projected to our moving points of observation. What doesn't change, however, is the relationship between angles that sit diagonally across from each other (the cross-ratio), and these uniquely specify a rectangular surface (and also a rigid one). Perception, then, is the activity by which animals and humans detect environmental invariants. (Barrett 2011, p. 105)

When we attend to the fact that looking is an activity, that part of seeing is moving, we see that the true data of perception are extremely rich, multimodal, and perfectly capable of revealing the higher-order invariants in our environment and uniquely specifying the shape of the world. Consider figure 5.2 based on figures in Gibson (1966, 1979). In addition to sensitivity

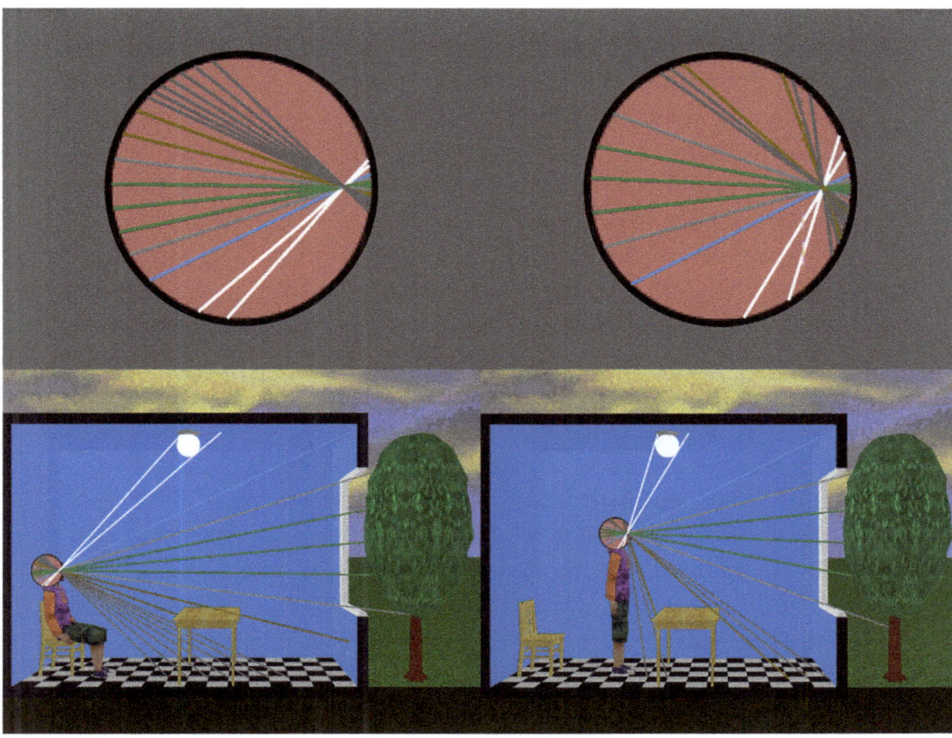

Figure 5.2
The retinal projection of objects in the environment while seated and standing. Perception is a matter of picking up on the changes in these projections as position and posture change. The changes uniquely specify the objects that cause them. From Proffitt and Linkenauger (2013), reprinted with permission.

to higher-order invariants such as the one noted above, perception also relies on the fact that movement causes characteristic patterns of *change* in the retinal projection given changes in posture and position. For the stationary figure, the retinal projections of objects in the world are potentially ambiguous; as illustrated in figure 5.1, each projection is potentially caused by any number of different objects. But it is *not* the case that each of these possible objects would cause the same set of *changes* in the retinal projection as the observer moves around. If the light on the ceiling weren't *that* shape in *that* position, then when the perceiver stands up, its retinal projection wouldn't go though *that* change.

Thus, as both Dewey and Gibson insist, no organism ever *starts* with sensory stimulation, for each and every stimulation was itself preceded (and is generally accompanied) by an action—action and perception are constant, ongoing, and intertwined. As Gibson writes, "The active senses cannot simply be the initiators of *signals* in nerve fibers or *messages* to the brain; instead they are analogous to tentacles or feelers" (1966, p. 5). If the organism does not start with a given stimulation, then no science of the organism or of its epistemic predicament should, either. When we recognize that perception and action are linked in this way—that the perceptual system is an exploratory and not an inferential system, and it is *because* one's view changes as eyes, head, and body move around the world that it is possible to know the world—we see that the presumed poverty of the stimulus disappears. There is no infinity of possible worlds to sort through, for the data of perception are not limited to the momentary two-dimensional image on the retina (or the surface of the skin) but consist rather in the much richer set of distinctive transformations of that image given changes in posture and position. These data are sufficient to reveal the shape of the world without the need for reconstructive inference. With no need for perceptual hypothesis generation, the argument that high-level, symbolic processing is *fundamental* to cognition begins to look unmotivated. Cognition is not a matter of algorithmic reconstruction but consists in maintaining a certain kind of environmental attunement by managing the perception-action coupling. There is no systematic problem of underdetermination here to solve.

Indeed, as Warren (2005) points out, it is a matter of ecological fact that things in the world have stable structures; that environmental change itself possesses higher-order stability; and that our perceptual system evolved precisely to pick up on such environmental invariances:

> Perceptual systems become attuned to informational regularities in the same manner that other systems adapt to other sorts of environmental regularities (such as a food source): possessing the relevant bit of physiological plumbing (whether an

enzyme or a neural circuit) to exploit a regularity confers a selective advantage upon the organism. Since the water beetle larva's prey floats on the surface of the pond and illumination regularly comes from above, possession of an eye spot and a phototropic circuit can enhance survival and reproductive success. But if illumination were ambiguous and prior knowledge were required to infer the direction of the prey, it is not clear how such a visual mechanism would get off the ground. Natural selection converges on specific information that supports efficacious action.

What the [traditional] view treats as assumptions imputed to the perceiver can thus be understood as *ecological constraints* under which the perceptual system evolved. The perceptual system need not internally represent an assumption that natural surfaces are regularly textured, that terrestrial objects obey the law of gravitation, or that light comes from above. Rather, these are facts of nature that are responsible for the informational regularities to which perceptual systems adapt, such as texture gradients, declination angles, and illumination gradients. They need not be internally represented as assumptions because the perceptual system need not perform the inverse inferences that require them as premises. The perceptual system simply becomes attuned to information that, within its niche, reliably specifies the environmental situation and enables the organism to act effectively. (Warren 2005, pp. 357–358)

This brings us to a second significant mistake that underlies the traditional approach and follows directly from the first. Insofar as one supposes that it is the fundamental job of the perceptual system to produce an accurate representation of the world, then it becomes natural to think of the fundamental content of perception as written in a perceiver-independent physics-influenced vocabulary of edges, colors, velocities, and weights (Gibson 1979). But if the purpose of perception is to guide action, then the more parsimonious hypothesis is that the organism will be primarily sensitive not to these sorts of properties but rather to action-relevant relationships between organism and environment (Anderson & Chemero 2009; Anderson & Rosenberg 2008).

The frog's visual system, for example, is tuned to particular patterns of motion that, in the restricted context of its niche, specify small edible prey and large looming threats. The fish's lateral line organ is tuned to pressure waves that specify obstacles, the movements of predators and prey, and the positions of neighbors in the school. Even the narwhal's tusk turns out to be a sense organ tuned to salinity differentials that specify the freezing of the water's surface overhead. The narwhal is thereby in perceptual contact with a property of its niche—the penetrability of the surface—that is critical to its survival. (Warren 2005, pp. 340–341)

It is the overall job of the organism's brain and nervous system to manage various organism-environment relationships; and the perceptual system keeps it in contact with the values of those relevant relationships (the closeness of the obstacle; the penetrability of the surface). Put differently, the world properties it is important to pick out for the purpose of *reconstruction* are not the same as those that best support *interaction*, and psychology has tended to (mistakenly) focus on the former class of properties to the exclusion of the latter.

5.3 The Vocabulary of Perception

Perhaps the biggest vice of the traditional view is its tendency to make accurate attunement to objective relational properties seem like inaccurate perception of intrinsic, observer-independent, reconstructive properties. Consider the size-weight illusion. First described by Charpentier over 100 years ago (Murray et al. 1999), the basic finding is simple and remarkably robust: given two objects of the same weight but different sizes, the smaller object will be judged heavier. The effect has been intensively studied, and there have been many different theories proposed to account for it. Most, however, center on some form of the idea that the effect is generated by problematic or inaccurate expectations, assumptions, or scaling functions that bias judgment. For instance, the sensory mismatch hypothesis suggests that the effect is due to the expectation that the smaller object will be lighter than it is; the contrast between expectation and experience then induces an inaccuracy of judgment (Ross 1969). In discussing the possible reason for us to have a weight-perception system "so liable to illusions," Ross writes:

> The term "weight-constancy" has been used in a proper sense by several authors to contrast with changing proprioceptive information, though the object lifted remains the same. Fischel (1926) and Katz (1935) used it to refer to the fact that an object feels much the same weight regardless of the manner of lifting. … We perceive the weights of objects, and not the systematic variation in stimulation due to the different positions of the arm or a mechanical apparatus.
>
> A weight-perception system which responded to the total sensory input would have disadvantages which are probably more serious than the occasional illusions which result from the existing system. One disadvantage is that it would not allow for weight-constancy. To do this the system must be able to discount expected changes in sensory input unconnected with the object itself. Another disadvantage is that the number of distinguishable rates of nervous firing is limited, and if the available rates had to cover all possible sensory inputs, discrimination between different intensities would be poor.
>
> The perception of weight seems to be greatly influenced by the contrast between the expected and received sensory inputs, though the value of the expected input is also taken into account. This system has the advantage that a large part of the sen-

sory capacity can be used for measuring contrast, thus enhancing contrast information at the expense of information about absolute value. (Ross 1969, p. 353)

Note the various ways in which the logic of the traditional view shapes this analysis. Given the assumption that the value of the environmental property in question (weight) must be designated by value of the momentary sensory input (for instance, via nerve firing rates), the variability in sensory inputs provides a potential source of error; changing sensations of weight—as when one curls a barbell and it requires less force to move it as the angle of the elbow dips below ninety degrees—can not possibly track an unchanging physical property of weight. Because much of sensory input must thus be discounted, the gaps must be filled by knowledge and expectations, which induce detectible errors in perceptual judgment because they will not be suitable in all circumstances. More generally, on this view, the perceptual system is a heuristic compromise that typically falls short of specifying the true nature of things in the world.

In contrast, on the ecological view that I am advocating here, and that I argue gives us a way to understand the brain that is more compatible with the findings reported in part I of this volume than is the traditional view, changes in sensory stimulation are often *precisely* the information that is needed to accurately specify the properties of objects; perception is generally accurate and rarely needs inferential "correction"; and we are very likely to have hit on nearly optimal solutions to perceptual-motor coordination problems. More generally (and here I am adopting one of the mantras of Andrew Wilson and Sabrina Golonka, who author the consistently useful and interesting blog *Notes from Two Scientific Psychologists*) the discovery of persistent perceptual (and cognitive) illusions—illusions that remain even in the face of experience and knowledge—is often a sign that one is asking the wrong scientific questions or bringing inappropriate concepts to bear on psychological phenomena.

In the case of the size-weight illusion recent work seems to bear this out. Given the robustness of the illusion in adult humans, its emergence in children as young as 18 months (Kloos & Amazeen 2002; Robinson 1964), and the fact that the illusion persists both when one is informed that the objects weigh the same and when—after repeated lifting—one uses identical amounts of finger force to lift the objects (that is, the illusion persists even in the face of two different ways of "knowing" that the objects in fact weigh the same; Flanagan & Beltzner 2000), Zhu and Bingham (2011) suggest that in perceiving these objects people may in fact be accurately tracking some property other than weight. And indeed, viewed from within the ecological perspective, weight does not seem a good initial candidate for one of the physical properties to which we would be natively or directly sensitive, for in what context do we actually care about the weight of things per se? We *weigh* things—and we have developed technologies for doing so accurately, primarily in the context of and constrained by the practices of trade. But the fact that we needed such technologies and the shared, externally validated standards they imply should already have suggested that weight evaluation was not to be found among our basic repertoire of perceptual capacities. Instead, what we care about—and what we have likely evolved to perceive—are ecologically relevant properties such as whether an object is edible, reachable, graspable, movable, liftable, or throwable.

Zhu and Bingham (2011) explored the last of these possibilities in understanding the origins of the size-weight illusion. They note that long-distance throwing is a uniquely human capability that is known to have played an important role in the survival and ecological adaptability of our species. Thus, perceptual and motor capacities favoring forceful, accurate long-distance throwing are likely to have been subjected to significant selection pressure. Given this, there is a strong case to be made that we are perceptually attuned to the throwability of objects in our environment. Describing some of their earlier work, they write:

> An object of graspable size and liftable weight affords throwing. Bingham, Schmidt, and Rosenblum (1989) investigated the perception of an affordance for throwing. In their study, spherical objects of different weights in a particular size were given to participants to judge the throwability, that is, the optimal weight for the size that could be thrown to a maximum distance. The task was intuitive and participants exhibited strong preferences in each of four graspable sizes of objects. They hefted objects and selected larger weights in larger sizes. When they were asked to throw every object as far as they could, the preferred objects were reliably thrown to the farthest distances. This result was recently replicated by Zhu and Bingham (2008). (Zhu & Bingham 2011, p. 290)

In the current case, Zhu and Bingham (2011) presented participants with a series of 48 objects of six different sizes and eight different weights for each size and asked them to select and order the top three best objects for throwing in each size class. Then, in a second part of the experiment, participants were asked to choose which of the 48 objects were the same weight as one of the objects they had chosen for optimal throwability (one of the small objects for one group of participants and one of the large objects for another group of participants). In each case the objects that participants thought were equal to their comparison object were in fact lighter (when the comparison object was large) or heavier (when the comparison object was small) than the comparison object—that is, they evidenced the size-weight illusion. But strikingly, the participants' choice of which objects weighed the same as the comparison object was very specific: the objects that they judged equal in weight to their comparison object were the *other* objects that they had previously judged optimal for throwing. That is: *participants reported all the optimally throwable objects to be the same weight,* despite the fact that the weights of these objects in fact increased with size. What this pretty clearly indicates is that the primary property of the objects to which the participants were sensitive in hefting them was their *throwability*—the "equally weighted" objects do not in fact weight the same but are in fact equally (maximally) throwable. Because we know from previous work (Bingham, Schmidt, & Rosenblum 1989; Zhu and Bingham 2008) that people track this (relational) property very accurately, this means that the observed inaccuracy in weight judgments exemplified by the size-weight illusion is in fact an artifact of asking participants to make a secondary extraction of information—weight—from their fundamental estimation of throwability. Put differently, throwability is a mixture of underlying physical properties of the object (weight, size, shape, density), the physical abilities of the agent (strength, hand size), and the relationships between these, in exactly the same way that weight is itself a mixture of mass and gravity-induced acceleration. It took a lot of hard science to deconvolve mass and gravity-induced acceleration, and nobody would think we would have (or maybe even could have) evolved native mass estimation capacities. Similarly, because it appears that we evolved a sensitivity to the deeply mixed property of throwability, we should not be at all surprised that we have trouble accurately extracting the weight component out of this mixture using just our native cognitive capacities—which is why we developed technologies and techniques to do it, instead.

Obviously I am using this case as something of a parable, for it is a specific instance of a moral we need to appreciate more generally: our brains are not primarily in the business of constructing observer-independent models of the world. We will only rarely have native neural sensitivities to the kinds of reconstructive properties (e.g., mass or weight) that would be best for model building. Instead, our brains are in the business of getting us around in and managing our interactions with the world. Our primary sensitivities, then, will be to the (often relational) properties of the world most relevant to this fundamental task. Just as the Narwhal maintains perceptual contact with the penetrability of the overhead surface, so we maintain perceptual contact with the throwability of objects in our immediate environment (and for the similar reason that this contact was crucial to our survival).

This, of course, is the thought behind the Gibsonian affordance-based theories of perception that have been widely influential in embodied cognition (Gibson 1979). Affordances are relationships between things in the world and an organism's abilities (Chemero 2009): for the average human the chair (but not the twig) affords sitting, the path (but not the wall) affords walking, and the rock (but not a tiny urticating bristle) affords throwing, but things are different for (and look different to) the bird and the spider. According to the view I am describing here—following Gibson, among others—perception is primarily perception *of* such affordances; the world is seen as a changing set of opportunities for action and interaction, and behavior is to be explained in terms of a combination of agent's purposes and environmental structures.

Early in her book, Barrett (2011) discusses the ever-present temptation to account for complex behavior with complex underlying mechanisms. The purposeful, collective behavior of ants in retrieving food, for instance, does not result from anything more complex than the interactions between the dispositions of individual ants (1) to lay down pheromones while returning to the nest with food and (2) to follow pheromone trails laid down by fellow ants. An observer able to detect pheromone trails, then, can see that complex behavior emerges from simpler interactions between the ants and their environment given their basic goal of food retrieval. Similarly, one might say that to an observer able to perceive human affordances it might well appear that our behavior could be largely explained by our dispositions to follow out the trails the affordances etch across our environment, in light of our goals. But of course *we are* such observers. It is just that we have come to explain our observations using the representation-rich vocabulary of beliefs, desires, and intentions and to account for our ability to predict in terms of our ability to know others' beliefs and desires. But this was a contingent choice, and alternate explanations exist, including of course developing a vocabulary centered around affordances and purposes (it would be correct to hear echoes of Sellars' 1997 "myth of Jones" in this assertion but incorrect to infer that what I am advocating here is a return to mindless behaviorism; we need to get beyond this simplistic dichotomization of psychology). One need not deny that it is indeed *possible* to predict and explain our behavior using such constructions to also recognize that reconstructing both folk and scientific psychology around the concept of an affordance could have consequences as significant and far-reaching as the shift from phlogiston to oxygen. Part of making this transition in our science of the mind will be learning to expect sensitivity to throwability rather than weight; reachability rather than distance;

climbability rather than slant or slope; and so on. And indeed, there is evidence for all of these.

Consider reachability. In the same way that systematic errors in weight perception indicate that the relevant underlying sensitivity is to the throwability of the hefted objects, so systematic errors in distance estimation in near space indicate that the underlying sensitivity is to the reachability of the objects around us. For instance, people with longer arms estimate target objects to be closer to them than do those with relatively shorter arms (Proffit & Linkenauger 2013). Similarly, when provided with a hand tool that extends their reach, people estimate target objects to be closer when the tool is present than when it is unavailable (Witt & Proffitt 2008; Witt, Proffitt, & Epstein 2005).

Proffitt and Linkenauger (2013) introduce the term "perceptual ruler" to capture this fact that we measure our world in units defined by our phenotype—our shape, abilities, and behavior. The idea is that one's capacity defines the unit of measure in question—thus the limits of the reachable for a person are perceived as something like one distance unit of reachability (let's call this one *ru*). When we compare a participant who can reach 100 cm (perhaps with a tool) to one who can only reach 75 cm, an object 50 cm away will be perceived as being 0.5 *rus* away to the first participant, and 0.67 *rus* away to the second—thus, it will seem closer to the first. For the same reason, the hand tool changes our perception because it changes our ability. Such "phenotypic reorganization" (Proffitt & Linkenauger 2013) occurs because as we prepare to act, we selectively attend to those aspects of the environment relevant to our preparation. We adopt a particular action stance, which in turn shapes our perception of the world. The same is true for graspable objects, where one's grasping ability alters estimations of object size in the same way (Linkenauger, Witt, & Proffitt 2011), and for acquired skills such as golfing and batting: better putters see the hole as larger, and better batters see the ball as larger (Witt et al. 2008; Witt & Proffitt 2005). The list of such effects is long:

> In addition to reaching and grasping, other action boundaries have also been shown to be used as perceptual scales. Jumping ability influences the perceived extent of jumpable gaps (Lessard, Linkenauger, & Proffitt 2009). Throwing ability influences the perceived extent over which one anticipates throwing (Witt, Proffitt, & Epstein 2004). The ability to pass through an aperture affects its perceived size (Stefanucci & Geuss 2009). The ability to balance on the thin beam of wood influences the beam's perceived width (Geuss & Stefanucci 2010). The ability to duck under a horizontal barrier influences the barrier's perceived height (Stefanucci & Geuss 2010). (Proffitt & Linkenauger 2013, p. 22)

Moreover, "ability" as measured in other ways has been shown to have systematic effects on perception and estimation. A series of studies has shown, for instance, that the perception of the slant of a hill systematically varies with the expected *effort* of the walk up it; a hill is perceived as being steeper when one wears a heavy backpack (Bhalla & Proffitt 1999), when one is fatigued (Bhalla & Proffitt 1999; Proffit et al. 1995), when one is less physically fit (Bhalla & Proffitt 1999), and when one has lower blood glucose levels (Schnall, Zadra, & Proffitt 2010). Similar effects are obtained for estimations of walking distance (Zadra et al. 2010). That such effects reflect an underlying perception-action calibration is shown by inducing a *recalibration* by, for instance, having participants walk on a treadmill, which pairs walking effort with no distance gain, and thereby induces an adjustment to the effect that effort is required to maintain one's location (as might be experienced in a more naturalistic setting when swimming against a current or walking into a strong wind). After experiences such as these, participants estimate distances to be substantially greater than they do prior to such recalibration (Proffitt et al. 2003). In these cases one is not misestimating distance but correctly estimating climb- or walkability in light of one's experience of moving around and acting in the world. That is, these various effects should not be seen as distortions of perception but as accurate reflections of the agent's underlying relationship to the environment. It is to the values of these relationships that our perceptual systems are primarily responsive.

Developmental work also underlines the more general point that action and ability educate perception. For instance, Linda Smith (2005) points out "babies, like birds, are confused by transparency" (p. 291); 9-month-old babies are better at retrieving toys from opaque containers than transparent ones because they tend to try to reach *through* the walls of the transparent containers (Diamond 1990a, 1990b). But giving children of this age transparent containers to play with for a while eliminates this problem (Titzer, Thelen, & Smith 2003).

> Why? These babies in their play with the containers—in the inter-relation of seeing and touching—had learned to recognize the subtle cues that distinguish solid transparent surfaces from no surface whatsoever and had learned that surfaces with the visual properties of transparency are solid. (Smith 2005, p. 292)

Something similar might be said about the visual cliff, an apparatus in which a steep drop is covered over by a thick sheet of glass suitable for walking (think of a laboratory version of the famous glass skywalk at the Grand Canyon). The interesting finding is that infants (and other animals) with little experience in self-locomotion will venture across the apparent drop, whereas those with extensive experience will avoid doing so (Campos et al. 2000; Gibson & Walk 1960). Actions alter and guide perceptions, and perceptions guide actions; what it is for the visual cliff to *look* different is to be understood in terms of the way we *act* differently. More generally, seeing is something we achieve by doing—by the activity of *looking*. Yogi Berra thus understated the case when he said "you can see a lot just by looking"; in fact, it is the only way to see anything at all. And we see things in terms dictated by action—the possible and the desirable opportunities that the world affords.

There is also some suggestive evidence that descriptors that are more distant from perception and generally considered the hallmark of central cognition, such as concepts, categories, and common-sense knowledge, are also organized along practical, action-oriented, and generally relational dimensions. For instance, one classic study of the semantic relations between animal terms (cat, rat, gorilla, wolf, elephant, and the like) used factor analysis techniques to specify the underlying structure of the semantic field (Henley 1969). The basic idea is that one can uncover the dimensions along which people take these animals to be similar and different, and that this will tell us something about how people conceive of these animals. The results will not be surprising in the context of the current discussion: animals are reliably categorized along dimensions of size and fierceness (and sometimes along the lines of what appears to be "similarity to humans"); that is, the underlying structure of the semantic space containing our animal terms is specified in relational, action-relevant terms.

More recent work uncovering the underlying structure of common-sense knowledge appears to paint a consistent picture. The Open Mind Common Sense (OMCS) database is a repository of "shared common sense" in the form of facts that are so obvious that they are only rarely explicitly formulated (Singh et al. 2002). As of 2008 it consisted of over 700,000 statements of general knowledge, including such things as "People don't want to be hurt" and "A trunk is a part of a car" (Speer, Havasi, & Lieberman 2008). Using matrix decomposition and factor analysis techniques similar to those used in the animal study above, Speer, Havasi, and Lieberman (2008) explored the underlying semantic structure of this database. It is worth pausing to offer a brief and simplified account of what they did because the techniques they used are—not coincidentally—similar to the ones described in chapter 4 that I have claimed will in fact allow us to better capture and characterize the true underlying functional profiles of brain regions. This methodological intersection is, I think, one of the

Table 5.1
Hypothetical feature matrix for three objects

	Solid	Animate	Mobile
Car	1	0	1
House	1	0	0
Cat	1	1	1

factors that will facilitate the convergence of the neuro- and biobehavioral sciences.

Speer, Havasi, and Lieberman (2008) began by extracting a matrix from the OMCS database having objects along one axis and properties along the other—essentially constructing a very large spreadsheet with all the objects in the database listed in the first column and all of their possible properties in the database in the first row. Information about objects in such a matrix can be captured by placing a number in each cell (x,y) that captures whether or to what degree a given object y has property x (they do not actually specify whether they built a binary or an analog matrix, but this does not matter for our purposes here). For instance, if we have three objects {car, house, cat} and three properties {solid, animate, mobile} our matrix would look like the one shown in table 5.1.

Each object is thereby characterized in terms of its location in a multidimensional space—in this case three dimensions (solid, animate, mobile). So cars are at location (1,0,1), houses at location (1, 0, 0), and cats at location (1, 1, 1). Applying dimensional reduction techniques to such spaces can help us to reveal and understand the semantically meaningful dimensions that capture the underlying relationships among objects and their properties that matter most to us. In the current case note that all of the variance in the matrix—all of the information that distinguishes among cars, houses, and cats—is captured by the last two dimensions. Thus, in the simplest form of dimensional reduction, we would remove the first dimension and leave the other two intact. Other possibilities for transformations of a matrix so it better captures or expresses the underlying variance would include collapsing two dimensions into one when they seem to capture the same information (they covary) or rotating the matrix (the space) when all of the variance in the descriptions of objects seems to be on a diagonal *between* dimensions rather than lie directly *along* one of the original dimensions.

The overall claim is that such transformation can reveal the properties that most matter to us or best capture the actual similarities and differences that we perceive between objects. If the tiny database constructed above had in fact captured a good deal of what and how people think about these objects, then the dimensional reduction would be said to reveal the underlying difference that we perceive and that helps us distinguish among these objects in thought and deed. That is, we might claim to have discovered that we care most about mobility and animacy—that these are the dimensions that do the most cognitive work for us.

Naturally, this tiny database captures no such thing. But the OMCS database, just by virtue of its sheer size, might plausibly claim to offer at least a significant glimpse into the primary dimensions that structure our "common sense" knowledge. It is thus interesting and significant that the two dimensions that capture the *most* variance appear to be desirability (things people want or avoid) and feasability (things people can do or not). Other dimensions of variance include urban versus rural; animate versus inanimate; and indoors versus outdoors (Speer et al. 2008). Unfortunately, the authors do not provide a more complete account of the dimensions they discovered (their concerns were more methodological than psychological), but the results nevertheless appear consistent with the story I am telling here. Future psychologically oriented work along these lines is clearly called for.

Measuring (and organizing) the world using a ruler denominated in action-relative (and action-relevant) units is a perfectly sensible thing to do when the primary purpose for gathering the information is to guide action, as these are the units that best serve the purpose. These facts call for a change in how we understand the fundamental job of the brain: *brains evolved to control action* (Anderson & Chemero in press), to manage the values of agent-environment relationships, and the perceptual system evolved to pick up on such action-relevant properties as these. Note this is not a matter of giving up on objective in favor of subjective perception or of pitting accurate against inaccurate perception; reachability (for instance) is an objective, relational, scalar property of objects in an agent's environment, and it is quite clear that agents generally perceive this property accurately. The point is rather that this reconceptualization of our epistemic relationship (and access) to the world requires a reunderstanding of the nature of our perceptual sensitivities, which in turn helps us better to see the true nature of the tasks that brains must master.

One interesting implication of this is that when we invent scales, rulers, clocks, and other measuring devices, along with the specific practices necessary for using them, we are not merely doing better with tools what we were doing all along in perception. Rather, we are constructing new properties to perceive in the world (Danziger 1997; Zeruvabel 1991)—properties that actually *require* these tools to perceive them accurately. One way of characterizing the nature of the mistake that I am claiming was made in the traditional cognitive sciences, then, is that they illegitimately supposed that the properties of the world that these technologies and techniques uncovered are the very ones to which we were imperfectly sensitive prior to their invention. But this does not appear to be so.

Heidegger (1962) famously leveled a version of this charge against Descartes: to suppose that the structures of mind revealed by isolated, logical reflection are more fundamental than, and in fact underlie and give rise to, our capacities for active engagement with the world was to fundamentally misconstrue the nature of human being (see Dreyfus 1990; Guignon 1983 for insightful commentary). In the next chapter I make a similar claim about computation and symbol manipulation. The invented technologies of language, logic, and mathematics should not have been taken to reveal what our brains had been doing all along. Rather, these formal systems gave us different ways of interacting with the world—offered new affordances—that we have been able to exploit to achieve ever more sophisticated goals. This reflection brings to the fore a fact about brains, bodies, and human beings that we will leave aside for now but return to in the next chapters: brains evolved to control action, but actions are not restricted to movements of the biological body. We are social environment-altering tool users. Tools give us new abilities, leading us to perceive new affordances, which can generate new environmental (and social) structures, which can in turn lead to the development of new skills and new tools in a virtuous cycle that, through a process that Clark (1997) calls "scaffolding," greatly increases the reach and variety of our cognitive and behavioral capacities. In this chapter I have primarily been interested in understanding our native capacities as a way of characterizing the nature of our basic epistemic relationship to the world and through that coming to better understand the nature of the brain's basic job. But to understand human cognition we will eventually have to explore the many ways in which our native capacities interact with elements of the built and social environment.

5.4 Perception and Control: Caching and Catching

The view being expounded in this chapter leads naturally to the further idea that the fundamental *cognitive* problem facing an organism—deciding what to do next—might best be understood not as choosing the right response in light of a given stimulus but as choosing the right stimulus in light of a given goal. Here knowledge of sensorimotor contingencies (Noë 2004)—how perceptions change with action—and the perception of affordances work hand in hand to allow an organism to follow environmental affordances to the right sensations, such as "the perception of a full stomach, not an empty one; the perception of safety, not fear" (Barrett 2011, p. 100), given an organism's purposes.

Paul Cisek (1999) developed this line of thought in an interesting and novel way, beginning with some reflections on the nature of life itself. What distinguishes living from nonliving things, fundamentally, is that living things actively, if temporarily, resist entropic dissolution. That is, they possess homeostatic mechanisms that keep certain biologically relevant variables such as temperature, pH, or chemical concentrations within some acceptable range. Some of these mechanisms are specialized for controlling internal variables (say, the amount of a certain metabolically important molecule) and work via reciprocal feedback mechanisms. But, as Cisek points out, some of these mechanisms serve to ensure that the *external* environment has certain characteristics. In the case where this can not be accomplished by changing the environment itself (giving off heat, excreting chemicals)—that is, where the local value of the variable is not within the organism's sphere of influence—then such homeostatic regulation must be accomplished by *going where the local value is appropriate*. The bacteria *E. coli* has developed a simple mechanism of just this sort: as the bacteria move about, turning rates are increased in inverse proportion to nutrient concentrations. The effect of this mechanism, given that "food sources are usually surrounded by a chemical gradient with a local peak" (Cisek 1999, p. 8), is that *E. coli* tends to move up nutrient gradients from lower to higher levels (Koshland 1980). Cisek comments:

> As evolution produced increasingly more complex organisms, the mechanisms of control developed more sophisticated and more convoluted solutions to their respective tasks. Mechanisms controlling internal variables such as body temperature or osmolarity evolved by exploiting consistent properties of chemistry, physics, fluid dynamics, etc. Today we call these "physiology." Mechanisms whose control extends out through the environment had to exploit consistent properties of the environment. These properties include statistics of nutrient distributions, Euclidean geometry, Newtonian mechanics, etc. Today we call such mechanisms "behavior." In both cases the functional architecture takes the form of a negative feedback loop, central to which is the measurement of some vital variable. Fluctuations in the measured value of the variable outside some "desired range" initiate mechanisms whose purpose is to bring the variable back into the desired range. ... The alternative "control metaphor" being developed here may now be stated explicitly: *the function of the*

brain is to exert control over the organism's state within its environment. (Cisek 1999, pp. 8–9, emphasis in original)

Indeed, as Cisek points out in more recent work (2007; Cisek & Kalaska 2010), if the brain originally evolved as a control system for managing interactions with the environment, then because brain evolution has been demonstrably conservative (and in fact we saw evidence throughout chapter 1 that the brain evolved largely by applying existing resources to new ends), we should expect the same to remain fundamentally true today. Thus, to put this insight into a vocabulary somewhat closer to the framework being developed here, organisms are primarily sensitive to, and act so as to select and maintain desired values of, salient organism-environment relationships, whether those consist in relative temperature differentials, surface penetrability, nutrient concentrations, predator or prey distances, object reachability and throwability, or emotional and social relations (which are discussed in part III of this volume). In this framework the problem of action selection is better understood as achieving the right perception(s) given a goal than it is as choosing the right response to a given stimulus.

Just before drafting this section I watched a Western scrub jay cache an acorn in the ground just outside my study at CASBS. It held the nut in its beak and shoved it into the ground. Then it looked at the nut—cocked over its head to put an eye directly over it—removed the nut, and tried again. It repeated this several times, sometimes pecking the nut deeper into the earth until—I would submit—it "looked" right, until the jay experienced the right perception. It then proceeded to cover the nut with leaves, again looking after each placement, and again stopping only when the perception met whatever visual criteria scrub jays use to signify successful caching (see Clayton, Bussey, & Dickinson 2003 for a scientific description of caching behaviors of these remarkable animals). It was perfectly clear to me that the animal was purposefully acting so as to achieve the right perception. As Barrett says, "the lovely thing about Gibson's theory is that it is a theory of perception that is automatically a theory of cognition, with no false separation between the two" (Barrett 2011, p. 110).

Of course, I am not even an *amateur* naturalist, so my observations of scrub jays are not to be taken too seriously. But there is a good deal of solid science behind the general idea. Consider just one example, solving the so-called outfielder problem (Fink, Foo, & Warren 2009; McBeath, Shaffer, & Kaiser 1995). The problem is simple to state, but it has proven somewhat hard to specify its solution: How does an outfielder put herself in a position to catch a fly ball? With the traditional view, the natural solution would be to measure the initial trajectory and velocity of the ball and, given knowledge that the ball will follow a parabolic path, predict its landing point and run there (Saxberg 1987). To put it bluntly, however, there is currently no evidence for this proposal and good reasons to doubt it: for instance, expert baseball players are both poor at predicting landing locations and good at catching fly balls (Shaffer & McBeath 2005). So it seems likely that there is some other mechanism or strategy at work.

From the perceptual control perspective, we should be looking for mechanisms that rely not on reconstructive perception and predictive modeling but rather on the management of organism-environment relationships and, more specifically, acting so as to achieve the correct perceptions. Consider first that a fly ball will look a certain way as it moves through the air—it will cause a particular pattern of perceptions. For instance, it will appear to accelerate into the air, decelerate, and then accelerate toward the ground. One proposal for how outfielders coincide with fly balls, known as optical acceleration cancellation (OAC), suggests that if a runner moves so as to cancel out the perceived vertical acceleration of the ball she will end up where the ball lands (Chapman 1968). Recent work by Fink, Foo, and Warren (2009) tested OAC against the traditional view by systematically perturbing the flight of the ball in a virtual reality environment and recording the compensatory movements of the participants. The results favor the OAC hypothesis: participants moved so as to cancel the optical acceleration and, interestingly, were roughly as accurate in catching perturbed fly balls as unperturbed ones—a result that would seem to further undermine the traditional view because the perturbed balls would not have been following a parabolic path.

It is important to underline how big a blow this apparently simple finding is for the traditional view. In order to solve this fairly complex perception-action coordination problem, people do not appear to be reconstructing or otherwise representing the flight path of the ball or generating predictive hypotheses. Instead they are acting continuously in real time so as to achieve a particular perception. To solve the problem, that is, the organism enacts a perception-action feedback loop in which perception is directly affecting action adjustments, and those adjustments are directly affecting perception. There is no need to posit any further epistemic mediators, symbolic or otherwise, to characterize the nature of this perception-action coupling. If perception is sufficient at least sometimes to drive behavior without heuristic reconstruction, then any global argument from the poverty of the stimulus to the necessity of perceptual reconstruction and symbolic processing of necessity fails. Any and all arguments that predictive modeling or other symbolic processing is *required* to solve a particular problem, then, must be made on a case-by-case basis and situated within (or alongside) the evidence that many of the basic problems of acting, interacting, and surviving in the world are treated as perceptual control tasks and *not* symbol processing ones. One upshot of this is that the more distant a cognitive mechanism (or its parts) appears to be from perception and action, the better argument we are owed for its existence; we no longer get amodal symbol processing as a scientific freebee. Instead, we need an account of how an organism that does not *need* symbol processing to manage its everyday affairs came to need and develop (evolve?) such capacities. I do not believe that there currently exists a satisfactory answer to this question in the traditional cognitive sciences. Note this is *not* to say that the ecological view has a fully satisfactory account of such vital higher-order behaviors as speaking, reading, and computing; it is rather to assert that these realms are now genuinely contested territory (Crowley 2011; Lakoff & Nuñez 2000; Landy & Goldstone 2007a, 2007b, 2009, 2010; see chapters 6 and 7 for further discussion).

What an organism's brain is fundamentally doing, then, is managing the relationships between the organism and its environment, and its perceptual apparatus is specially suited for facilitating that task. Recall from the last part of this volume that I was urging us to understand the differential activity in and functional activation of regions of the brain to be a sign of the different loadings of each region on a set of common neuroscientifically relevant psychological (NRP) factors. I argued that we should understand these loadings as representing each region's differential tendencies to help manage some particular aspect of the organism-environment relationship and thus to help determine the character of the organism's interaction with its environment. In this chapter I have been focusing on elucidating some of the more basic relationships that brains detect and manage, but of course there are also important higher-order relationships that will eventually need to be addressed—this is one of the central research agendas of the framework I am advocating here. I will not be in a position to fully close this loop until the reflections of the next chapter, but I do hope it is beginning to become clear how all this could fit together: the organism perceives the values of salient organism-environment relationships and, in light of some goal, acts so as to perceive the right changes in those relationships. The brain that manages this process is structured in such a way that its various parts have different dispositions to manage the values of these perceived relationships, and the summed cooperation and competition among the active dispositions of these regions in a situation both determines the current goal and structures the control loop that facilitates the required behavior.

5.5 Embodiment and Symbolic Processing

In this section I offer a brief recontextualization of some of the work discussed in chapters 1 and 3, for the ecological framework may allow us to cast these efforts in a somewhat different light and will also serve to recapitulate the central argument I have offered here. A central tenet of the ecological framework as presented is that the temporally extended stream of perception in fact generally contains sufficient information to specify and guide interaction with an organism's world. Insofar as this is true an important upshot of this is that *stored* traces of perceptual experience would be content carriers that automatically possess two elements crucial to the implementation of representation: (1) causal decoupling from the immediate environment and (2) intentionality grounded in action guidance (Anderson & Rosenberg 2008; Clark 1997; Glenberg 1997). Thus, even within the ecological framework, it strikes me as perfectly reasonable to hypothesize, in addition to our primary capacities to enact perception-action feedback loops with elements of the present environment, an ability to enact such a loop *as if* an element of the environment were present. Given the centrality of the notion of representation to most theorizing about cognition, it is worth taking the time to consider the issue directly as well as what the argument offered here does and does not show about the potential role of representations in cognition.

The classic experiments of Shepard and Metzler (1971) suggesting that our ability to compare 3-D shapes for congruity is supported by a mechanism for imagined rotation have been largely taken to support the existence of some such capacity to simulate physical interactions with perceived objects—that is, to *imagine* the interaction and its effects. And there has been a great deal of similar work since that suggests that during what Clark (1997) has called offline cognition our brains are working, not with the amodal symbols posited by the PSSH but with stored traces of perceptual experiences (Kosslyn, Thompson, & Ganis 2006). For instance, Barsalou, Solomon, and Wu (1999) asked participants to name properties associated with generic nouns such as grass or watermelon. They hypothesized that if visual simulation plays an important role in property recall, then when asked to name properties of "watermelon" versus "half watermelon," participants in the latter condition would tend to retrieve more internal properties ("seeds," "red"), and those in the former condition more external properties ("green," "oval"). This is indeed what they found. Similarly, Solomon and Barsalou (2001) showed that responses in a property verification task ("Do ponies have a mane?") are faster in a target trial when an earlier trial asked about a visually similar property ("Do horses have a mane?")

than when the earlier trial asked about a visually dissimilar property of the same name ("Do lions have a mane?"). The suggestion here is that if part of the verification process in the earlier trial involves the construction of a perceptual simulation, then visually similar properties will be easier to retrieve in the later target trial. Rounding out this very brief sketch, Pecher, Zeelenberg, and Barsalou (2003) discovered that during a property verification task modality switches induce performance decrements. Thus, for instance, if asked to judge whether pillows are soft, participants will be faster when an earlier trial asked about a property of the same modality ("Is sandpaper rough?") than when asked about a property in a different modality ("Are lawnmowers loud?"). Because this echoes an effect seen in online perception—for example when judging whether two successive stimuli come from the same spatial location, participants are faster when the stimuli are in the same modality than when the modality is mixed across trials—this suggests that the "conceptual" property verification task may activate the same modal systems used in online perception. And recall that the claim that there will be neural overlaps between the "perceptual" and "conceptual" systems is part of the backdrop of the general hypothesis of concept empiricism for which a great deal of other evidence was already presented in chapter 1.

Examples like these appear to show that stored traces of perceptual experience do indeed play a role in offline cognitive processing. But it is important to recognize that although such examples are generally taken to be arguments against the existence of amodal symbols, in fact they do *not* (and probably cannot) prove that there are no amodal symbols or systems implemented in the brain. Not only is it difficult if not impossible to prove a negative, many of the findings can in fact be accommodated by amodal theories that posit frames or other context effects that might affect retrieval times (see Machery 2007 for a nice discussion). Indeed, Machery's analysis suggests that a global question such as "does the brain work with modal or amodal representations?" can never be definitively resolved as such, and we should be scientifically content to test specific amodal models against specific modal models of various phenomena until there is but one model left standing. I certainly agree that in the end the science will come down to testing specific claims in specific contexts about the underlying processes and mechanisms of cognition. But I also think that the increasing strength of the ecological approach to psychology outlined here (see also Barrett 2011; Chemero 2009) shifts the debate on symbolic processing—along with the burdens of proof—in ways that are important to appreciate.

As Machery suggests, the debate is currently stuck in what might be called a war of amodal attrition: many theorists largely accept the basic tenets of reconstructive perception but are arguing on a case-by-case basis over the nature of the symbols employed in various tasks and processes. Consider the following case in point. Larry Barsalou (1999) famously posited that simulation mechanisms could serve as the basis for implementing a *perceptual symbol system*, a cognitive system with all of the features and capacities of amodal *physical* symbol systems but built instead on modal foundations. In my view Barsalou's framework should be understood as an attempt to address certain important shortcomings of the PSSH while holding on to the basic logic of reconstructive perception. Barsalou did the field a great service by showing how one could implement abstract computational/cognitive processes using native perceptual resources. But from the standpoint of the ecological approach I am advocating here, he did not break sufficiently with the traditional framework, with the result that he is able to tell only half of the new story of the brain: he rightly highlights the centrality of perception and perceptual embodiment to cognition, but he largely neglects the reciprocal importance of action and active embodiment.

Admittedly, Barsalou (1999) does not discuss the fundamental nature of our perceptual relationship to the world explicitly, so it would take more work than I can devote here to definitively establish his adherence to the tenets of reconstructive perception. But the logic of his presentation of perceptual symbol systems closely follows the logic motivating the PSSH: he takes perceptual processing to be extractive and abstractive, resulting in the generation of image schemas and perceptual symbols; his perceptual symbols are componential and amenable to many of the same sorts of transforms as the amodal symbols of the PSSH, such as arbitrary recombination and transformation by logical rules; and he takes his theory to solve certain conundrums such as the symbol-grounding problem and the problem of meaning (Harnad 1990; Searle 1980) that arise uniquely from within the logic of reconstructive perception. Thus, I feel fairly confident that I have put his theory in its proper context.

This being said, the accuracy of this contextualization and its attendant *explanation* of the neglect of the role of the active body in cognition is less important than the *fact* of that neglect and the impact it had on the development one of the more prominent theories of embodied cognition and thereby on the development of the field over the past 15 years. For because Barsalou imagined cognitive processing to centrally involve *symbols*, albeit symbols with a structure, syntax, and semantics dictated by the nature and content of perception, he imagined the basic mechanisms of perception and cognition to resemble classic *symbol-processing* mechanisms. In contrast, although I believe the ecological view that I am advocating here can *also* take aboard the possibility of a cognitive role for stored traces of perceptual experience in support of offline cognition, the most natural processing mechanisms for such perceptual simulations will be the very ones that underlie the online perceptual control loops described in section 5.4, above (and indeed, you may recall that many of the examples discussed in chapter 1, such as "shifting spatial attention along an imagined number line," have just this flavor of applying sensorimotor processing to stored traces of perceptual experience). Moreover, *both* symbol systems hypotheses ignore the potential cognitive role of the active body interacting with tools—including *external* symbols—in the environment.

Thus, I believe the debate over the constituent mechanisms of cognitive processing must be broadened beyond the debate over symbols in at least three ways. One set of mechanisms often neglected when the focus is exclusively on symbols and symbol structures consists of iterated and nested perceptual control loops. We have already seen that these can play important roles in generating adaptive behavior. Another set of mechanisms involves offline motor processing of simulated perception—a possibility that could involve applying sensorimotor processes to various sorts of perceptually grounded and action-oriented representations (Clark 1997) including guidance representations (Anderson & Chemero 2009; Anderson & Rosenberg 2008), indexical-functional symbols (Agre & Chapman 1987), pushmi-pullyu representations (Millikan 1995), or emulator-based representations (Churchland 2002; Grush 1997, 2004), all of which are best understood as control structures with intentional content and not (for instance) as arbitrary symbol structures. A third set of cognitive mechanisms harnesses our capacities for interacting with *external* resources, including symbolic ones (see chapters 6 and 7 for discussion). Once we include the active, interactive body in our repertoire of resources that can be brought to bear in solving cognitive problems, I believe we will find that it is only the rare case that will require the particular kinds of central computational resources envisioned by the two symbol systems hypotheses. Notice I do *not* claim that there are no such cognitive processes. I insist only that cognition is supported by a broad diversity of constitutive mechanisms.

One important example involving the second set of mechanisms—and one that points to another way that bodily interaction with the environment gets neglected when the debate is reduced to one about symbols—is the case of causal knowledge derived from and generated for the guidance of our interventions in the world (Gopnik 2012; Gopnik & Shulz 2007). A good deal of evidence has shown that children are able to acquire from experience sophisticated, structured knowledge of causal relationships, sufficient for them to predict the effects of actions or events not previously experienced (e.g., Sobel & Kirkham 2006). This should remind us that "stored traces of perceptual experience" are not limited to stored *visual images* or other merely qualitative aspects of perception; nor is learning limited to the association of stimuli explicitly paired in experience. Rather, because experience itself has sensorimotor structure, and because we integrate and consolidate our experiences over time, stored experience can capture sophisticated sensorimotor and other relational contingencies and thus guide actions in circumstances not previously experienced (Anderson & Chemero 2009; Anderson & Rosenberg 2008). This being said, it is important to observe the following caveat: the work on causal learning is typically presented as supporting the analogy that I rejected above between perceiving and acting in the world and scientific reasoning and hypothesis testing (e.g., Gopnik 2012). I think, however, that the empirical evidence does not in fact support this gloss. What the evidence shows is (1) we learn best through interacting with and intervening in the world, (2) what we learn guides future action, (3) we are sensitive to mismatches between expected and actual perception, and (4) these mismatches alter learned sensorimotor contingencies (and thus behavior) in systematic ways. The precise neural mechanisms that underlie these capacities are not well understood, but it is at least clear that they *do not* necessitate the explicit, syntactically structured symbols that are involved in scientific hypothesis testing, nor do they necessarily point in the direction of reconstructive perception, the generation of neutrally specified models of the world. Woodward (2007) offers the following summary:

> [B]oth instrumental learning in rats and human judgments of causal strength (as well as actions based on this) behave as though they track the perceived degree of control or manipulative efficacy of the instrumental action over the outcome, which is what one would expect on an interventionist account of causation. In addition, phenomena such as sensitivity to contingency, backward blocking, and causal discounting show that at least some causal representation and judgment are sensitive not only to information about the rates of occurrence of cause and effect and the processes that connect them but also to information about what would or does happen in the absence of the cause and under the occurrence of potential alternative causes of the effect. (Woodward 2007, p. 27)

As we perceive and act in the world, we are learning to see what the world affords and where and how to intervene to generate preferred outcomes, and we are at the same time inducing the neural structures that make such control possible. The causal knowledge we acquire appears to be best understood as a guide to action, written primarily in the vocabulary of sensorimotor contingencies. In this sense the literature on causal learning appears to be solidly within the pragmatist tradition despite the cognitivist (structuralist) gloss applied to it by many of its leading proponents.

5.6 Knowledge and Practice

This brings me to one final bit of reconceptualization, with which I end this chapter. Again because the classic debate over cognitive processing has focused on the nature of the symbols and symbol transformation processes, the debate between the PSSH and alternative approaches—parallel distributed processing (PDP) models in particular—has largely focused on whether and to what degree the symbols involved in cognition are "distributed" (Rumelhart & McClelland 1986; McClelland & Rumelhart 1986). My own view is that our understanding of the underlying nature of PDP models was overly constrained from the outset by the example set by the PSSH, which drove the field to focus on the nature of the symbols in PDP architectures. In fact, the brain networks of active beings are perceptual control systems, and the transformations of activation values through layers of the network should have been understood in this context.

At this stage, however, it must be acknowledged that even those readers who find this direction compelling—and especially those who do not, should any remain who have read this far—will no doubt be asking: Yes, but *how* does the brain do this? It is all well and good to focus on control over computation, but given the underlying capacities of the brain, don't the mechanisms of control in fact have to be computational and therefore representational? The brain may have evolved to control action, but didn't it develop computational mechanisms to do this?

As emerges over the course of the next chapter, my answer to this question is: "yes, but." I believe that the notion of computation, construed as the calculation of mathematical functions—mappings between input values and output values—is indispensible for the brain sciences. But I also think that CCTM drew from this fact the wrong moral—abstracted to the wrong class of explanatory models. To see what I mean, it is worth considering one of the classic exemplars in the debate between CCTM and dynamic systems explanations of the brain: the Watt centrifugal governor.

A significant problem faced early in the Industrial Revolution was how to maintain a consistent driving force for machinery such as the automated loom, given that the speed of a steam engine depends on both the workload placed on the engine and the pressure of the steam driving it, which variables are constantly fluctuating. The Watt centrifugal governor is the solution to this problem. The mechanism consists of a pair of weights at the end of some scissored arms, coupled both to the engine's rotating flywheel and to its throttle in such a way that as the engine speeds up, the weights will move further apart, driven by the centrifugal force of the spinning flywheel; this motion causes the throttle valve to close somewhat, reducing the speed of the engine, which in turn causes the weights to move closer together, opening the valve, and so on, thus maintaining speed at a consistent level (figure 5.3; the description and analysis of the Watt governor offered here are based on those in Shapiro 2011; van Gelder 1995).

In his 1995 paper van Gelder introduces the Watt governor as a candidate exemplar for modeling cognition by first describing the sort of solution it is *not*. For from a certain perspective it might seem natural to turn

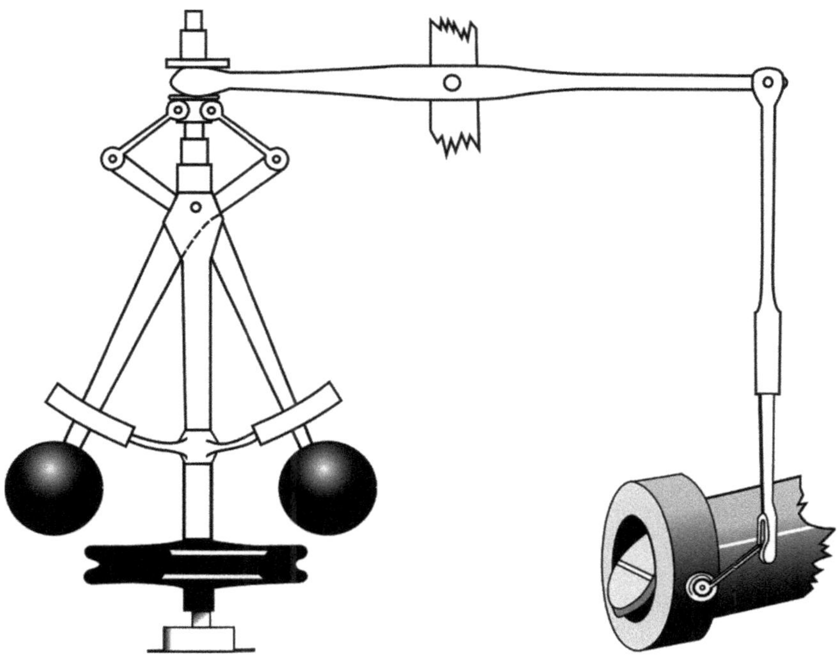

Figure 5.3
Watt's centrifugal governor. As the flywheel spins, the weights move closer or further apart depending on the speed; this movement pulls on the throttle lever, changing the amount of steam driving the engine. The coupling between these parts ensures that the engine maintains a steady speed. Picture from the Wikipedia Commons, made available under the Creative Commons CC0 1.0 Universal Public Domain Dedication.

this *control* problem into a *measurement* problem: monitor the engine speed; when there is a discrepancy between actual and desired speed, measure the steam pressure; calculate the necessary change in steam pressure; calculate the required throttle valve adjustment; and make the adjustment. After all, if one knows what adjustment is required, making it is mechanically trivial; all the hard work is in knowing what adjustment is needed. Thus, one solution to this problem is to focus on building the right measuring and calculating devices. If this is the approach taken, then, as van Gelder writes:

> There must be some physical device capable of actually carrying out each of these subtasks, and so we can think of the [imagined alternate] governor as incorporating a tachometer (for measuring the speed of the flywheel); a device for calculating the speed discrepancy, a steam pressure meter; a device for calculating the throttle valve adjustment; a throttle valve adjuster; and some kind of central executive to handle sequencing of operations. This conceptual breakdown of the components of the governor may even correspond to its actual breakdown; that is, each of these components may be implemented in a distinct, dedicated physical device. (van Gelder 1995, p. 348)

The reader will of course recognize this approach as analogous to that motivating CCTM in general and the symbol systems model in particular: given a control problem, start by measuring and recording the values of some relevant set of objective properties in the world and use these values to calculate the appropriate response. Adapting such a framework makes it natural to suppose that there might be dedicated (neural) devices for representing and calculating, and it becomes the job of cognitive neuroscience to identify and describe these components. As van Gelder puts it:

> Perhaps the most central of the [imagined alternate] governor's distinctive properties is its dependence on representation. Every aspect of its operation, as outlined above, deals with representation in some manner or another. The very first thing it does is measure its environment (the engine) to obtain a symbolic representation of current engine speed. It then performs a series of operations on this and other representations, resulting in an output representation, a symbolic specification of the alteration to be made in the throttle valve; this representation then causes the valve adjusting mechanism to make the corresponding change. ... Finally, notice that the governor is homuncular in construction. Homuncularity is a special kind of breakdown of a system into parts or components, each of which is responsible for a particular subtask. Homuncular components are ones that, like departments or committees within bureaucracies, interact by communication (that is, by passing meaningful messages). (van Gelder 1995, pp. 350–351)

In contrast, van Gelder argues, the Watt governor does *not* operate via the manipulation of representations, nor do its parts interact via communication. There are no measurements and no steps where the result of the measurements are subjected to computational transformations into other representations. Rather, the parts of the system are dynamically coupled, and their states are continuously codetermining. For instance, the change in the angle of the weight arms over time is given by the following equation:

$$\frac{d^2\theta}{dt^2} = (n\omega)^2 \cos\theta - \frac{g}{l}\sin\theta - r\frac{d\theta}{dt}$$

where θ is the angle of the arms, n is a gearing constant, ω is the speed of the engine, g is a constant for gravity, l is the length of the arms, and r is a constant for friction and the hinges (van Gelder 1995, p. 356). A similar equation describes the change in speed ω over time as a function of arm angle θ and various other parameters. The point here is that the system works not because there are representations and ordered transformations of one kind of information into another but because the relevant variables are dynamically coupled.

The Watt governor implements something quite similar to the perceptual control processes described above: engine speed (the "perception") directly guides throttle adjustment (the "action"), which itself partly determines ongoing engine speed ("perception"), and so on, thus enacting an ongoing coordination that, given the design of this particular system, tends to maintain a steady engine speed. Recall the nature of the mechanism we apparently employ to solve the outfielder problem—the problem of running to where a fly ball will land. In order to manage this feat we do not appear to be reconstructing or otherwise representing the speed, position, or flight path of the ball, nor calculating its likely landing point. Instead we act continuously in real time so as to maintain the perception of zero vertical acceleration. We enact a perception-action feedback loop in which perception directly determines our action adjustments, and those adjustments determine ongoing perception; to catch a fly ball, that is, we instantiate a special kind of Watt governor between our eyes and our legs.

For current purposes, there are three points I need to make about this example. First, I think van Gelder is clearly correct in his analysis of the nature of the Watt governor and of the differences between it and the measurement-driven representation-manipulating solution to the same problem. The Watt governor does not operate by generating, maintaining, or transforming representations; rather, it maintains an ongoing reciprocal coordination between input and output. Van Gelder is also right to insist that because of those differences, the Watt governor is a much better exemplar for the fundamental mechanisms of cognition than is the symbol system.

Second, I do not see any good reason to deny that the Watt governor computes in the sense that it implements a mathematical function that in this context solves the required control problem. As is made clearer in the chapters that follow, I think we can understand brains—neurons and networks—in much the same way, as implementing various kinds of mathematical functions (multiple regression, Bayesian updating, and the like) that help solve various control problems. Put differently, the solutions to the fundamental problem(s) of achieving desired perceptions and maintaining organism-environment relationships within desired bounds almost certainly involve the computation of mathematical functions and feedback loops that implement the necessary dynamic coupling between perception and action. Knowing what these functions are and which neurons and networks help instantiate them are interesting questions that should certainly be pursued in the cognitive neurosciences. But the evident importance of computation should not have led (and should no longer lead) us to focus exclusively on what is represented and where. In the current case, for instance, it would appear to be a mistake to ask where (or in what) the system represents ω and where it is transformed into the required angle of throttle adjustment. This is not because we *can not* do it—for certain purposes, it might actually be useful and illuminating to think about the circumstances in which the angle of the arms *would* be used as a representation of the speed of the engine—but because doing so leads us to abstract away from properties that are in fact crucial to understanding the system and its operation and so leads us to the wrong class of guiding models for cognition. For instance, when we think about computation as the transformation of representations, we are led to think in terms of computational steps (measure the speed, compute the discrepancy ...) and to wonder about the storage and communication of intermediate results. We focus on the timeless algorithmic structure of the computation and abstract away from the temporal dynamics of the sensorimotor coordination.

Third, and following directly from the second point, it was perhaps in retrospect unfortunate that van Gelder framed the contrast between the systems in terms of the presence or absence of representations per se. For although he allows for the possibility that dynamic systems could incorporate some form of representation without thereby becoming symbol systems, this notion was left largely undeveloped. Thus, much of the debate between symbolic approaches to cognition, on the one hand, and dynamic systems approaches, on the other, has centered around whether or not a given system could be construed as having or using representations (Bechtel 1998; Chemero 2009; Prinz & Barsalou 2000; Shapiro 2011). In fact, the otherwise important debates between CCTM and dynamic systems approaches frequently devolve into arguments simply over whether or not a system is representing. The result has been to reinforce the perceived equivalence between representational systems and classical computational ones such that it is often assumed that establishing that a system uses representation is sufficient to prove that it is classically computational. Indeed, anticipating just this sort of turn to the debate, van Gelder makes a rather shrewd prediction.

> [A]lthough many connectionists are thoroughly dynamical in their general vision of the nature of cognitive systems, many others attempt to combine their connectionist dynamical substrates with an overall conception of the nature of cognitive systems, which owes more to the computational worldview. ... A classic exemplar is David Rumelhart and James McClelland's past-tense learning model. In this kind of work, underlying systems that are basically dynamical in nature are configured so as sequentially to transform static input representations into output representations. They retain much of the basic structure of the computational picture, changing some of the ingredients (in particular, the nature of the representations) but retaining others. Connectionism of this kind can be regarded as having taken up a half-way house between the computational and dynamical conceptions, combining ingredients from both in what may well turn out to be an unstable mixture. If this is right, we should expect as time goes on that such connectionist models will increasingly give way either to implementations of generically computational conceptions of cognition, or to models that are more thoroughly dynamical. (van Gelder 1995, p. 374)

As we have already seen, this *is* right, and contemporary connectionism has been moving back toward a componential computationalism that differs from classic symbol-system-inspired cognitive models such as ACT-R only in the nature of the underlying implementation (see section 3.2). Thus, although I do think that the debate over representations and their role in cognition was a necessary one, I think the overly strong association it created between representations and classical computationalism is an unfortunate development for the field. As I noted in the opening pages of this chapter, and emphasized in section 5.5, control processes can often usefully incorporate representations, and symbols have an important place in human cognitive skills including language and mathematics. The real quarrel that enactive views have with CCTM is against the notion of reconstructive perception and the cognitive architecture that follows naturally from that approach. This debate is not (or should not be) about the existence of representations but rather is about the kinds of models that best capture the underlying nature of cognitive processes. The fact that representations play a major, central role in the exemplars of CCTM and often no role or possibly a minor, peripheral one in the exemplars of dynamic systems theory is an important thing to notice, but this is at best a heuristic marker for the differences between the approaches.

Rather than organize the field around the distinction between representational and nonrepresentational systems, then, I would like to suggest a contrast orthogonal to the issue of representation that may serve a similar and ultimately more useful purpose. The contrast comes from Andrew Pickering's (2010) analysis of the nature and history of cybernetics. [Not incidentally, the word cybernetics, Pickering tells us, comes from the Greek *kybernetes*, meaning "governor," which seems appropriate given the current context, for the field was named by Norbert Weiner precisely in honor of the Watt governor (Pickering 2010, p. 6)]. Pickering writes:

> To put it very crudely, there are two ways to think about the brain and what it does. The way that comes naturally to me is to think of the brain as an organ of *knowledge*. My brain contains representations, stories, memories, pictures or the world, people and things ... If I know something, I have my brain (and not my kidneys, say) to thank for it. [In contrast,] the cybernetic brain was not representational but *performative* ... (Pickering 2010, pp. 5–6; emphasis in original)

So far, that is largely just a restatement of the distinction I am drawing between computation and control, and Pickering's use of "representation" as allied with knowledge but not performance reproduces the problem I am hoping to get beyond. And yet I think that this particular contrast between knowledge and performance points us in a useful direction, in part because it should be obvious that we cannot do entirely without either perspective if we hope to fully understand the brain. The brain is an organ of knowledge *and* an organ of performance; the questions are how to understand the nature of each and their relations.

The answers are important in part because they allow us to situate the current perspective in a long tradition of post-Cartesian thinking about the mind and behavior. One legacy of the Cartesian framework reflected in CCTM is the notion that knowledge is fundamental and prior to practice, in the sense that it both grounds and explains it. One aspect of this mistake is the supposition that knowledge how is driven by knowledge that. As van Gelder writes: "[a] fundamental Cartesian mistake is, as Ryle variously put it, to suppose that practice is accounted for by theory; that knowledge how is explained in terms of knowledge that; or that skill is a matter of thought" (van Gelder 1995, p. 381). But the failure to understand that skills comprise an autonomous form of knowledge is only one aspect of this mistake; the other is the foundationalism itself—the supposition that one and only kind of knowledge provides the fundamental ground for everything else. Thus, although I sympathize with the tendency exhibited by some of my colleagues (and sometimes even by me) to turn the equation on its head and make practice the ultimate foundation, I think that this is an unstable and untenable position in the end. Consider the following from van Gelder:

> In the Cartesian framework, the basic stance of mind toward the world is one of representing and thinking about it, with occasional, peripheral, causal interaction via perception and action. It has been known since Bishop Berkeley that this framework had fundamental epistemological problems. It has been a more recent achievement to show that escaping these epistemological problems means reconceiving the human agent as essentially embedded in, and skillfully coping with, a changing world; and that representing and thinking about the world is secondary to and dependent upon such embeddedness. (van Gelder 1995, p. 380; for discussion see Dreyfus 1990; Guignon 1983; Ryle 1984)

It is most certainly true that human agents are essentially performative; that, as one might say, our *telos* is action. We need to absorb that lesson to be able to understand the brain. It is also true that cognition is largely a matter of taking real-time advantage of our embeddedness, of acting in and on the world in ways that rely on its temporal and physical structure; for this reason, explanations of cognitive phenomena must generally be spread over brain, body, and environment. And it is finally true that representing and thinking about the world is dependent on such embeddedness, if only because our physical interactions with the world are a necessary condition for intentionality (Anderson 2006; Anderson & Rosenberg 2008; O'Donovan-Anderson 1997).

But what, in the end, does it mean to say that thinking is "*secondary to* and dependent upon" practice? Here it becomes too easy to overinterpret this valid post-Cartesian insight as establishing practice as the ultimate foundation for all aspects of mind, and I think this would be a mistake both substantively and rhetorically. It would be a *substantive* mistake because practice is not always and everywhere prior to thought (not temporally, nor metaphysically, nor epistemically). We do not always act first and think second any more than (to paraphrase Dewey 1896) we sense first and act second. What we do depends on (follows from, is motivated by, is shaped by, is partially explained by) what we know and think; and what we know and think depends on (follows from, is motivated by, is shaped by, is partially explained by) what we do. Knowledge and practice are in this sense intertwined and coconstructing. To apply Erikson's (1986) creepy and visceral metaphor for social interaction ("climbing a tree that climbs you back"): knowledge and practice climb each other. It is crucial to the science we are developing to keep this firmly in mind, for it will help us understand both knowledge and practice.

Consider again the Watt governor analogy for the mechanism that allows us to coincide with fly balls. It certainly appears correct to assert that the mechanism is best understood as implementing a sensitive feedback loop allowing the agent to couple perception to behavior, and because information about the state of ball and body is continuously available and sufficient for the purpose, there is simply no role for reconstructive perception to play in this capacity. But clearly this does *not* mean that the mechanism requires no *knowledge*. It is, after all, a learned skill. Neurally speaking, the mechanism is a *structural achievement* that involved an initial search of the set of possible neural pathways and partnerships that would support the behavior, the consolidation of the partnerships that worked, and the subsequent fine-tuning of connections and other parameters to allow the mechanism to work well. These aspects and parameter values must be retained—stored—in such a way that the skill is retained over time. This is a necessary condition of knowing how—and being able—to catch a fly ball. It is probably the case that most of these connections, weights, and parameters are best understood as simply necessary parts of the coupling mechanism—conduits for the causal forces at play, no different from the pins and hinges and welds of the Watt governor. But might it turn out that one of more of these values represents some (dynamic or stable) property of the agent, world, or both? Of course! Discovering this would be extremely interesting! Yet it would hardly undermine the facts that this particular mechanism implements a perception-action coupling; that neurocognitive mechanisms are generally best understood as control structures; and that models based on reconstructive perception and classic componential computationalism generally offer a poor fit for human cognition. For even as this would make the skill dependent on knowledge, it would remain true that the knowledge depended and depends on practice, is learned in the course of practice, and is useful only in the context of a given skill. Knowledge and practice climb each other.

It is for this reason that adopting a performative foundationalism would also be a *rhetorical* mistake: no one should be surprised, nor should anyone's global theory of the mind be either refuted or vindicated, should a particular skill or behavior turn out to require some knowledge—a representation or a stored parameter or some other bit of information whatever its form—that cannot be reduced to or otherwise analyzed away in favor of practice. The performative framework being presented and developed here, in insisting that we take action and not knowledge to be our primary *telos*, most certainly points us to a new class of models and mechanisms for understanding brain and behavior, but it should not be a framework inimical to all kinds of representation. Defining dynamic and enactive and performative accounts of cognition by their rejection of representations risks building an unwarranted fragility into the argument for these approaches, as if merely discovering a representation would be enough to refute them and reinstate CCTM. But this is foolishness. It is time to admit that simply discovering a representation in a cognitive system tells us virtually *nothing* about the nature of the underlying architecture in which the representation plays a role. I am hoping this point will be easier to take on board if we stop contrasting control with representation and stop equating representation and knowledge and organize our thinking on this matter instead around the concepts of knowledge and performance. For in the end, the human agent is both a doer and a knower, and we can accept the point that our cognitive architecture is organized such that we generally *know* in order to *do*, and still recognize that the relative importance of doing, knowing how, and knowing that will vary with the skill to be explained as well as across time and circumstance.

This returns us to Pickering's general point, and the broader reason that the contrast between knowledge and performance is useful to us here, for the performative perspective has been long neglected not just in the cognitive sciences but across the modern sciences more generally (Hacking 1983). It is well past time to emphasize it and reshape our science to account for it.

> The modern sciences invite us to imagine that our relation to the world is basically a cognitive one—we act in the world through our knowledge of it—and that, conversely, the world is just such a place that can be known through the methods and in the idiom of the modern sciences. One could say that the modern sciences stage for us a modern ontology of the world as a knowable and representable place. And, at the same time, the product of the modern sciences, scientific knowledge itself, enforces this vision. Theoretical physics tells us about the unvarying properties of hidden entities like quarks or strings and is silent about the performances of scientists, instruments and nature from which such representations emerge. … [T]he performative aspects of our being are unrepresentable in the idiom of the modern sciences. … I want to say that cybernetics drew back the veil the modern sciences cast over the performative aspects of the world, including our own being. (Pickering 2010, pp. 20–21)

Another way to express Pickering's insight here that brings it closer to the focus of the current work is: we need to keep in mind that brain science is a natural science, but also a life science and, in part, a human one. In developing a biologically grounded science of the mind, we need to be sensitive not just to the empirical properties of the object of our study but also to the underlying nature of and the assumptions that structure the science itself. The idea that physics offers a poor model for the psychological sciences in particular is so commonplace to be cliché, and yet here we have a very specific example of why and how this might be so: it is harder within such a framework to appropriately capture the world of *doers*.

> It has become difficult for us to recognize that much of our being does not have a cognitive and representational aspect. I suppose I could figure out how my doorknob works, but I don't need to. I established a satisfactory performative relationship with doorknobs long before I started trying to figure out mechanisms. A science that helped us thematize performance as prior to representation might help us get those aspects of our being into focus. And of course, beyond the human realm, most of what exists does not have the cognitive detour as an option. It would be good to be able to think explicitly about performative relationships between things, too. (Pickering 2010, p. 23)

It is in part this difficulty that has led connectionism—the single most plausible and flexible class of dynamical models of the brain yet developed—both to a largely cognitive and representational set of self-interpretations (McClelland 2009) and in some cases to the embrace of a kind of componentiality that I am arguing is better left behind (Jilk et al. 2008).

In the next chapter I outline an account of dynamic computation in the brain that, in contrast, fits better with the performative perspective being developed here and that treats the brain primarily as a control system in general and as an affordance processing system in particular (Cisek 2007; Cisek & Kalaska 2010). My account attempts to capture the fundamental feature of the functional brain model I am advocating, whereby regions of the brain are differentially responsible for managing organism-environment relationships. This responsibility is distributed across the brain but not evenly, leading to functional differentiation without functional specialization. According to this schema, interaction with an environment offering multiple affordances causes regions of the brain to be differentially activated—as determined by the differential weights between nodes in the brain network and as captured in the differential loadings on the common set of NRP factors. A situation posing several possible courses of action, then, will cause multiple distributed patterns of neural activation across the brain, and the behavior of the organism in this situation will be ultimately determined by competition among the patterns of regional brain activity. Such competition was described in classic PDP networks by Smolensky (1986) but was interpreted by him to be a sign of competition among representational states. In fact, I believe that the observed competition in the brain should be understood to reflect tension among various behavioral control loops that could be enacted; pattern competition indexes affordance competition (Cisek 2007). If I am successful, the account will show how the brain harnesses computations to behavioral control without requiring the implementation of a full-blown internal symbol system.

Interlude 5 Network Thinking

It ain't so much the things we don't know that get us into trouble; it's the things we do know that just ain't so.
—Artemus Ward

One of the things that everybody knows about motor cortex is that it is organized somatotopically, with each small region of the cortex responsible for managing the movements of each small part of the body. The idea is illustrated in one of the most iconic images of neuroscience: the motor homunculus (Penfield & Rasmussen 1950). That grotesque humanoid is shown draped over the motor cortex like a lounge singer across her piano in probably every Introduction to Psychology textbook on the planet, where it is presented as a paradigmatic exemplar of brain organization. Both the explicit claim it makes and the implicit advice it offers could hardly be clearer: good systems neuroscience means discovering the crisp, clean mappings between cortical structures and behavioral and cognitive functions.

The trouble, as the epigraph from Artemis Ward foreshadows, is that it "just ain't so." This is not a radical claim from the fringe and will not be news to any researcher who studies the neural bases of motor control. We have known for some time that there is no such simple one-to-one mapping of cortical region to muscle or body part, neither anatomically (Scheiber 2001) nor functionally (Graziano et al. 2002a, 2002b; Schieber & Hibbard 1993; see Anderson 2007c for further discussion). Schieber (2001) summarizes the matter this way:

> 1) *Convergent* output from a large M1 [primary motor cortex] territory controls any particular body part, joint, or muscle. 2) *Divergent* output of many single M1 neurons reaches multiple spinal motoneuron pools. 3) *Horizontal connections* interlink the cortex throughout a major body part region. 4) *Widely distributed activity* appears in a major body part region whenever any smaller body part is moved. 5) *Partial*

inactivation of a major region affects multiple smaller body parts simultaneously. 6) *Plasticity* limits the degree to which control of a specific body part can be assigned to a particular piece of cortex. (Scheiber 2001, p. 2125)

This of course begs the question: If we know better, then why do we keep telling the same misleading story about the organization of the brain? In the case of primary motor cortex, Graziano (2011) suggests that credit goes to the continuing influence of a classic study of functional anatomy. In 1890 Beevor and Horsley discovered that if you dissect *away* the cortex to uncover the pyramidal tract, those descending connections to spinal motor neurons are indeed organized as a kind of muscle map of the body. From this observation sprang the idea that the cortex developed to individually target these pyramidal neurons; or, as Graziano put it, that each region of the cortex stands at the head of a cable that joins it to the body, muscle by individual muscle. This picture of cortical function encourages us to focus on the individual actions of each region—when and how it tugs on its cable—over the interactions between the regions. It offers a clear, simple, easy-to-understand model for cortical organization that is easy to generalize and has proven hard to shake. But it is also a mistake we need to move beyond.

In fact, motor cortex, as with the rest of the brain, is dominated by recurrent connections in which the evolving patterns of activity determine functional outcomes (Capaday et al. 2011; see chapter 6 for further discussion). In an architecture such as this, interactions dominate, and so if we want to understand motor control—or any other function of the brain, for that matter—we need to give up on our cable-oriented focus on point control and attend instead to interactive network dynamics. Indeed, once we start to engage in "network thinking" (Mitchell 2006, 2009), we come to see that the interactive perspective makes much better sense of brain function than a cortical point-control model ever could, for behavior (and thought!) is itself a dynamic, evolving phenomenon in which each moment remains intertwined with the past and anticipates the future. Even simple paw movements illustrate this fact:

A basic tenet of [the Jackson-Walshe] thesis is that activity at a cortical focus represents the leading part of a movement (e.g., paw) and that of parts further removed (e.g., wrist and elbow). Thus, the state of neural activity at any one cortical point in the recruited area and that of the area itself are anticipations of what additional muscles may need to be recruited as the movement evolves or is perturbed. Sequencing, or anticipation, is a property of recurrent neural networks (Dayan & Abbott 2001). Central to the Jackson-Walshe thesis is the idea of overlapping and graded movement representations as a mechanism underlying integrated control of the musculature. ... Thus, cortical points are not functionally independent, they cannot operate in isolation. (Capaday et al. 2011, p. 2525)

The crucial sentence here is the last. Odd as it is to say, the realization that neurons are parts of networks and cannot be functionally understood in isolation has still not penetrated as deeply as it needs into neuroscience's collective unconscious. Instead, as the cable metaphor suggests, we tend to assign simple purposes to individual cells and regions and use this to guide our hypotheses for the functions of higher-level structures. Thus, this cell stimulates that muscle, so this region controls that body part. Or this cell responds to contrast at that orientation, so this region is responsible for edge detection. But, as we have seen throughout this volume, the matter is not that simple, not in perceptual and motor areas nor anywhere else.

So, what would it mean to get beyond this localist approach? Network thinking offers something like a family of heuristics for guiding one's scientific attention to global properties and overall system dynamics. It involves focusing on the relationships between entities, as much as or more than on the entities themselves. It means attending to global dynamics over local activity. Consider in this light the following two principles of network thinking.

First, rather than approach a complex system by breaking functions into subfunctions and assigning functions to proper parts—a heuristic that has been quite successful across a broad range of sciences (Bechtel & Richardson 1993, 2010)—network thinking suggests one should look for higher-order features or patterns in the structure and behavior of complex systems and advert to these in explaining the functioning of the system. The paradigm exemplars for this sort of approach come from the discovery of common, functionally relevant topological structures in various kinds of networks, ranging from human and insect social networks to the phone grid and the Internet, and from foraging behaviors to the functioning of the immune system (Barabási & Albert 1999; Barabási et al. 2000; Boyer et al. 2004; Brown, Liebovitch, & Glendon 2007; Huneman 2010, Jeong et al. 2000; Newman, Barabási, & Watts 2006).

Here is a simple example: the brain network is organized such that there are relatively dense local interconnections, with sparser connections to other more distant regions. This organization—called a "small world" architecture—is one of the reasons the brain can efficiently process a great number of signals and yet still fit into a skull small enough to allow for childbirth (Sporns 2011). Moreover, small-world structure facilitates neural synchronization (Mitchell 2009), which as we have seen is crucial to overall function. Importantly, small world structure is not a property of any single part of the brain; it is a feature of the whole. Similarly, some brain regions such as anterior insula play the functional role they do because of the strength and number of the *relationships* they have to the rest of the brain; they are the functional "hubs" in the brain's network (Uddin et al. 2014).

Second, network thinking means being aware of the constant interplay between bottom-up and top-down, feed-forward and feedback processing. A signal from an individual neuron contributes to the overall pattern of activity, but it is at the same time shaped by that pattern. Its signal can be enhanced or degraded, its activity modulated in myriad ways, and what the signal "means" depends on that overall context (see interlude 3).

One might be tempted to respond: Of course, who ever denied it? We have always known that the parts interacted to produce overall function. But this response underestimates the depth of the reconsideration that is required here. For it is not as if we can identify the one fixed function of an element of the system and then worry about the effect of interactions later. Rather, the interactions are often precisely what fix local function—what the neural (and genetic! see Jablonka & Lamb 2004) element is *for* and is *doing for* the organism at any given moment is determined by the relationship between that element and its functional partners (see chapter 3 and interlude 4).

The value of attending to evolving network dynamics and functional interactions has, of course, been a theme that runs throughout this book. But in the next chapter I hope to showcase more fully how such a focus helps us make better sense of the brain and what it is doing for the organism. The aims are both to situate this emerging picture of brain in the context of the evolutionary, embodied approach to cognition developed in chapter 5 and also to demonstrate how the dynamic, interactive brain facilitates such cultural achievements as mathematics and dialogue.

6 Embodiment, Computation, and Control

To this point I have been asserting that the componential computational theory of mind (CCTM) offers an inadequate architectural framework for understanding the brain. More particularly, I have argued that the kind of local functional specialization and static componentiality assumed by CCTM does not do justice to the dynamic complexity seen in the functional organization of the brain. In fact, as we saw in chapter 2, the case of sensory substitution and the possibility of dynamically establishing novel neural partnerships to support new behavioral capacities suggest that a scientific focus on a perception-oriented modal organization for the brain might be better supplanted by an exploration of the brain's (meta-modal) *task-oriented* organization (Engel et al. 2013). Consistent with this suggestion we have seen that CCTM—as represented, for instance, in the symbol systems hypothesis—fundamentally misconstrues our epistemic position in the world, thus misunderstands the nature of the problems brains evolved to solve, and so is led to posit a metaphysically possible but by no means empirically necessary sort of reconstructive, representational mechanism to solve those problems. In its stead I have suggested that we need to adopt and adapt a framework for understanding the brain that centers around the notion of *control*. The brain is a dynamic control system that modulates the sensorimotor coupling at multiple timescales. More specifically, I have been arguing that the fundamental behavioral control problem should be understood in terms of achieving particular values for salient organism-environment relationships and that behavioral goals should be cashed out in terms of achieving particular perceptions in given circumstances.

In light of this action-oriented framework, I have argued that local activity in regions of the brain should be interpreted as reflecting that region's differential propensities to contribute to managing some set of these organism-environment relationships (and have outlined an empirical approach to understanding regional activity that can help us specify these relationships in the form of NRP factors). Under this interpretation, then, *patterns* of neural activation across the brain reflect the functional coalitions between regions that actually implement the control processes and pathways determining an organism's dynamic response to a situation—its acting to achieve a particular perception. (Here it is perhaps useful to recall the simple bubbler analogy from chapter 2, where interactions between bubblers implement the currents that move the cargo around the system.) Finally, I have suggested that dynamic systems approaches, properly understood, offer a more promising class of exemplar models for understanding the neural solutions to such behavioral control problems.

From the dynamic systems perspective, the scientific focus in the cognitive and behavioral neurosciences should be on how activity in and across regions of the brain impacts the behavioral trajectory of the organism. As Randall Beer puts the matter:

> [T]he focus is on how the system's behavior changes over time and the various "forces" that shape this trajectory. The importance of the states that the system passes through lies not so much in any content that they may be assigned, but rather in their sensitivity to subsequent inputs and the future behavior that they make possible. (Beer 2003, p. 212)

The main task for the current chapter, then, is to begin to describe how the model for brain function outlined in the first four chapters fits in with the model for understanding perception and behavior outlined in chapter 5. That is, in this chapter I begin to outline a framework for understanding the neurobiological foundations of behavior in light of the observed functional complexity of the brain. I do that in three steps. First, I offer a general overview of dynamic computation in the brain and the way it gives rise to adaptive behavior. Second, I ground these theoretical (in principle) mechanisms in actual neurobiology. And finally, I show how these mechanisms can be applied to cognitive processes that do not appear to naturally lend themselves to the perception-action coordination framework. I begin that last task at the end of this chapter, with a discussion of mathematics, and continue in chapter 7 where I offer an interactive account of natural language.

6.1 Connectionism, Pattern Competition, and Control

Abstractly, a neural network (McClelland 2013; McClelland & Rumelhart 1986; Rumelhart & McClelland 1986) consists of a set of processing units (nodes), each connected to the others with a certain strength (weight). Each node has a varying degree of activation determined by its current activity and any activation inputs, that is, the degree of activation in the nodes connected to it and the weight of those connections. Physical limitations of the nodes dictate that activation remains within a certain range (usually represented as a number between 0 and 1); thus, the summation function that translates inputs to activation must be nonlinear, a point to which we return shortly.

The neural network of the brain is massively recurrent (Sporns 2011). Activation does not just feed forward from the inputs of sensation to the outputs of motor control. Instead, feedback connections at multiple levels of organization ensure that sensory inputs are influenced by and integrated with ongoing activity—an architecture that should not be surprising in active *perception-seeking* organisms like us. Sporns notes:

> Even in regions of the brain such as primary visual cortex that are classified as "sensory," most synapses received by pyramidal neurons arrive from other cortical neurons and only a small percentage (5 percent to 20 percent) can be attributed to sensory input (Douglas et al. 1995). Cortical areas that are farther removed from direct sensory input are coupled to one another via numerous mono- and polysynaptic reciprocal pathways. ... [These] recurrent or reentrant processes make an important contribution to the shaping of brain responses and to the creation of coordinated global states . . . essential for the efficient integration of multiple sources of information and the generation of coherent behavioral responses. (Sporns 2011, p. 150)

As has been discussed in earlier chapters, the brain consists of a neural network, an underlying genetic network, and an overlying chemical gradient field. We are far from understanding the dynamic interactions of these three elements. Here, I am assuming that, at least to a first approximation, we can treat brain operations on the timescale of milliseconds to seconds as being governed by the neural ("wired") network, which is modulated and remodeled on the timescale of seconds (and longer) by the other two layers. That is, I am assuming that the main effect of genetic modulation and volume transmission is to change the weights between the nodes, modulating the effective connectivity of the network.

At any given moment in a quiescent network, the effective connectivity in the wired network would dictate the evolution and time course of any induced pattern of activity. But of course the brain network is never quiescent. It is always active to some degree, whether due to the purposeful activity of the organism or the endogenous dynamics of the brain at "rest" (Raichle et al. 2001). The effect of perceptually induced neural activation patterns, then, will depend not just on the connectivity of the network but also on the *ongoing* activity resulting from past patterns, memory, feedback connections, and so on. These overlapping patterns index the multiple dynamic states of the brain or, if you prefer, the analyzable contributors to the brain's overall state, and they reflect the functional coalitions between regions that actually implement the control processes and pathways contributing to an agent's ongoing behavior. These evolving patterns serve to manage the organism's interaction with its environment, and ongoing perceptual experience along with the recurrent connections described above serve to reinforce some and disrupt other patterns, in a continuous process of biased pattern competition.

How does all this work? A result from the classical parallel distributed processing framework is especially relevant here, as it will help us to better understand the underlying rules governing pattern expression and evolution in neural networks. Here I describe the situation qualitatively; the reader should consult Smolensky (1986) for the mathematical details (my discussion follows Smolensky's quite closely). One of the first things to realize about neural networks is that because their state at any given time is determined by the activation values of each node, that state can be captured as an ordered set $\{a_1, a_2, a_3 \ldots a_n\}$ where a is the activation level of the node and n represents the number of nodes in the network. That is to say, the state of a neural network can be expressed as a multidimensional vector or a point in a multidimensional state space. The space of possible network states has a particular shape, and the range of the activation values available for each node, then, imposes constraints on the extent of that space. For instance, in a three-node network, where the activation value for each node must be in the range $\{0,1\}$, the space of possible states is described by a cube having sides of 1 unit. The state of the network at any given time is a point somewhere in that cube. Similarly, a network of n nodes with the same range of activation values will be defined by an n-dimensional hypercube where each side equals 1. It is important to note that this global constraint on the shape and extent of the state space is in fact imposed locally at each node by the nonlinear summation function that determines the moment-by-moment activation value of the node as a function of its current state and the inputs it receives.

Now, it is natural to think of the state of a neural network as consisting of patterns with particular meanings. For instance, Smolensky (1986) speaks of different patterns of activation values as representing concepts or hypotheses. Under this sort of interpretation the changing activation values—more precisely the evolving patterns of these values—represent such things as inference (evolving conceptual structures) or changes in the degree of confidence in a set of hypotheses.

Alternatively, as I took pains to demonstrate in chapter 5, one can emphasize not knowledge but performance; neural networks *do* things, from causing wheels to turn or legs to move (Braitenberg 1986 Reeve & Webb 2002) to sorting mail (LeCun et al. 1989) to producing speech (Sejnowski & Rosenberg 1987) to classifying and differentially responding to objects (Beer 1996, 2003). In this case different patterns of activation values might index different possible behaviors, and the dynamic evolution of those patterns would track the convergence on a particular action in light of the current system state and sensory inputs. Although both knowledge-oriented and performance-oriented interpretations can be useful under different circumstances, from here forward I take the position that—especially when networks are embedded as part of the control architecture in a perceiving, acting system—the performance perspective is the preferable global stance to take. It is also worth emphasizing that in no case should we expect or require that every momentary state of the network admit of some high-level interpretation; one hazard particularly of knowledge-oriented interpretations of network evolution is the tacit assumption that if *some* states can be usefully understood as corresponding to a certain stage in a cognitive process—a logical deduction, for instance—then there must be some mapping of network states to *every* stage of the process. Similarly there is a temptation to understand control processes in terms of discrete steps that can be mapped into network states. But this is to use the algorithmic structures of classic computationalism as a lens through which to view neural network evolution, and this will generally be a mistake. Neural networks are dynamic systems and not algorithmic ones (Spivey 2007).

In any case, from the pragmatic, action-oriented approach we are developing here, let us imagine that the pattern of activation $p_1 = \{0, 0, 0, 1, 1, 1\}$ in a simple network indicates part of the preparation for throwing a stone, whereas the pattern $p_2 = \{1, 1, 1, 0, 0, 0\}$ indicates preparations for running away. For such a network it might be natural to interpret the state $s = \{0.3, 0.3, 0.3, 0, 0, 0\}$ as "maybe run?" Smolensky (1986) points out there is a concise way to express such partial activations mathematically. If the pattern $p_2 = \{1, 1, 1, 0, 0, 0\}$, and the network state $s = \{0.3, 0.3, 0.3, 0, 0, 0\}$, then $s = 0.3p_2$. He calls this the "pattern view" of the network. In general, a pattern can be expressed perfectly, but also *partially*. Partial activation is captured by weighting the pattern, $w \times p$.

One of the crucial features of neural networks is their ability to express more than one pattern simultaneously. This is called superposition, and it is captured mathematically by adding together the weighted vectors representing the patterns, such that $s = w_1p_1 + w_2p_2 + \ldots + w_np_n$. In a simple case imagine the network described above in a state that might be interpreted as "throw stone and run away." Assuming each pattern is fully expressed, that is, has a weight equal to 1, if $s = p_1 + p_2$, then in this case $s = \{1, 1, 1, 1, 1, 1\}$. That state s is in one of the extreme corners of the state space but clearly still is one of the possible states of the system. In general, it is possible for a neural network to fully express (superpose) any number of nonoverlapping patterns simultaneously.

What about the case of overlapping patterns? This question is crucial for us here because I have been arguing in some detail that the different functional states of the brain are instantiated by *highly overlapping* patterns of activation. A node (here imagined as a neuron or small brain region) is activated across many different task contexts but has different activation partners in each; the brain as a whole expresses a different pattern of activity in its various task states, but the same nodes are active across multiple states. Sticking with our simplified example, we can introduce a new network pattern $p_3 = \{0, 0, 1, 1, 1, 0\}$, implementing "eat." It quickly becomes apparent that this network *cannot* in fact be in the state "throw and eat" nor "eat and run"—that is, it cannot fully express these patterns *simultaneously* although it clearly could *sequentially*—because those states lie outside of the state space of the system; $p_1 + p_3 = \{0, 0, 1, 2, 2, 1\}$ and $p_2 + p_3 = \{1, 1, 2, 1, 1, 0\}$. These vectors specify a point outside of an n-dimensional hypercube with 1-unit sides. The states $s = 0.5p_1 + 0.5p_3$, and $s = 0.3p_2 + 0.7p_3$, and many others, are all possible, but it will not in general be possible to superpose two (or more) fully expressed overlapping patterns.

This restriction is determined by the shape and extent of the state space, itself imposed (or, enforced, if you will) by the local summation functions that determine activation values. Smolensky (1986) calls this situation "natural competition," and he shows that the nonlinear summation function naturally results in the amplification of the weight differences between strong overlapping patterns but not between weak ones.

That is, the nonlinear summation function enforces a dynamics on the system such that the stronger of two strong, overlapping patterns naturally wins out and comes to dominate the state of the system. Weak overlapping patterns, in contrast, can superpose without competition. Interestingly, Smolensky also finds that the nonlinear summation function tends to *reduce* the differences between the weights of *nonoverlapping* patterns, which suggests that any natural competition between neural states might in fact be a sign of their overlap—another indication of the fundamental nonmodularity of the brain.

With this understanding of the significance of neural patterns as background, there are a few things worth observing about the phenomenon of natural competition in neural networks. First, if we grant that neural activation patterns generally overlap in the brain and that neural networks capture something important about the underlying implementation of real neural processing, then we get competition between strong high-level activation patterns *for free*, simply as a consequence of the nonlinear summation and activation ceiling that characterize the response properties of individual nodes. That is to say, insofar as patterns of neural activity index different control processes, then *competition between different behavioral priorities or response options is built into the functional architecture of the system.* Second, and following directly from the first observation, the phenomenon of natural competition would appear to make the control interpretation of neural networks (and organic brains) rather more natural than the conceptual interpretation. This is because one would presumably want a conceptual/representational system to be generally able to simultaneously express *many* different concepts or hypotheses; true incompatibility between these will be the exception rather than the rule, and thus a system that imposed competition as a matter of course would seem a poor choice. But the opposite is true for a control system: there incompatibility between response options will be the rule, and the possibility of doing two or more things at once the exception. Thus an architecture that generally forces competition between strong—that is, salient or pressing—behavioral options makes good sense. Finally, this is one of the ways in which the neural overlaps that we observe in the brain actually make functional sense; the overlaps are not *just* the result of evolutionary and developmental pressures to conserve neural resources but offer functional features that may have been taken advantage of by and integrated into the overall functional architecture of the brain.

Any given circumstance will naturally contain or be partially defined by multiple organism-environment relationships and opportunities for action. In my model the detection of these will activate regions of the brain according to their native dispositions, as modulated by their interactions with other regions. When this is combined with the neural expression of the organism's other relevant states—temperature, glucose levels, general arousal, and so on—the result is a particular pattern of neural activity jointly determined by the organismal and environmental circumstances. The patterns that overlap naturally compete in the way outlined above, and this process is biased by the organism's ongoing evaluation of future reward and continuing interaction with its environment: sensory inputs and reward estimates will reinforce some patterns and disrupt others, changing their weights. Naturally, the process can be indirect as well if, for example, online experience triggers the recall of stored experience, which might itself tend to reinforce or disrupt some set of neural patterns and partnerships (Platt 2002).

It is worth noting here that I am using "biased competition" in a fairly generic sense to index the various mechanisms whereby distributed patterns of activity in the brain interact to change both an organism's behavior and also the underlying network state of the brain (which in turn changes behavior, activity patterns, and so on). That is, the model I am outlining is also compatible with the "guided activation" framework (Miller & Cohen 2001) and with "flexible hub" theory (Cole et al. 2013), which are developments of the original biased competition account (Desimone & Duncan 1995).

The basic idea is beautifully and simply illustrated by Sarah Newmann (1999). Newmann reports a finding that will not be at all surprising in the context of the present volume—that each node in the "social behavior network" contributes to more than one social behavior (e.g., sexual behavior, parental behaviors, aggression) and that each social behavior is supported by activity in multiple nodes (see figure 6.1). Differential degrees of activation in each node, together generating overlapping patterns of activation for two different social behaviors, compete in this network to determine the animal's (initial) behavior in its environment. She writes:

> [B]orrowing an important concept from our colleagues in cognitive neuroscience, we envision that a particular social behavior, for example, male sexual behavior, is an emergent property of the pattern of activity across the network (Mesulam 1990).

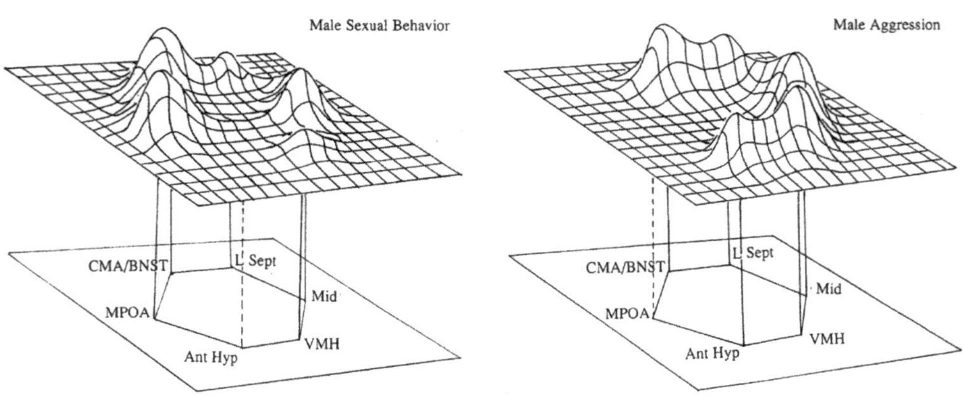

Figure 6.1
Hypothetical representations of the patterns of activity in the mammalian social behavior network at the outset of different behavioral possibilities. Adapted from Newmann (1999) and reprinted with permission.

> It is not an action produced by the "on" or "off" state of any one of the nodes, such as the medial preoptic area, but a sequence of multiple behaviors (e.g., sniffing, mounting, ejaculating, grooming) that is initiated by and emerges from a temporal pattern, and therefore a dynamic pattern, of activity across the network. (Newmann 1999, p. 248)

An animal's behavior will naturally change its perceptions in the environment—as, for instance, *other* animals in its environment react to the behavior—leading to changing patterns of activation in the relevant structure, further responses by the animal, and so on. Thus, the animal implements a perception-action feedback loop that depends on its evolving brain-network state but also, crucially, on the responses elicited in other animals. The social component of such feedback loops is something to which we return in chapter 7.

In the language of dynamic systems theory, the brain network charts a continuous path through a very high-dimensional state space, driven by the underlying attractor landscape and the biasing inputs that themselves change the brain's overall state via the competition between patterns. The rapid establishment of various functional partnerships—TALoNS at multiple physical scales—through learning and neuromodulation dynamically reconfigures the underlying attractor landscape and allows the different sets of biasing inputs to play their causal roles. In turn, the winning patterns can produce further changes in the attractor landscape by inducing new configurations of TALoNS (Cole at al. 2013). Note that such a system does not implement computation in the classic sense of applying discrete operations to stable representations (Spivey 2007). Nevertheless, it will sometimes be the case that the dynamics implement a *function*—that is, a characteristic mapping between inputs and outputs—and so in this sense it will often be appropriate to speak in terms of computation. Learning in such networks will often be best described as a process of the discovery and implementation of the right mapping—the right control equation—mediating inputs and outputs. One of the main claims of this book, of course, is that a fundamental functional principle of the brain is the deployment of basic control loops for multiple purposes, aided by the fundamental architectural principle of neural reuse.

6.2 Action Selection as Affordance Competition

Once one adopts the global perspective that local brain activity indexes local neural dispositions to shape ongoing behavior given the values of particular organism-environment relationships and the opportunities for action afforded by the circumstances, and that neural *patterns* index both the emerging behavioral decisions and the functional coalitions that will carry them out, it is a fairly small conceptual step to the idea that neural pattern competition of the sort described is also at the same time a process of affordance competition; what the brain is doing is determining which affordance(s) the organism will exploit. Precisely this is the basis of an intriguing model of action selection offered by Paul Cisek (2007; Cisek & Kalaska 2010).

As Cisek (2007) points out, prevailing models of decision making have generally inherited the assumption from the reconstructive perception framework that our fundamental epistemic approach to the world involves building an objectively specified world model (perception), which we then use to generate possible action plans (cognition), decide among them (action selection), and determine how the motor system will enact them (action specification). But we have seen in chapter 5 that it is better to think of perception largely in terms of the assessment of the values of salient organism-environment relationships and the detection of opportunities for changing those values through action. This shift in perspective regarding our basic epistemic situation naturally also impacts our understanding of the nature of the problem of deciding what to do. In particular, if we perceive the world in terms of the opportunities for action it affords, then deciding what to do might be thought of as largely a matter of choosing which perception-action path to follow by enacting one of the many available sensorimotor feedback loops. Cisek writes:

> The proposal made here is that *the process of action selection and specification occur simultaneously* and continue even during overt performance of movements. That is, sensory information arriving from the world is continuously used to specify currently available potential actions. … From this perspective, behaviour is viewed as a constant competition between internal representations of the potential actions which Gibson (1979) termed "affordances." Hence, the framework presented here is called the "affordance competition hypothesis." (Cisek 2007, p. 1586)

Although it is imprecise to talk of "representing" affordances—technically, an affordance is the perceivable relationship between an organism's abilities and features of the environment (Chemero 2003, 2009)—clearly it is the notion of internal (neural) competition among possible courses of action that is the center of Cisek's account. In the view being developed in this volume, which I take to be compatible with and supported by Cisek's hypothesis, brain regions have different dispositions to manage sets of organism-environment relationships in light of the opportunities for action afforded by circumstances. What are competing, then, to determine action outcomes are not representations of affordances but the neural patterns that are triggered by the perception of affordances and that evolve to implement the different control systems that would allow the organism to follow out or take advantage of one or another of the afforded opportunities.

There are two central tenets of Cisek's hypothesis, both of which fit in nicely with the overall framework being developed here:

1. The process of action selection and specification occur continuously and in parallel. Because the organism's brain evolved to support interactive behavior, and perception is to be understood in terms of the detection of opportunities for action, it stands to reason that the process of selecting and specifying actions is a continuous, ongoing part of simply perceiving and acting in the world. Cisek and Kalaska write:

> Schmolesky et al. (1998) showed that neural responses to simple visual tasks appear quickly throughout the dorsal visual system and engage putatively motor-related areas such as FEF [frontal eye fields] in as little as 50 ms. This is significantly earlier than some visual areas such as V2 and V4. ... In a reaching task, population activity in PMd responds to a learned visual cue within 50 ms of its appearance (Cisek & Kalaska 2005). Such fast responses are not purely visual because they reflect the context within which the stimulus was presented. For example, they reflect whether the monkey expects to see one or two stimuli (Cisek & Kalaska 2005), reflect anticipatory biases or priors (Coe et al. 2002; Takikawa et al. 2002), and can be entirely absent if the monkey already knows what action to take and can ignore the stimulus altogether (Crammond & Kalaska 2000). In short, these phenomena are compatible with the notion of a fast dorsal specification system that quickly uses novel visual information to specify the potential actions most consistently associated with a given stimulus (Gibson 1979; Milner & Goodale 1995). (Cisek & Kalaska 2010, p. 285)

2. For any given behavior, both processes occur in the same regions of the brain. Regional differentiation in the brain reflects different capacities for managing certain classes of actions, rather than (for instance) specializations for perceptual discriminations, decision making, and action execution. Cisek and Kalaska write:

> [S]tudies on the neural mechanisms of decision making have repeatedly shown that correlates of decision processes are distributed throughout the brain, notably including cortical and subcortical regions that are strongly implicated in the sensorimotor control of movement. Neural correlates of decision variables (such as payoff) appear to be expressed by the same neurons that encode the attributes (such as direction) of the potential motor responses used to report the decision, which reside within sensorimotor circuits that guide the online execution of movements. These data and their implications for the computational mechanisms of decision making have been the subject of several recent reviews (Glimcher 2003; Gold & Shadlen 2007; Schall 2004). (Cisek & Kalaska 2010, p. 270)

In fact, recent evidence suggests that such "mixed selectivity" may be the norm for brain structures underlying cognitive processing and task management (Rigotti et al. in press). Such functional overlaps have functional implications that can be discerned in the characteristics of the behavioral responses observed in the experiments and that are hard to explain within the classical framework. For instance, one body of evidence shows that the trajectory of reaching movements depends on the amount of separation of the targets in space: subjects move directly to the chosen target when the options are far apart but initially reach between the targets when they are close together and only later veer to their selection (Ghez et al. 1997; see also Howard & Tipper 1997). Such behavior is naturally explained as a side effect of the similarity of the response options in the case of the close targets: nearly identical patterns will tend to reinforce one another, perhaps even merging for a time, and it is only when the differences in the reach trajectory between the options become more pronounced as the hand approaches the targets that the two possibilities become distinguishable and competitive (Cisek 2006; Erlhagen & Schöner 2002). Similarly, when subjects in two alternative forced-choice tasks are asked to indicate their decision by moving a hand or a cursor to the chosen location, the subject's confidence in his or her choice predicts aspects of the movement including endpoint and peak velocity (McKinstry, Dale, & Spivey 2008). Cisek and Kalaska conclude:

> These findings are difficult to reconcile with the idea that cognition is separate from sensorimotor control (Fodor 1983) but make good sense if the continuous nature of the representations that underlie the selection of actions has been retained as selection systems evolved to implement increasingly abstract decisions. (Cisek & Kalaska 2010, p. 283)

Cisek (2006, 2007) offers a computational model of visually guided reaching decisions to make concrete the theoretical proposal outlined above and to compare the performance of the model with known neurophysiological performance data. The model consists of multiple interconnected populations of simulated neurons. Each neuron in the population is broadly tuned for some preferred value of a movement parameter, such as direction, and thus the neural populations can—through superposition—encode multiple distributed simultaneous possibilities for the eventual direction of the movement.

> The model suggests that sensory information in the dorsal visual stream is used to specify the spatial parameters of *several* currently available potential actions in parallel. These potential actions are represented simultaneously in frontal and parietal

> cortical regions, appearing as distinct peaks of activity in the neural populations involved in sensorimotor processing (Cisek & Kalaska 2005; Cisek, Michaud, & Kalaska 2004; Platt & Glimcher 1997). Whenever multiple peaks occur simultaneously within a single frontal or parietal cortical region, they compete against each other through mutual inhibition. (Cisek 2007, p. 1590)

It is worth pausing here to acknowledge that, in Cisek's model, neurons with similar tuning have excitatory connections between them, and neurons with dissimilar tuning inhibit one another. Thus, pattern competition of this sort—which I call structural competition to distinguish it from the natural competition described by Smolensky—is driven primarily by the existence of these antagonistic causal relationships. In contrast, when outlining the biased competition approach to action selection in the last section, I emphasized pattern competition due to overlaps in the patterns themselves. Strong overlapping patterns are forced to compete when full expression of both would take the system outside of its state space; the nonlinearity of the summation function ensures that in such cases the stronger pattern comes to dominate the weaker.

In reality, of course, both mechanisms are likely to be important. Certainly it will be the case that when the neural partnerships that support a particular behavioral capacity get established, inhibitory connections will be established along with excitatory ones. In some cases these connections will reflect aspects of the control system itself—the implementation of a negative feedback loop, for instance—but in other cases they may reflect the necessity to reduce activity in a competing control loop, without which the capacity in question could not be properly expressed (or learned). In this latter case a strong expression of a given pattern—a given potential control loop—would structurally inhibit activity in competing control loops, just as Cisek describes.

In addition to competition induced by mutual inhibition and nonlinear summation of overlapping patterns, an important aspect of the model is the bias induced from multiple sources. Here bias is understood simply to be activity in the system that reinforces one pattern over another; such activity might be induced by perception or result from feedback or other ongoing dynamics. When a given pattern is reinforced (strengthened) or a competing pattern is disrupted, natural competition and inhibition will tend to further strengthen the stronger pattern, enabling it to eventually emerge over competing options (which, of course, may also be receiving positively biasing inputs). In this model, then, "decisions are proposed to emerge as a 'distributed consensus,' which is reached when a competition between representations of potential actions is unbalanced by the accumulation of evidence in favor of a given choice" (Cisek 2007, pp. 1593–1594). Although I would prefer to talk about competition between potential control systems rather than between "representations of potential actions," this semantic preference should not be allowed to obscure my fundamental agreement with Cisek's framework. As he writes: "[i]n summary, instead of viewing the functional architecture of behaviour as serial stages of representation, we view it as a set of competing sensorimotor loops" (Cisek 2007, p. 1594).

The work reviewed above primarily highlights the role of biased pattern competition in action selection, but it is worth reminding the reader that the mechanism is universal and known also to play a role in perception, visual attention, and memory, among other things (e.g., Deco & Rolls 2003; Desimone 1998; Desimone & Duncan 1995; Mather & Sutherland 2011). The potential sources of bias are multiple in the brain and well documented in the decision-making literature. For instance, for actions largely modulated by the frontoparietal system modeled by Cisek, there is evidence for the importance of inputs from prefrontal cortex (PFC) (Miller 2000; Tanji & Hoshi 2001), orbitofrontal cortex (Schultz, Tremblay, & Hollerman 2000), and basal ganglia (Redgrave, Prescott, & Gurney 1999). Basal ganglia and orbitofrontal cortex in particular are thought to bias action selection by helping predict the rewards associated with various response options (Schultz et al. 2000). The neural sources of bias will vary from system to system and between different functional coalitions in the brain; orbitofrontal circuits may help bias patterns subsisting in PFC but not elsewhere in the brain, for instance. This is not the place to comprehensively review the relevant literature to provide a list of potential sources of biasing input (but see Gold & Shadlen 2007; Middleton & Strick 2000). Rather, the point to emphasize here is the control-oriented framework within which I am arguing these well-documented features of the neural bases of perception, attention, action selection, and decision making should be interpreted. Cisek captures this sentiment well:

> The affordance competition hypothesis … differs in several important ways from the cognitive neuroscience frameworks within which models of decision making are usually developed. Importantly, it lacks the traditional emphasis on explicit representations which capture knowledge about the world. For example, the activity in the dorsal stream and the frontoparietal system is not proposed to encode a representation of objects in space, or a representation of motor plans, or cognitive variables such as expected value. Instead, it implements a particular, functionally motivated mixture of all these variables. From a traditional perspective, such activity appears surprising because it does not have any of the expected properties of a sensory, cognitive, or motor representation. It does not capture knowledge about

> the world in the explicit descriptive sense expected from cognitive theories and has proven difficult to interpret from that perspective. ... However, from the perspective of affordance competition, mixtures of sensory information with motor plans and cognitive biases make perfect sense. Their functional role is not to describe the world, but to mediate adaptive interaction with the world. (Cisek 2007, p. 1594)

Put differently, it is an explicit and important feature of the project being pursued in this book that I rarely if ever question the veracity and usefulness of any body of neuroscientific data. There is, for instance, overwhelming evidence that basal ganglia play an important role in action selection. There is nothing in the present treatment that could or should be construed to deny these observations. At issue, instead, is how to understand the nature of that role in light of the broad synthesis of results that I am engaged in here. Thus, for instance, rather than conceive of basal ganglia as a gatekeeper in a central, specialized action-selection circuit or as a central conflict resolution device (e.g., Prescott, Redgrave, & Gurney 1999; Redgrave et al. 1999), according to the model I am advocating it is arguably better to think of basal ganglia as one important source of biasing inputs that can influence ongoing pattern competition between different response opportunities. Michael Arbib (2010; Bonaiuto & Arbib 2010) offers a model of action selection as affordance processing in which basal ganglia play just such a biasing role, updating estimates of the "desirability" of various possibilities for action. This seems to me a better interpretation of the observations because not just basal ganglia but in fact *most* of the brain is continually involved in action selection and guidance, for this is what it evolved to do. The coordination of behavior emerges from natural and structural competition modulated by multiple sources of internal and external bias—in which process basal ganglia do indeed play an important role.

6.3 Toward an Interactive Account of Higher Cognition

What I have argued so far is that it may be more scientifically fruitful—and more ecologically valid—to approach the brain as a *control system* and that from this perspective neural mechanisms capable of implementing input-output mapping functions and perception-action feedback loops are likely sufficient for both choosing between and executing basic behaviors. Indeed, I have argued that because the brain implements dynamic causal relationships between inputs and outputs—the latter change as a strict function of the former and the current and past state of the network, and the former change as a result of the latter because the agent's actions alter its perceptions—the brain is always simultaneously "selecting" and "implementing" actions in real time as the patterns of activity across the network change. The evolution of patterns across the brain network, then, *is* the ongoing process of deciding what to do next (and doing it). As was noted before, this perspective highlights less any individual decision points and more the ongoing behavioral trajectory of an agent, along with the forces that bias that trajectory.

However convincing this picture may be for simple sensorimotor behaviors such as reaching, it is less likely to seem sufficient as an account of the mechanisms responsible for higher-order cognition. This is in part because we have a deeply ingrained tendency to think of tasks such as language use and mathematics in knowledge-oriented, and thus *symbolic, representational,* terms leading us to imagine they must be supported by the usual suite of computational mechanisms that have proven themselves not just capable of but in fact specialized for such uses as mathematical calculation, planning, and logical reasoning (Russell & Norvig 2009). Extending the control-oriented approach to these domains, then, will require two things: a reminder that such behaviors as speaking, reading, and calculating are in fact also *processes*; and an in-principle account of how perception-action feedback loops—iterated interactions with an environment—could underlie such processes. For this account we turn to what might be considered an unlikely source: Alan Turing, the twentieth-century British mathematician who is largely credited with spurring the cognitive revolution in psychology. As Fodor writes:

> The cognitive science that started fifty years or so ago more or less explicitly had as its defining project to examine a theory, largely owed to Turing, that cognitive mental processes are operations defined on syntactically structured mental representations that are much like sentences. . . . Turing's theory was thus a variant of the Representational Theories of Mind that had been familiar for centuries in the British Empiricist tradition and elsewhere. What RTMs have in common is the idea that mind-world relations (or mind-proposition relations if that's your preference) are mediated by mental particulars that exhibit both semantic and causal properties. ("Ideas" in Hume's terminology; "concepts" and "mental representations" in the vocabulary of thoroughly modern cognitive psychologists.) From this point of view, Turing's suggestion was that the mental particulars in question are syntactically organized was crucial; it opened the possibility of treating their causal interactions as computational rather than associative. (Fodor 2000, pp. 4–5, 105n5)

It is certainly true that Fodor accurately characterizes one crucial thread of the cognitive revolution. But what I argue here, following Wells (2006), is that the cognitive revolution got Turing wrong in one crucial respect: he did not propose that thinking is based on the manipulation of, nor do his famous computing machines actually manipulate, mental representations.

Fixing this error, and seeing clearly what Turing in fact suggested, will get us an important step closer to understanding our brains, what they do, and how *we* do such things as speak and calculate.

The main issue that concerned Turing in his famous 1936 paper was the notion of computability—the question of which numbers (specifically; but which answers more generally) it is possible to compute. Turing's analysis of this problem (arrived at independently by Alonzo Church 1936) considered computable numbers to be those for which it is possible to specify a series of steps or actions that, when completed, would produce the number in question. Computable answers, by analogy, are those for which there is an *effective procedure* for reaching them. A Turing machine is a simple device that can carry out effective procedures; and the Universal Turing Machine is, as its name implies, a machine that can carry out *any* such effective procedure and thus can compute any computable answer. Before I describe the nature of Turing machines and how they work, it is important to note that I am taking for granted that Turing machines *can* in fact compute anything computable and thus that any mechanism with the power of a Turing machine is a candidate for the mechanism whereby *we* do such things as compute answers to cognitive questions, whether arithmetic, linguistic, or otherwise. Because CCTM also accepts this conclusion, I think it is fair game for me to assume it as part of my argument *against* CCTM.

A Turing machine consists of two parts: a finite-state machine and a tape divided into squares from which the finite-state machine reads and on which it writes, one square at a time. A finite-state machine, as the name suggests, is simply a machine can be in one of some finite set of possible states. Figure 6.2 represents a turnstile as a finite state machine. The

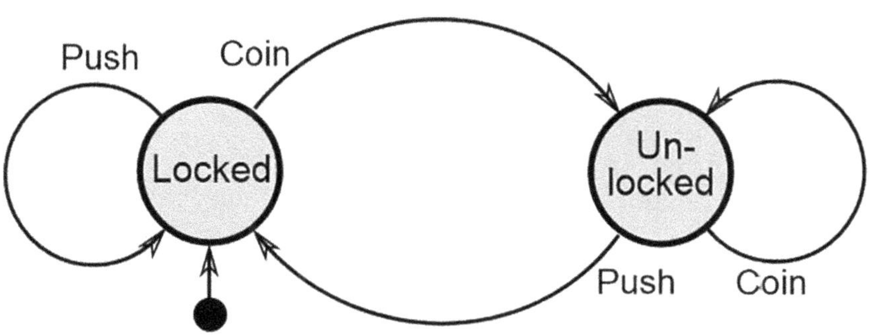

Figure 6.2
A turnstile represented as a finite-state machine. Picture from the Wikimedia Commons, made available under the Creative Commons CC0 1.0 Universal Public Domain Dedication.

turnstile can be in one of two states, locked and unlocked, and can take two kinds of inputs from its environment: coins and pushes. Depending on its state and the input, the turnstile moves between its states, as the arrows in the diagram indicate.

In his excellent discussion of Turing's analysis and its relevance to psychology, Wells (2006) calls the finite-state machine portion of a Turing machine a "mini-mind" and treats the tape as the special kind of environment in which mini-minds operate. I adopt those conventions here. A mini-mind has only some very basic capacities: it can perceive its immediate environment; it can change that environment by erasing or inscribing a symbol on it; and it can move to a different location in that environment. What a mini-mind does at any given time depends both on what it perceives and on the state it is in when it perceives. Naturally, the state it is in can change over time. These changes—the state transitions of the mini-mind—also depend on what it perceives and the state it is in when it perceives. A mini-mind is completely described by its states and transitions.

Wells (2006) illustrates the concept with a mini-mind of turnstile-level simplicity (figure 6.3). It has two states: happiness and sadness, and transitions between them are triggered by good and bad news.

As noted above, the mini-mind is only one component of a Turing machine; the other crucial component is its environment. In Wells's analysis, the most psychologically fruitful way to understand Turing machines is as mini-minds moving and acting in a highly specialized environment, consisting of a series of places (the squares of the tape) of indefinite extent.

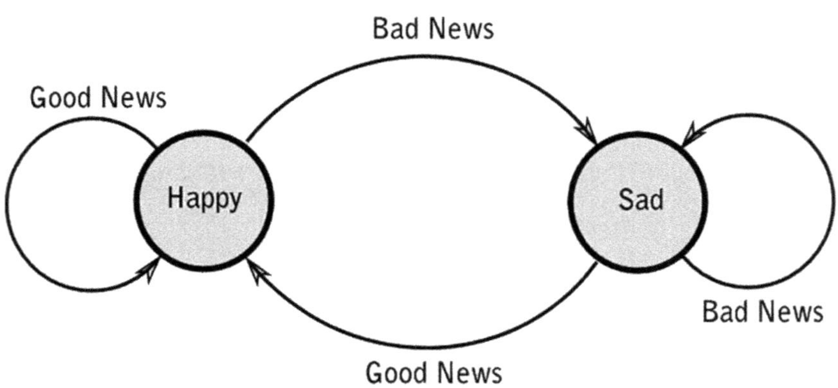

Figure 6.3
A simple emotional mini-mind with two states, happy and sad, and transitions defined for good and bad news. After Wells (2006), p. 29.

Each place in the environment may contain at most one of the stock of available symbols, which the mini-mind may perceive. Depending on the state of the mini-mind and what it perceives, it will remove a symbol from the environment, place a symbol in the environment, or neither; move to one of the adjacent places, or not; and make a state transition, or not.

Although calling the tape of a Turing machine an "environment" may seem forced to those used to the way Turing machines are generally discussed in the cognitive science literature, Wells (2006) demonstrates quite convincingly that this is in fact consistent with what Turing had in mind. In analyzing the problem of computability, Turing was intentionally building an abstract model of how a human—a *computer*, as a person who did calculations for a living was called in Turing's day—works with pencil and paper to provide mathematical answers:

> Computing is normally done by writing certain symbols on paper. We may suppose this paper is divided into squares like a child's arithmetic book. In elementary arithmetic the two-dimensional character of the paper is sometimes used. But such a use is always avoidable, and I think that it will be agreed that the two-dimensional character of the paper is no essential of computation. I assume then that the computation is carried out on one-dimensional paper, i.e., on a tape divided into squares. (Turing 1936, p. 249; quoted in Wells 2006, p. 78)

Interestingly, the justification for abstracting away from the two-dimensional character of the relevant part of a (human) computer's environment—"[it] is no essential of computation"—indirectly implies that paper of *some* form in the environment *is* essential. Turing also makes clear that the characteristics of the machines he describes are determined in part by the physical and perceptual limitations of human computers working with pencil and paper. For instance, he writes: "I shall also suppose that the number of symbols which may be printed is finite. If we were to allow an infinity of symbols, then there would be symbols differing to an arbitrarily small extent" (Turing 1936, p. 249; quoted in Wells 2006, p. 79). What Turing seems to have in mind here, argues Wells (2006), is the fact that as one increases the number of symbols available for use in computation one will eventually reach a point where the differences between them can no longer be perceived when they are printed in one of the squares of the tape. Given our perceptual limitations, we are only able to distinguish between some large but nevertheless finite number of symbols. Naturally, one might make the squares larger, which would allow the possibility for a larger number of distinguishable symbols, but this, too, is a self-limiting option, for eventually the squares become too large to be useful or to be perceived in a single glance. In a similar vein Turing notes:

> The behaviour of the computer at any moment is determined by the symbols which he is observing, and his "state of mind" at that moment. We may suppose that there is a bound *B* to the number of symbols of squares which the computer can observe at one moment. If he wishes to observe more, he must use successive observations. (Turing 1936, p. 250; quoted in Wells 2006, p. 81)

The bound *B*, like the finite collection of available symbols, is a reflection of Turing's sensitivity to the constraints and conditions under which human computers operate: they perceive, write, and erase symbols in locations in their environment; which action they take depends on what they perceive and their current state; and they perceive and act on different locations in their environment serially and successively until their mathematical task is completed. Although the exact constraints Turing places on his machines and their environments—such as perceiving and acting on *one* square containing at most *one* symbol at a time—are not biologically realistic, that there are *some* such constraints on human computers is certainly true. Thus, if a machine working under the stricter constraints Turing imposes can compute a given number or function, then machines with more powerful faculties for perception and action can clearly do so as well.

The reason Turing machines came to be so important to cognitive science—and the main finding of relevance to us here—is the proof that for any computable number or sequence, there is a Turing machine that can produce it. Moreover, there is a kind of machine, a Universal Turing Machine, that can simulate the function of any other Turing machine and thus can itself compute any computable result. I do not discuss the substance of these proofs here but take the conclusion as a given. An appropriately structured Turing machine, by simply responding to and changing its environment in accordance with its own changing states—by, that is, following out the appropriate effective procedure—can compute any computable result. Turing machines are at once mechanistic, relatively simple, and extremely powerful, and they have thus come to be seen as something of an existence proof for how the human mind might be realized in the physical universe (J. R. Anderson 2007). Indeed, it seems fair to say that CTM is founded on the hypothesis that any system functionally equivalent to a Universal Turing Machine has sufficient capacity to implement general intelligence. Newell and Simon (1976) offer perhaps the most well-known such assessment:

> *The Physical Symbol System Hypothesis.* A physical symbol system has the necessary and sufficient means for general intelligent action. ... A physical symbol system is an instance of a universal machine. Thus the symbol system hypothesis implies that

intelligence will be realized by a universal computer. (Newell & Simon 1976, pp. 41, 45; quoted in Wells 2006, pp. 19–20)

I think there is reason to doubt that we are universal machines of quite the sort that Turing imagines—for instance, the Universal Turing Machine has a fixed architecture, whereas ours evidently changes as the result of learning. But I am happy to accept two general points: (1) a system functionally equivalent to a Universal Turing Machine has the capacity to implement crucial aspects of high-level general intelligence; and (2) we are functionally equivalent to a Universal Turing Machine in the sense that we can *learn* to compute any computable number or sequence. Having accepted these two points, however, I now want to argue, in light of the analysis of Turing machines offered above, that the functionally equivalent system is *not* the human brain, but the brain + body + environment system, which operates via iterated perception-action feedback loops. As Wells puts the point:

The activity of the mini-mind consists of a sequence of perception-action cycles. Each cycle consists of an observation of the scanned square and action, a movement and a state transition. … The moral is that explanations of behaviour patterns must make reference to both internal and external structure. This means that neither the environment alone nor the organism alone can provide a complete basis for psychology. … One might expect, therefore, that a psychological theory based on Turing's analysis of computation would be ecological in character and would seek to understand how functional states are instantiated in the brain, which environmental inputs they are attuned to, what the structure of those inputs is and how behaviour unfolds in a continuous sequence of perception-action cycles. (Wells 2006, pp. 57, 65, 161)

As I hope is abundantly clear by this point in the volume, I am attempting to offer just such a theory. Much of our behavior can be adequately explained without reference to symbols. But explaining some of our behavior—certain kinds of calculation, linguistic interaction, and the like—very likely requires reference to symbol structures of various sorts. What I am suggesting, following Wells and Turing, is that *when* we implement a symbol system, we generally do so by coupling brain structures implementing the appropriate input-output mappings to *external*, perceivable, manipulable symbols (see Clark 1997, 1998b for a nice discussion of the importance to cognition of such scaffolding via interaction with external structure). The point bears emphasis: the mini-mind in a universal machine does not need to be, in and of itself, a symbol system. The symbol system being described here is a combination of the mini-mind and its environment. If we accept the CCTM claim that a system functionally equivalent to a

Universal Turing Machine has the capacity to implement crucial aspects of general intelligence, and if an appropriately configured mini-mind + a structured environment is functionally equivalent to a Universal Turing Machine, then an appropriately configured mini-mind + a structured environment has the capacity to implement crucial aspects of general intelligence. No reconstructive perception, mental symbols, or any of the other standard accoutrement of CCTM is required.

It will come as no surprise (but see Cleermans 1993 for the proof) that recurrent neural networks have the capacity to implement arbitrary mini-minds—to emulate finite-state machines with the required input-output mappings. Thus Wells's (and Turing's) analysis definitively shows that abstract computation of any computable result can be achieved by coupling the appropriate connectionist mini-mind with a suitable environment including a repertoire of readable and writable symbols. Just as with the other capacities and behaviors described here, our ability to master abstract cognitive tasks such as mathematics ultimately depends on our ability to exploit the possibilities for perceiving, acting in, and changing our environments. Our apparent facility with the abstract symbolic processing exemplified by speaking, reading, and calculating can be explained by the application of our native interactive behavioral control processes to externally available symbol systems. The *brain* in this framework needs only the ability to implement finite-state machines—mini-minds—that respond to environmental circumstances and produce output changing the environment and/or the organism's relation to it. The fact that the relevant part of the environment for some behaviors is a shared system of symbol structures does not change this fundamental fact about brains and what they do. Contra CCTM, our brains are not Turing machines. Rather, our brains are controllers capable of marshaling various internal (memory for abstract contingencies), physical (pencil and paper and the skills to use them), and sociocultural (formal and natural language) resources that together can generate Turing machines (or their functional equivalents) on the fly.

It is worth noting that the founders of the parallel distributed processing approach reached essentially the same conclusion, albeit from within their version of the information-processing framework:

> If the human information-processing system carries out its operations by "settling" into a solution rather than applying logical operations, why are humans so intelligent? How can we do science, mathematics, logic, etc.? How can we do logic if our basic operations are not logical at all? We suspect the answer comes from our ability to create artifacts—that is, our ability to create physical representations that we can

manipulate in simple ways to get answers to very difficult and abstract problems. (Rumelhart et al. 1986b, p. 44)

From the standpoint of this general theoretical perspective, the research program going forward is not to discover what is represented where in the brain, which is a natural question to ask within the CCTM framework. Rather, we must ask: What are the nature and range of environmental and bodily structures that, when coupled with appropriately tuned neural networks, can implement our celebrated capacities for planning, abstract thought, and reasoning (Anderson 2003; Barrett 2011; Chemero 2009)? What precise role do perception and action play in human computation, and what are the functional side effects of deploying perception-action loops to accomplish abstract semantic (rather than physical) tasks? When external symbols are not available for perception and manipulation, what internal resources (e.g., memory, imagination) can be deployed in their stead? Under what circumstances is this substitution possible, and what limits does it impose on human computation? What kinds of networks are there, and what are their functional structures? Which networks are more suited for interacting with the natural environment, which for interacting with the social environment, and which for interacting with the artificial symbolic environment offered by language and mathematics? What are the differences among these networks, if any? How do they interact? Are (entire) networks developed for one purpose often reused for other purposes? If so, under what circumstances, and with what functional and structural effects? Or are the networks themselves relatively selective and specialized but composed of parts that are marshaled in the service of multiple networks and operate across these different targets of interaction (natural, social, symbolic)? These and many similar questions await the concerted efforts of the field.

Before I close this chapter with a review of some suggestive work on the role of perception and action in mathematical calculation, one final point about embodiment and cognition is worth mentioning. I have so far been emphasizing the various ways in which perception-action feedback loops—or, more generally, iterated interactions with a structured environment—lie at the heart of cognition. This is because that perspective allows me to tell a single consistent story that unites brain, body, and environment in the service of both rudimentary and higher cognitive processing. But such a story, although consistent, is not by any stretch a complete account of the importance of the body and environment to cognition. Among other things, the body plays a role in structuring experience (Lakoff & Johnson 1999; Lakoff & Nuñez 2000), in reducing the dimensionality of perception

(Hochner 2012), and in limiting the degrees of freedom possible in action (Thelen 1995; Thelen & Smith 1994), all of which simplify the job that the brain needs to do. As this work has been extensively reviewed elsewhere (Anderson 2003; Barrett 2011; Chemero 2009; Clark 1997; Shapiro 2011), I do not provide further details here. But I do want to underscore a point that this work illustrates: it has always been tempting to presume that the brain had to do all of the cognitive work for an organism and that where external structure could be shown to matter, it could be cognitively efficacious only when that structure was itself represented in and by the brain. This tendency to ascribe to native capacities of the brain what are in fact the contributions of bodily and environmental context is the neural version of the fundamental attribution error. This error appears to be part of the reason that, in the history of the cognitive sciences, Turing's external symbols migrated into the head—and may explain why even as committed a theorist of embodiment as George Lakoff sometimes evinces the tendency to unnecessarily re-present and reproduce bodily structures as neural ones. But this is a mistake that distorts both our understanding of the problems that brains must solve and the resources it has to solve them. To put the matter in the language being developed for this volume (this picks up on a brief discussion from chapter 4), the psychological does not depend solely on the neural, and thus not every psychological difference needs to manifest itself in a neural one. Bodily and environmental differences can make no detectible difference to brain activity—the situation can express or fall under the identical set of NRP factors—and yet the neural process can result in different cognitive or psychological outcomes. Cognitive work can be done by bodily and environmental structures where they are. Once we understand that cognitive processes are temporally extended, iterative, and interactive in nature, we see that the brain possesses the native resources to take advantage of such structures in causally efficacious ways without reproducing them as elements of the neural system. Brains can focus on what they do best: managing interactions with the world. Cognitive processing emerges from—is indeed identical to—these iterated interactions.

6.4 Mathematics as Symbol Pushing

I would like to close this chapter with a summary of a growing body of work that has begun to address one of the research questions offered above and thus offers both a nice example of embodied cognitive science and some preliminary evidence that mathematics really is, in part, a sensorimotor skill characterized by the iterative interaction with external symbols. For if our capacity to compute generally depends on our capacities to perceive and interact with external symbols—is a matter of coupling perception-action loops to an appropriately structured symbolic environment—then we should expect to see traces of this underlying mechanism in the observable processes of problem solving. Here I discuss three sets of such results. Note the evidence is *not* for the proposition that we implement Turing-machine-like operations; our perceptual and motor capacities are rather different from those of a mini-mind, and our environment is more complex. Rather, the evidence showcases the particular role for perceptual and motor capacities and practices in performing calculations and other forms of mathematical reasoning and begins to specify the sensorimotor operations that we might actually use in solving mathematical problems.

I start with what will be an uncontroversial observation to anyone who has witnessed students engaged in solving math problems, whether on their own or at the chalkboard for a classroom demonstration: people interact with mathematical symbols in myriad ways. They point at them, gesture over them, move them, and strike them out, among many other things. These external actions serve myriad purposes: they direct spatial attention (both one's own and that of witnesses); they aid memory for one's place in the problem-solving procedure; and, as we see below, they communicate and reveal something about the nature of the underlying processes enabling people to solve mathematical problems (see Radford 2010 for an interesting review). It is important to emphasize that such interactions are not peripheral to the process but are part and parcel of it; Goldin-Meadow (2003), for instance, discusses the role of gestures in both indexing and *aiding* mathematics comprehension.

One possible interpretation of this observation is that people often treat symbolic expressions not as abstract indicators of some mathematical statement to be reproduced in an inner language of thought but rather as icons (Radford 2010). In this case manipulations of the spatial properties of the image would be at the same time manipulations of its mathematical content. Or, as Landy and Goldstone (2009) put the matter:

> Recently it has been suggested that formal languages, and mathematical languages in particular, often serve as diagrams whose analog physical structure relates systematically to mathematical or formal truths. Therefore, treating the formal notations as images directly suitable for perceptual motor processing provides a way to implement abstract cognition in perceptual-motor systems (Dörfler 2002; Landy & Goldstone 2007a). (Landy & Goldstone 2009, p. 1072)

In one simple demonstration of the importance of spatial format in mathematical expressions, Landy and Goldstone (2007a) asked participants to write out simple arithmetic equations by hand. They found that the spacing that participants used in the equations indexed mathematical content such as order of precedence. For instance, by order-of-precedence conventions, multiplications are performed before additions in equations that have both. Landy and Goldstone found that in such equations participants grouped numbers to be multiplied closer together than numbers to be added. From the perspective being advocated here it is natural to interpret this as a spatial strategy that makes it easier to see what to do with the equation. Equations, like other physical objects, have affordances, and this form of spacing makes those affordances easier to see. Doing math is affordance processing, too!

If that perspective is correct, it should be possible to manipulate these affordances in ways that will cause people to do the (mathematically) *wrong* things with equations. Landy and Goldstone (2007b) did just that. They asked participants to make validity judgments of equations containing both multiplication and addition. In some equations—called the sensitive set—incorrectly doing addition before multiplication will lead one to make the incorrect validity judgment. In other equations—the insensitive set—one can come to the correct validity judgment even when applying an incorrect order-of-precedence rule. For both sets of equations Landy and Goldstone systematically manipulated the spacing around the operators such that in some equations spacing was consistent with order of precedence—that is, the grouping around multiplication sign was tighter than that around addition signs—in others it was inconsistent, and in still others spacing was even and thus neutral. What Landy and Goldstone found was exactly what one would predict if spacing is used as an action-guiding cue: first, spacing only affected performance on equations from the sensitive set; and second, participants gave correct answers for only 55% of the sensitive-inconsistent equations but over 80% of the sensitive-neutral equations and well over 90% of the sensitive-consistent equations. Consistent spacing appeared to facilitate performance, whereas inconsistent spacing induced significant errors. Interestingly, they got similar results with other forms of perceptual grouping. For instance, in one experiment they manipulated similarity of form, such that an expression like $(d^*d^*d) + (4/q) * (9/m)$ would represent a consistent perceptual grouping whereas $(d^*d^*d) + (q^*q^*q) * (9/m)$ would represent an inconsistent grouping because the operation with mathematical precedence would be performed on visually similar elements of the expression in the first case but not in the second. The authors conclude:

> The current experiments show an influence of perceptual grouping on mathematical problem solving. The results are noteworthy for several reasons. First, they demonstrate a genuine cognitive illusion in the domain of mathematics. ... Our participants were systematically influenced by the grouping variables of physical spacing, alphabetic familiarity of variable names, and notational form. The normative formalism of mathematics does not include these factors. ... The second impressive aspect of the results is that participants continued to show large influences of grouping on equation verification even though they received trial-by-trial feedback. ... This suggests that sensitivity to grouping is automatic or at least resistant to strategic, feedback-dependent control processes. ... The third impressive aspect of the results is that an influence of grouping was found in mathematical reasoning. Mathematical reasoning is often taken as a paradigmatic case of purely symbolic reasoning. ... (Landy & Goldstone 2007b p. 730)

Other results point in the same direction. A natural extension of the view that equations have spatial affordances is that they invite us to act *on and with* them, manipulating and moving their parts around as part and parcel of the problem-solving process. In this view, to learn algebra is to acquire a sensorimotor skill, and acting in accord with the rules of algebra is a matter of learning to see and act in accordance with the transformations that the equations afford. "Learning the practice of mathematical proof is a matter, then, of learning physical constraints on the way parts of an equation can move and transform, akin to learning physical constraints on the motions of real objects" (Landy & Goldstone 2009, p. 1073).

Landy and Goldstone (2009) investigated the possibility that people may employ such a motion strategy to solve algebraic equations in a clever set of experiments in which equations were displayed against an apparently moving background. The motion was either neutral, congruent, or incongruent with the direction symbols would have to be "pushed" from one side of the equal sign to the other in order to solve the equation (see figure 6.4 for an illustration).

For one set of equations—the sensitive set—an incorrect ending location for the moved symbol would lead to the wrong answer because, for instance, $(8-2)$ is not the same as $(2-8)$, whereas for an insensitive set of equations there would be no such order effects $(8+2) = (2+8)$. As with the experiments reported above, their results were entirely consistent with the sensorimotor interaction view of mathematical reasoning. Performance was facilitated by congruent motion and impaired by incongruent motion— but only for the sensitive set of equations. Interestingly, Landy and Goldstone found that this effect was actually *stronger* for more advanced mathematics students, a finding that underlines the point I am pushing for here:

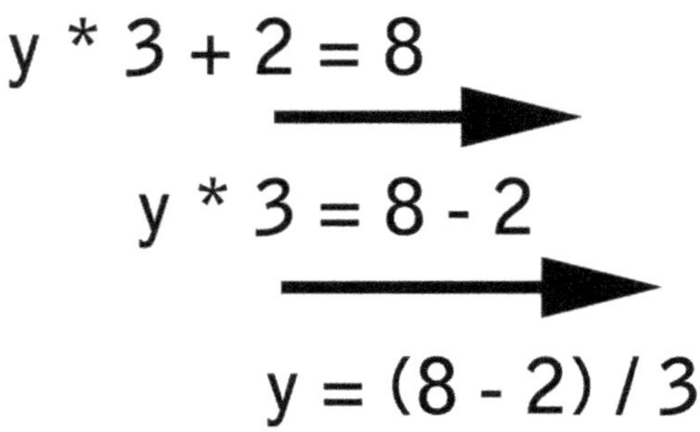

Figure 6.4
Illustration of the motion strategy. The 2 and 3 need to be moved to the far right to get the correct answer. The authors hypothesized that background motion might interfere with this movement, inducing more errors in equations where ordering mattered (had the left side of the initial equation been, for example, $y \div 3 - 2$, then placement would not matter). Based on figure 1 from Landy and Goldstone (2009).

what we are uncovering in experiments like this are not multiple instances of odd and anomalous interference by irrelevant neural systems on those specialized for mathematical reasoning but, rather, evidence of *what it is to learn to do mathematics*. Advanced mathematics students have learned to rely even more strongly on their sensorimotor systems in doing mathematics because these are the systems that we marshal to accomplish even such abstract cognitive tasks as solving equations.

Reflecting on such results Landy and Linkenauger (2010) suggest that we might also expect to see some of the same effects on distance estimation that we see in the case of physical actions, such that actions that are easier or more available will tend to make the action targets appear closer—recall our review of this literature in chapter 5 of this volume (Lessard et al. 2009; Proffitt & Linkenauger in press; Proffitt et al. 2003).

> One clear prediction is that symbols that afford action appear proximal relative to those that do not. … Therefore, if solving a mathematical equation requires actions on the part of the solver, high-precedence terms—those most available for actions—should generally seem most proximal in arithmetic expressions. Put simply, years of experience acting first on multiplications in expressions [such as] $3 + 4 \times 5 = 23$ will lead the multiplication to appear closer than the addition. (Landy & Linkenauger 2010, p. 2164)

Incredibly, this is just what they found. They showed participants displays like that illustrated in figure 6.5, asked them to solve the expression, and then asked them which baseball was nearer to them (the participants).

Figure 6.5
An example display screen shown to participants in Landy and Linkenauger (2010). Participants were asked to solve the expression and asked to judge which baseball was closer to them. Reprinted with permission.

The vast majority of the time, the baseball beneath the times sign appeared closer. The researchers got similar effects when the baseballs were moved beneath variables to be multiplied or added, as in expressions such as $c = a * x + b$. In such a case, a baseball beneath the a would appear closer than one beneath the b.

Clearly, many questions remain, and more research is called for. But overall, these results strongly suggest that we interact with symbolic equations using the same strategies and even perhaps some of the same perception-action systems that are used when we interact with physical objects. In general agreement with Turing's analysis of mathematical reasoning, "episodes of formal reasoning are indeed typically organized by attention-based interactions with external environments" (Landy & Linkenauger 2010, p. 2168). Abstract mathematics is generally possible for us not in spite of the fact but *because* we have affordance-exploiting perception-action processors that can be used to interact with objects of various types—even abstract objects such as algebraic equations.

The implications of this discovery are profound, not just for our understanding of what it is that brains fundamentally do (a subject that I have been continually returning to here) but also for any account of what formal systems do, why they are useful, and how they develop. As Landy and

Goldstone observe: "proofs are often physically designed to appeal to processing systems typically used for dynamic events" (Landy & Goldstone 2009, p. 1072), a suggestion that echoes the arguments of Christiansen and Chater (2008; Chater & Christiansen 2010) to the effect that natural language developed to fit the capacities of the human brain, even as it extended those capacities (Clark 1998b). It is to this subject, then, that we turn in the next chapter.

Interlude 6 Is Our Brain as Good as It Gets?

Evolution is typically presented as the ultimate Horatio Alger story: up from the primordial muck we came, with each generation and successive species fitter—better, faster, stronger, smarter—than the one before. Tales of brain evolution offer few exceptions to this script: mammals are smarter than reptiles, primates are smarter than other mammals, and humans smarter than other primates (politics notwithstanding). Popular science (and science fiction) abounds with speculation about what heights we might scale next.

But a recent article from *Scientific American* (Fox 2011) showers a bit of acid rain on this ticker-tape parade. If you take a look at the facts—the coldest, hardest facts of physics—there doesn't seem to be much room left for improvement. This may be as good as it gets.

Consider: the human brain already jams 100 billion neurons into our skulls, giving us the largest brain-to-body ratio of any species on the planet. To achieve this inspired feat of packing that should be the envy of vacation goers everywhere, evolution had to make a series of trade-offs that may have boxed our brains, and us, into a corner—what theorists call a *local maximum*. Being at local maximum is like scaling a lesser peak. There are taller mountains, somewhere, but for you there's no place to go but downhill. In our case the local maximum is defined by the following constraints.

We could try to improve our brains by making our neurons faster—able to carry and process signals with greater speed and efficiency. The only known way to do this is to increase the diameter of our axons, the impulse-transmitting parts of neurons. But thicker axons need more energy and more space, and to fit the neurons (and any nutrient-carrying supports) into our skull, we would require either that we use *fewer* neurons (defeating the purpose) or grow bigger heads. Putting aside the question of whether we could give birth to babies with even bigger heads (ask your mother), a bigger brain forces neurons further apart, and so it requires longer connections, which in turn require more energy and—you guessed it—makes signaling slower, once again defeating the purpose.

How about utilizing more connections between the neurons we already have? Well, more wires likewise take up more space (see above), and longer wires take longer to transmit signals (ditto).

Finally, what if we try to implement one of these options by making neurons smaller, so we can fit more, or more connections, or both, into the skulls we already have? Thinner neurons slow down signal transmission (as we noted). But more importantly, they also tend to fire more randomly—and noise is signal processing's worst enemy.

You see the evolutionary bind we are in? Biologically, we aren't really in a position to be adding new cells or changing their fundamental properties as a way of improving function. Instead, we evolve and develop by using the resources we have in multiple ways, mixing and matching the functional pieces in novel combinations to achieve novel capacities.

But perhaps even more crucially, these biological limitations highlight the importance of culture to achieving human potential—as we see in detail in chapter 7. The most important moral of this article is this: it is up to *us* to design the social systems such as schools and work teams, and technologies such as encyclopedias and smartphones that will *enhance* human intellectual capacities rather than deaden them (Lanier 2011). Are we up to this challenge? I certainly hope so. I'd hate to think we will collectively stay exactly this dumb.

Of course, fully embracing this challenge will require a much better understanding of how our brains actually do their job—and although the article by Fox (2011) is a fine piece of reflective science journalism, it did miss some opportunities for education in this area. In particular, in the discussion of the structural compromises that allow us to fit so many neurons into so little space, the article suggests that our brain is organized into localized, functionally specialized regions dedicated to such things as speech comprehension or face recognition. But as we have been seeing throughout this volume, it is not so. Indeed, given Fox's own set of points, it almost *can't* be so. Without the ability to use and reuse neurons and larger pieces of the brain in a variety of circumstances, we would have exhausted the brain's functional capacity long ago.

It is certainly true that our brain is organized into network modules consisting of regions of dense local interconnections, with sparser connections to other more distant regions. This organization—known as a "small world" architecture—allows for fast communication across the whole brain without the need for wires connecting everything to everywhere, which would take up too much room. But these network modules are not functionally specialized; they are put to many different uses under different circumstances. And they are not static: functional connectivity is modulated at multiple physical and multiple temporal scales to support learning and adaptive behavior.

Still, Fox makes a very deep fact about the brain quite clear: intelligence is less about local processing than about *cooperative connectivity*. Brain function is about establishing and coordinating neural partnerships across the brain; and adaptive behavior is a matter of establishing functional partnerships both inside and outside, within and across the body-environment boundary.

Part III Beings

7 Languaging with an Interactive Brain

Questioning and answering, requesting and commanding, declaring, promising, and apologizing are all things we do largely with and through language, through speaking and listening. Indeed, one might say in general that language is something we *do* (Austin 1975).

> [T]o speak and to listen and respond to talk, that is to indulge as talk-in-interaction, and to write and read, and to use language in modern hybrid media, imply involvement in actions, in acting in and through language. Such a perspective will highlight dynamic processes; as several authors have claimed quite emphatically, discourse is a *process*. Thus, for example, Potter et al. (1990) contend that "discourse is a verb." (Linell 2005, p. 3)

In a similar spirit, Becker (1988, 1991) employs the term "languaging" to underscore the fact that language is better thought of as an *activity* than as a *thing* (for good reviews of this perspective see, e.g., Gumperz 1982; Linell 2009; Tannen 2007). As emerges in the course of this chapter, I believe that this is the natural perspective to take when thinking about language and the brain. Natural language is beyond any doubt a deeply important realm of human action and interaction, and insofar as our brains are specialized for managing action in and interaction with the world—including especially interactions with one another—then it seems we would need very strong justification to approach the study of one of our most important *sorts* of interaction from any different perspective. Gumperz argues:

> Communication is a social activity requiring the coordinated efforts of two or more individuals. Mere talk to produce sentences, no matter how well formed or elegant the outcome, does not by itself constitute communication. Only when a move has elicited a response can we say communication is taking place. To participate in such verbal exchanges, that is, to create and sustain conversational involvement, we require knowledge and abilities which [sic] go considerably beyond the grammatical competence we need to decode short, isolated messages. (Gumperz 1982, p. 1)

For Gumperz in particular, and sociolinguistics in general, the capacities that establish and maintain conversational involvement—the extended, goal-oriented linguistic interactions between individuals—are what constitute the key bases for language: "understanding presupposes conversational involvement" (Gumperz, 1982, p. 2). That is, our context-creating interactions provide the grounds for understanding, and not the reverse. And yet the study of language and the brain, as well as the history of linguistics more generally, has been largely dominated not by this pragmatic, functionalist perspective but rather by a more formal, structural approach—an approach that is at this point standing in the way of a better understanding of the brain: "[A]s far as our ways of conceptualizing language [are] concerned, this has not been the dominant tradition. Instead, we have become used to saying that language is an *inventory of forms*, and rules for generating forms" (Linell 2005, p. 3, emphasis in original).

The focus on language as a structured system of abstract forms has an ancient history and was bolstered in more recent times by structuralism in psychology (Titchener 1898), by structural linguistics (De Saussure 2013), and most especially by the sort of "Cartesian" linguistics that brought us such notions as the universal grammar (Chomsky 1965, 1966; note this blanket claim of course covers up the important differences between Saussure's structural linguistics and Chomsky's generative grammar). The cognitive neurosciences inherited this focus on linguistic rules and forms, along with a tendency to organize itself around the investigation of psychological faculties, from cognitive psychology and theoretical linguistics (Gazzaniga, Ivry, & Magnum 2008; Miller 2003; see Anderson et al. 2012 for discussion). Together, faculty psychology and linguistic structuralism lead to the position widely embraced in the cognitive neurosciences that language mastery relies on the possession of unique, specialized, and complex neural and cognitive resources, including internal symbol systems in general and such items as a "biological specialization for grammar" in particular (Pinker & Bloom 1990, p. 707). As Pinker and Bloom write:

> All modern students of language agree that at least some aspects of language are due to species-specific, task-specific biological abilities, though of course there are radical disagreements about specifics. A prominent position outlined by Chomsky (1965, 1980, 1981, 1986, 1988), Fodor (1983), Lenneberg (1964, 1967), and Liberman (Liberman et al. 1967; Liberman & Mattingly 1989) is that the mind is composed of autonomous computational modules—mental faculties or "organs"—and that the acquisition and representation of language is the product of several such specialized modules. (Pinker & Bloom 1990, pp. 707–708)

This view, along with a characteristic emphasis on the complexity of the linguistic forms to be mastered, has even led to controversy over the very possibility that language is an adaptation—with Pinker and Bloom (1990) on the adaptationist side and Chomsky (1988), Fodor (1990, 2000), Bickerton (1995), and Lewontin (1998), among others, on the antiadaptationist side. Interestingly, the controversy echoes the debate between Darwin and Wallace (1870) about the limits of natural selection in general and of its ability to explain language in particular, and although the current debate is being waged in a more thoroughly naturalistic, evolutionary framework (but see Fodor 2007), this fact might already suggest the presence of some conceptual trouble, for if our capacity for language as we generally construe it cannot have evolved, we had better be very sure that this construal is the right one.

As noted above, the debate is driven largely by the supposed unique complexity of human natural language—the fact, for instance, that syntax theoretically permits the production of an infinite number of well-formed sentences as well as sentences with multiple modifiers and nested recursions of indefinite depth: "The groceries that the boy who went to the store that carried the correct brand, which was recently advertised in the paper that comes to the house, bought, spoiled." Such claims, along with the apparently intimate relationship between human thought and human language, have provided some of the strongest supporting pillars for the componential computational theory of mind (CCTM) and the inner symbol systems view of human cognition. Chomsky (2005), for instance, appears to endorse the view that the language faculty may have emerged initially as the biological underpinnings for human symbolic thought, such that the role of language in communication was a secondary development made possible by the first (see Fodor 1975; Jacob 1982; Tattersall 1998 for similar views). From within this cognitivist framework, that is to say, there is a tripartite equation among human thinking, human language, and an abstract formal system of rules and symbols: insofar as thought is language, and language is a system of abstract forms, then thought is the manipulation of abstract forms (alternately, one could just as well say: insofar as thinking is the manipulation of abstract symbols in a formal system, and thought is language, then language is the manipulation of abstract symbols in a formal system).

Finally, it has been claimed that there is insufficient information in experience for people to *acquire* grammar without a great deal of innately specified structure that biases the learning. This is the so-called "poverty of the stimulus" argument, which leads to the postulation of a "biological specialization for grammar" mentioned above (Chomsky 1965, 1988; Crain & Nakayama 1987). This nativist argument is somewhat orthogonal to the claims of most interest to us here, but it does reflect the same heavy bias toward explaining behavior with reference to complex, specialized, and domain-dedicated inner structures that this whole volume has been urging us to resist.

From this perspective it seems clear that language cannot have emerged from the development and application of preexisting capacities, if only because such a system appears to require a unique and powerful blend of computational resources (Chomsky 1956). In this way the idea of a language faculty supported by specialized, dedicated neural modules arising by genetic mutation comes to make the most architectural sense, and thus do the various strains of cognitivism mutually sustain a commitment to CCTM, perhaps nowhere more strongly than in the cognitive neurosciences of language. Indeed, this constellation of views still serves as a research-guiding frame for much work on the neural bases of language, with the result that researchers typically approach the subject by searching for the biological underpinnings of particular linguistic forms or rules, exemplified for instance in a focus on finding the neurophysiological substrates for the processing of words in particular formal or grammatical categories (e.g., Bedny & Thompson-Schill 2006; Devlin et al. 2004; Kaan & Swaab 2002). As Nusbaum (2011) notes:

> When this approach to the scientific study of language is imported into the neuroscience of language, putative linguistic structures become reified as neural structures. For example, in consideration of the neural mechanisms of language processing, recent debates focus finding the neural locus for syntax (e.g., Grodzinsky & Santi 2008) rather than understanding the complexity by which neural systems implement communication. (Nusbaum 2011, p. 669)

In fact, as Margoliash and Nusbaum (2009) point out, in most experimental work on the neurobiology of language, participants are asked to make metalinguistic judgments (e.g., of whether a particular string of symbols is a word or not; or is a noun or verb; or of whether a sentence is grammatical) rather than to engage in more natural forms of language use, such as storytelling or conversation. This not only reflects the grammar-first bias that infects the field but also means that current knowledge of the biological underpinnings of natural language is seriously impoverished.

The case I hope to make in this chapter is that, however valuable and interesting such work may be in its own right, when it comes to understanding the essence of language and the nature of its neural supports, the cognitive neurosciences have in fact grasped the wrong end of the talking stick. Language should be understood first and foremost as a realm and form of action and social coordination, as has long been argued by specialists in discourse and conversation analysis (Ochs, Schegloff, & Thompson 1996; Schegloff 2007; Tannen 1982, 1993). Seen in this light it seems quite plausible that, biologically speaking, language could have emerged largely from the redeployment of preexisting capacities and neural structures involved in managing relationships, action, and social interaction. As with *all* our cognitive abilities, we achieve the miracle of language by mixing and matching the very same set of resources and interactive capacities that we bring to bear in other circumstances. All this is to say that when we come to understand the true nature of language, and the importance of the interaction between evolution and culture, we will see that it does not in fact represent a significant challenge to adaptationist approaches to the mind (Christiansen & Chater 2008), nor does our linguistic proficiency require the postulation of specialized internal symbol systems (and thus should not be taken as a piece of evidence for CCTM). Far from being an obstacle to an evolutionary, action-grounded, interactive account of the brain, language may well turn out to be an exemplar of the approach.

This is not to say that language is not special; it certainly is. But its specialness comes from the leverage it offers the cognitive system in solving complex problems and not from the special challenges it itself presents. This is especially so in the case of *written* language. As we saw in chapter 6 in the case of mathematical symbols, once we are presented with external symbols to manipulate, our native cognitive capacities are remarkably enhanced (Clark 1997, 1998a, 1998b). Indeed, it would be fair to say that much of the evidence for the complexity of language and the specialized and powerful neural supports it therefore requires is actually derived from an investigation of linguistic behavior in the context of written languages (Linell 2005). We are indeed capable of some impressive cognitive gymnastics when our native capacities are coupled with external structures in general and with external symbol systems in particular. But to suppose that our brains must be capable of such gymnastics in the *absence* of such scaffolding is yet another instance of the fundamental attribution error that I have highlighted repeatedly in this volume. Our cognitive abilities are the result of a synergistic combination of internal neural mechanisms, bodily capacities and constraints, and environmental and social context; if we are ever to understand our brains, we must thoroughly absorb this lesson.

For the remainder of this chapter I lay out a picture of language and its evolution that emphasizes its origins as a *social* system for managing and structuring action and interaction. Ironically, this is a widely accepted view within linguistics itself; thus, my job in this chapter is largely to try to bring the news from sociolinguistics to those interested in uncovering the neural underpinnings of language, many of whom are still overly influenced by the form-first approach. I end with a brief outline of a possible research program for a more social, interactive neurolinguistics going forward.

7.1 Language Is Social

The view of language that I am urging us in the cognitive neurosciences to take on board has its roots in the pragmatics-focused ordinary language approach to discourse—that is, language in use, in context—pioneered by such figures as Ludwig Wittgenstein (2001) and J. L. Austin (1975). In contrast to an approach to language that begins with the question of what words mean (and do), the ordinary language school focused first and foremost on what *we* mean and what *we* do with words. If one could express the central insights of this school as maxims, there might be three, one methodological and two descriptive: (1) attend to what people actually say; (2) language is action; and (3) meaning is use.

That the study of language ever strayed from the first maxim might be surprising, but this is another instance where theoretical commitments arguably got in the way of empirical progress. The background against which the study of ordinary, everyday language use came to seem revolutionary was shaped by the (structuralist and Cartesian) notion that, because perception is reconstructive and language expresses thought, the primary function of language is to picture the world. Insofar as this is so, the reasoning went, it should be possible to specify for each sentence in the natural language the ideal, unambiguous factual expression—the logical proposition—that the sentence is meant to encode (Russell 1918; Wittgenstein 1995). When one takes this attitude toward language, the gap between what people say and the logical propositions that might be thereby expressed begins to loom large; natural language as embodied in folk usage is messy, imprecise, ambiguous, polysemous, and even in some cases apparent nonsense. Faced with this fact, one response might be to urge a reform of natural language, purifying and clarifying it by expunging from the language any expressions not demonstrably factual or logical, as the logical positivists did (e.g., Carnap 1937). Alternatively, one might question whether the theoretical frame (and the picture of cognition that motivates it) was in fact adequate to the object of study.

The ordinary language school took this latter approach. When a surgeon says "scalpel" or a police officer yells "stop!" do her listeners wonder at the logical proposition thereby expressed? Of course not: they hand her the proper instrument or cease moving. The surgeon and the officer were in each case *doing* something with the word—obtaining a scalpel, causing someone to stop—that they might have also chosen to do in other ways (pointing; picking the instrument off the tray; tackling the suspect).

There are a few lessons that might be drawn from this simple observation. One is that, despite the fact that the logical proposition such short utterances might express would likely be a matter of dispute, there was in these imagined cases no miscommunication, nor is there in the vast majority of natural language exchanges. Language, as spoken and used by ordinary people in everyday situations, is perfectly adequate to its task. The second lesson follows from reflection on the first, for insofar as this is so, it seems unlikely that the expression of logical propositions ought to be considered the central function of language. Instead, as these cases illustrate, language is a form of action, and those actions can take a variety of forms with a variety of effects. Finally, in reflecting on *why* these utterances are effective, we come quickly to the realization that it is not because scalpel-getting or movement-ceasing is what these words *mean* in the language per se, that is, in some unsponsored lexicon codified apart from its users—for after all, in a different context saying "scalpel" might have the effect of teaching an assistant the name of the instrument, or reminding a young physician of the right tool to use at a given step in the procedure—but is rather because of a shared pragmatic context. It is because the assistant who hears that word in an operating room will produce a scalpel that uttering that word is how to generate the desired effect in the context. "Scalpel," this is to say, is a token in the surgery language-game, in the naming language-game, in the teaching language-game, and in other language-games as well, and the effect it produces in each case will depend on the conventions of the game (Wittgenstein 2001). What the words mean—what we do in using them—is a matter of how (and when, and by whom) they are used.

Such reflections might serve to cement the general point that language is a tool for practical ends, a system that individuals learn to wield to achieve their various goals, but it is equally crucial to notice that to achieve these goals *with language* is for the most part to achieve them *with and through others* (setting aside for the moment the uses of speaking to oneself, such as repeating a phone number as an aid to memory). To get a scalpel with "scalpel" depends on the fact that the assistant will produce one. To put it in Grice's (1969, 1989) terms, it depends on the listener's ability to understand

and respond to the *speaker's intent* in producing an utterance, whatever particular form the utterance might take. It is for this reason that the query "is there any salt?" is likely (although, depending on the listener's degree of impishness, not guaranteed) to lead someone to hand you the salt—for in the context of dinner, that utterance will be taken to express that intent.

Language, then, is irreducibly social—and in at least two ways. First, unlike in the picture theory of language, which holds that an utterance instantiates an objective, dyadic relationship between a logical proposition and the world, in this view utterances are in fact intersubjective and triadic: they are attempts by the *speaker* to get a *listener* to do *X*; language is in this sense a form of *joint action* (H. H. Clark 1996; Sebanz, Bekkering, & Knoblich 2006). Second, and following directly from the first, language should be understood as a coordinating structure "created for the purpose of enabling or facilitating ... cognitive and social interactions" (Tomasello 1999, p. 95).

Given that language is *action* and that language is *social*, one might predict that language would find its neural niche among regions of the brain also involved with motor control and social interaction. And indeed, it appears to be so. In one simple demonstration Pulvermüller (2005) showed that words such as "lick," pick," and "kick" activate regions in primary motor cortex involved in mouth movements, hand movements, and leg movements, respectively. Moreover, a number of findings coming out of Art Glenberg's lab have linked the neural supports for motor processing and language understanding more generally (Glenberg & Kaschak 2002; Glenberg et al. 2007, 2008a, 2008b; Guan et al. 2013; see section 1.2 for a review).

Even more intriguing is the evidence for the involvement of mirror neurons in language processing (Arbib 2010; Pulvermüller & Fadiga 2010; Rizzolati & Arbib 1998). As is well known, mirror neurons—discovered originally in the brains of monkeys—possess very special receptive fields: they respond both when the monkey performs an action, such as grasping a peanut, and also when the monkey observes someone *else* performing the action (Rizzolati et al. 1988). For this reason the mirror system is understood to be an important component in supporting both action understanding and social interaction; and thus its involvement in language comprehension fits very naturally into the story I am developing here (see also Christiansen & Müller in press).

In fact, in chapter 1 we already reviewed evidence relevant to this claim. Broca's area, which has long been strongly associated with language processing, is in fact also involved in many different action- and imagery-related tasks, including movement preparation (Thoenissen et al. 2002), action sequencing (Nishitani et al. 2005), action recognition (Decety et al. 1997; Hamzei et al. 2003; Nishitani et al. 2005), imagery of human motion (Binkofski et al. 2000), and action imitation (Nishitani et al. 2005; for reviews, see Grodzinsky & Santi 2008; Hagoort 2005; Tettamanti & Weniger 2006). Broca's area, then, represents not a specialized neural adaptation for language but a region that likely evolved in the service of action and interaction that has since been redeployed to support communication-related tasks (Müller & Basho 2004).

In the linguistic examples offered above I focused on *directives*, as these best illustrate the thesis I am defending in this chapter and most obviously conform to the triadic formulation involving speaker, listener, and action *X*. But there are, of course, other kinds of speech acts: assertions of facts, declarations such as judicial findings that *make* facts; expressions of attitudes; and commissives such as promises or oaths that commit the speaker to some future act. Although these categories of speech acts are importantly distinct, they are all nonetheless social in the way outlined above: declarations and promises, for instance, depend in part on recognition by others, and assertions generally have a practical purpose (such as eliciting agreement) other than the mere depiction of some state of affairs. Thus, for our purposes, the nature of the action *X* can be left largely open: it can be as simple and practical as passing the scalpel or the salt or as abstract as coming to recognize and respect a given state of affairs or to adopt a particular belief or attitude.

And in fact, dialogue is known to subserve multiple purposes, often simultaneously, from facilitating practical activities to signaling status to building and maintaining relationships. For example, as Gordon (2013) summarizes a central thread in Deborah Tannen's work, "all conversations balance connection and hierarchy"; these relational dimensions are constantly being negotiated in all our linguistic interactions. Through what we say, how we say it, and how we respond to what is said, we both signal and instantiate our relative status and our degree of closeness, even in such domains as narrative, where the focus would seem to be rather on depiction:

> For instance, research has shown that storytelling in family discourse is a means for exerting social control (e.g., Langellier & Peterson 1993), reinforcing a family's hierarchical structure (e.g., Erickson 1990; Ochs & Taylor 1992a, 1992b, 1995), building and maintaining solidarity (e.g., Byers 1997), socializing children (e.g., Blum-Kulka 1997), helping children develop a sense of well-being (e.g., Fivush et al. 2004), constructing cultural or ethnic group identities (e.g., Blum-Kulka 1997), creating a

family's beliefs and values (e.g., Ochs, Smith, & Taylor 1996), and constructing a unique family culture (e.g., Langellier & Peterson 1993). (Gordon 2009, p. 111)

As already noted above, utterances are pragmatically effective in these ways only because both speaker and listener share a relevant context within which the speaker's intentions are interpreted. In the sociolinguistic study of discourse that developed in part out of the ordinary language school and Grice's pragmatic approach to linguistic exchanges, these contexts have come to be analyzed using the concept of a "frame." A frame is a jointly established characterization of the context of a linguistic interaction that helps discourse partners formulate effective utterances and interpret the (illocutionary and perlocutionary) intentions of their interlocutors—that is, what the listener is expected to understand and to do in response. Although the term "frame" is the same as one often used in artificial intelligence (Minsky 1975) and cognitive psychology (Schank & Abelson 1977) to refer to what Tannen and Wallat (1987) call "knowledge schemas," in sociolingusitics a frame is created by and shapes interaction:

> Goffman considers frames to be primarily social and situational; as he explains, they are "definitions of a situation" that interlocutors establish in interaction (Goffman 1974, p. 10). Thus a "real experience" differs from a "theatrical performance"; it is framed differently, and participants treat what transpires differently. (Gordon, in press; see Tannen & Wallat 1987 for further discussion)

Interestingly, and crucially, interlocutors can agree on the relevant frame governing their interaction but nevertheless take different attitudes toward it. Gordon (in press) offers the example of an undercover narcotics agent posing as a college student interested in buying illegal drugs for a party. While her "fellow students" and she may jointly establish that they are having a casual conversation and thus agree of the governing frame for their interaction, they will nevertheless establish different "footings" within that frame.

> As people create frames, they also construct footings, or alignments between one another as well as between themselves and what is said. In Goffman's words "[a] change in footing implies a change in the alignment we take up to ourselves and the others present as expressed in the way we manage the production or reception of an utterance "... (Goffman 1981, p. 128). (Gordon in press)

Although interlocutors may jointly construct a frame, they need not for this reason share goals for the outcomes of the conversation. Whereas the college students in this example may be primarily interested in establishing a friendship or in being perceived as cool or in simply passing the time in conversation, the narcotics agent intends to gather information useful in her investigation. These goals will be reflected in what gets said and how and in how the linguistic interaction is steered and managed by each party.

From the perspective I am developing here of a brain specializing in tracking and manipulating the values of salient relationships (naturally including such interpersonal dimensions as status, closeness, etc.) and in managing the organism's interactions with the world, the notion of taking up a footing within a frame appears to be very naturally analyzed in terms of the "phenotypic reorganization" (Proffit & Linkenauger in press) of bodies preparing for action, discussed in chapter 5. One's goals are partially expressed in the "stance" one takes up in and toward one's environment, including the other people in it; enacting the stance prepares us for and attunes us especially to a certain range of interactive possibilities. That is, being a "walker" or a "thrower" or a "caregiver" or a "friend" reconfigures the perception-action system in such a way that certain affordances are enhanced while others recede from attention. Moreover, depending on whether the frame is "hunting" or "skipping stones," the particularities of one's reconfiguration as a "thrower" will differ somewhat. This changes what one perceives and how in general one evaluates the attended elements of one's environment. In the case of language, then, this reorganization instantiates the cognitive aspects of a footing within the context of a frame; it limits and biases the way (physical, social, and linguistic) events will be interpreted and responded to. And these may differ for each member of the interacting group, depending on the footings each member instantiates, as is quite clear in the case of the undercover narcotics agent: what will seem to her to be most important and salient in the conversation will be quite different from what seems so to the student.

These aspects of discourse are so crucial that many theorists have argued that the manipulation of frames and footings is *the* defining function of language. For instance, Bruner writes that we have "a set of predispositions to construe the social world in a particular way, and to act upon our construals" (Bruner 1990, p. 73), and he takes it to be the central job of language in general and narrative in particular to manipulate and adjudicate between competing construals of the world. Similarly, Tomasello writes that "linguistic symbols are social conventions for inducing others to construe, or take a perspective on, some experiential situation" (Tomasello 1999, p. 118). We have already seen that language and discourse do more than this—are capable of a greater variety of effects. But it is nevertheless likely that the establishment and manipulation of frames and footings, through the biasing of perception and action that this achieves, constitute the crucial pathway whereby language can *have* its variety of effects. For brains like ours, attuned to see and respond to the opportunities for interaction that the world affords, the ability to bias this attunement is a powerful social tool, indeed. Later in this chapter I review some empirical evidence that suggests that discourse does in fact have this effect.

For all the reasons outlined above, I would like to suggest that we think of sounds and words and phrases and various chunks of language as possessing constructed social and cultural affordances. As Scollon (2010) argues, narratives and other forms of discourse afford some kinds of actions and interactions and impede others (see also Scollon & Scollon 2001, 2003). In fact, utterances typically offer multiple affordances; they invite us to use and respond to them in various ways (and sometimes in very few, in the face of which we are at a loss: the word "fine" offered in response to a query about one's day intentionally offers very little conversational grip). Depending on our frame and footing, some of these affordances—and thus some of our response options—will be more salient than others. What is gentle, humorous teasing in a play frame may be hurtful in a different context, and its effect on our behavior will be likewise different. And in fact, as Gordon (2009) clearly demonstrates, utterances are sometimes meant to be effective in multiple frames simultaneously; the humorous teasing about being clumsy that builds rapport in a play frame also serves as a reminder to replace the broken lamp and offers another opportunity to apologize for breaking it.

Of course, it must be admitted that some technical conceptual work would be needed to definitively extend and adapt the notion of an affordance to the case of language. An affordance is a perceivable relation between an organism's abilities and the properties of the environment—that is, it is a dyadic relationship, whereas we have seen that the linguistic relation is triadic. A *cultural* affordance, then, would be an intersubjective relationship between two agents and an artifact (whether a physical tool or a bit of language); one perceives what one *can* do with it but also what this use would mean to the other—how she is likely to respond—in light of its history of use and the nature of the current interaction. One important question to be worked out in this context is whether we should model this awareness of cultural affordances as direct perception or as the modulatory influence on directly perceived affordances of our praxis-based representations of cultural conventions (Anderson & Rosenberg 2008; Bruner 1990). Indeed, the relationship between affordances and cultural conventions has been left largely unexplored (but see Anderson & Chemero 2009; Turner 2005 for related discussions); this area offers opportunities for future advances.

But whatever the ultimate fate of this proposal about cultural and linguistic affordances specifically, for our purposes the crucial point is that language is a medium for social and cultural interaction, and acquiring a language is therefore in part learning what we can do with and in response to utterances and what our utterances will afford to others. As with learning any other practical skill, this process involves learning to see and to hear and to respond as do other practitioners of the skill: "dialogue rests on perceptual skills that develop in the service of action" (Cowley 2011, p. 188). The focus of a language learner, then, is on coming to act effectively with words; it is the acquisition of an interactional competence with and through others that also involves coming to see and hear and act like others. To acquire a language, this is to say, is to enter into a cultural practice (Bruner 1990). As Stephen Cowley argues:

> In learning to talk ... babies integrate vocalizations with activity and, only later, with wordings. It is by learning to listen as a member of a community that they discover the value of verbal resources. ... This depends on linking bodily coordination and wordings as, together, we act under the constraints of a cultural tradition. While grounded in moving bodies, wordings become integral to feelings, attitudes, beliefs, and displays of who we are. We learn to hear—and hear ourselves "repeating" patterns that call up response. (Cowley 2011, pp. 187, 188–189)

Not surprisingly, given the social nature of the tool in question, acquiring a language depends crucially on such imitation. Repetition of verbal patterns is crucial to learning in part because it is essential for the child to experience what *she* does (effects) when wielding wordings in that way and what she does when using them in a novel manner or context. But it is also an inevitable result of the conventional nature of discourse. This is not so say that language learners *only* repeat what they have heard—the child who says "runned" has inferred and then mistakenly universalized a rule for constructing the past tense—but is rather to emphasize the fact that acquiring a language is primarily a matter of mastering a social practice. There is no other way to speak than to use language—to access, that is, the stock of conventions and shared history that these sounds and symbols embody. Gordon comments:

> Indeed, research in discourse analysis has illustrated that repetition—of words, phrases, syntactic structures, ideas, and so on—is prevalent in conversations across a variety of contexts, in literary discourse, and in the media. Such repetition serves to generate links not only within but also between various written and conversational texts; it creates what Julia Kristeva (1980) calls *intertextuality*. ... As constructing a mosaic is a creative process in which bits and pieces from here and there are selected, adjusted, and fitted together to create something "new," so too is repeating. In A. L.

Becker's (1995) words, "prior text" is continually "reshaped" in interaction in various ways; moreover, this is what comprises language use of what he calls "languaging." ... Current uses of language always hearken back to those prior, giving all discourse an intertextual (or in Bahktin's words, dialogic) dimension. (Gordon 2009, pp. 7–8)

This illustrates yet another way in which discourse is interactive, social, and cultural: we communicate with one another by using and reusing a shared stock of sounds, words, and phrases that have been given their power to facilitate and coordinate by others; and moreover, as we acquire these tools (only) through interaction, our shared practices and shared linguistic tools together both create and reenforce a shared culture. Interestingly, but not surprisingly in light of the evidence recounted above, actual social interaction appears *necessary* for language acquisition, both for us and for vocal learning in other animals, and not just because that is the usual way that we gain experience with language:

[Z]ebra finches have been ... shown to need social interaction with a real-bird tutor (most often, the father) in order to learn from him and typically do not learn well from tapes (Eales 1989; Mann & Slater 1995). If, however, the experimental setting is such that a subject bird is required to activate a tape by pressing a key to elicit the zebra finch song model, the song is learnt. Yoked controls who can hear the same playback, however, do not learn the song (Adret 1993). ... The commonality between a live tutor and controllable playback is response contingency. ... [T]here is now ample evidence from human studies to indicate that response contingency is crucial to infant learning of advanced vocal forms (Bloom, Russel, & Wessenberg 1987; Goldstein & Schwade 2008). In addition, response contingency in the form of turn-taking interactions is an early feature of parent-infant face-to-face interactions (Papousek, Papousek, & Bornstein 1985), and human infants respond to social contingency from a very young age and show sensitivity to familiar contingency levels based on the responsiveness of their caregivers (Bigelow 1998; Bigelow & Rochat 2006). It has been suggested that social contingency aids learning by focusing infant attention on relevant features of stimulus information (Kuhl 2007). It is thus a property of the interaction itself, and not sensory signals or motor output that gates learning, an observation that is in fact fairly pervasive. (Syal & Finlay 2011, pp. 423–424)

Language skill emerges out of action and interaction because it is in fact a form of action and interaction. And the fact that language rests on a broad set of interactional competences can often be seen when communication fails. For instance, Gumperz (1982, chapter 8) offers an extensive and careful analysis of an unsuccessful encounter between an unemployed Pakistani teacher and a British job counselor. The conversation is awkward, both rhythmically and prosodically disjointed; each participant in the dialogue appears to be operating according to a different set of interactional conventions for how to use words, timing, and intonation to direct the attention and alter the behavior of the other. For instance, indirect requests for information fail to have their desired results, and pauses apparently intended to emphasize the importance of what is *about* to be said are instead interpreted as signaling a turn change, thus causing unwanted interruptions. As a result, "the speakers never achieve a joint understanding of where the interaction is going at any given time" (Gordon 2011, p. 70).

What is interesting here is that the failure of communication is revealed in, but also is in part *caused* by, the failure of interactional coherence and not, for instance, by any failure in the individual capacity of either speaker to produce well-formed English sentences. In such a situation two individuals with perfectly adequate mastery of a common language can nevertheless fail to communicate because of "different expectations about what needs to be said in the session and ... different uses of prosodic and paralingusitic features like rhythm and intonation" (Gordon 2011, p. 70). Communication is not about competence in the formal encoding of messages but rather is about the ability to successfully manage interactions to achieve joint (and individual) goals. And in fact, interactive competence can allow for successful communication even in the face of significant disfluency in formal encoding (consider in this light the story of the patient Chil who is able to successfully communicate and interact despite having a poststroke vocabulary of only three words, which he can deploy in various contexts via prosodic manipulation [Goodwin 2010]).

7.2 Language Evolved

I have been emphasizing the pragmatic, interactive aspects of language *not* to deny that there is a great deal of knowledge—formal and otherwise—that lies behind successful communication but instead to make the point that linguistic competence need not necessarily represent a mismatch with our native perception-action capacities and thus require the postulation of specialized, dedicated, hard-to-evolve neural hardware. Interactive brains like ours may in fact be especially suited to interactive languages such as ours. As Nusbaum has recently suggested:

> Rather than think about language as simply a signal for transmitting information, we can think about the communicative interaction itself as a psychologically significant act that was part of the basic force shaping the evolution of the brain. By this construal, the listener's goal may not be to interpret the linguistic message but to interact with the interlocutor in a way that satisfies specific social goals and motives.

> This would suggest that communicative behavior—broadly construed—should be affected by a conversational partner's behavior, even beyond the simple process of interpretation. (Nusbaum 2011, p. 676)

That is to say, insofar as language emerged as a refinement of existing capacities for social engagement and interaction (Tomasello 1999), then the *supports* for language will be largely existing supports redeployed in the service of this new form of interaction. This implies two related but subtly different effects. First, if linguistic coordination is achieved using many of the same mechanisms as social coordination more generally, then we should see signs of this on specifically linguistic behavior; the way people "language" should be affected by their efforts to socially coordinate. Second, if dialogue is a form of coordinated action, then people in conversation should show signs of acting in coordination.

Each of these effects appears to obtain, as demonstrated in a broad range of experimental work. For instance, interlocutors converge on a shared vocabulary (Brennan 1996; Brennan & Clark 1996) even when no miscommunication would occur if each interlocutor continued to use different words. Thus, for instance, dialogue partners might come to jointly use "glasses" to refer to eyeglasses even when one partner generally says "spectacles" and both are familiar with various ways are referring to eyeglasses. Although there is some evidence that such lexical entrainment improves understanding in dialogue, the main purpose is arguably to signal social cohesion. The effect is powerful enough that *refusing* to entrain itself carries a social message. For instance, half of my extended family refers to my niece with one name, and the other half with a different name (her middle name), even when speaking with one another ("Is Jane [not her real name] enjoying college?" "Oh yes, Sarah is having a great time."). In ordinary conversation speakers would converge on a single name; in this case the refusal to do so constitutes a meta-message (Bateson 2000) communicating and reinforcing the social estrangement of these two halves of the family. (Note that communication is not thereby impaired; no one is ever confused about who the subject of conversation is.) Likewise, apparently for largely social reasons, people pronounce things more similarly during a conversation than they do when speaking separately (Pardo 2006). They also tend to match prosody when engaged in conversation and joint action, and the degree of similarity predicts how positively the interaction is perceived by the interlocutors (Lee et al. 2010).

The second effect is evidenced by various forms of increased physical coordination between dialogue partners. Shockley, Santana, and Fowler (2003) found that participants engaged in conversation adopt similar bodily postures, even when they cannot see one another; and a series of studies has demonstrated coordination of the eye movements of dialogue partners, even when they are not looking at the same scene (Richardson & Dale 2005; Richardson, Dale, & Kirkham 2007; Shockley, Richardson, & Dale 2009). It is even the case that if one interlocutor in a conversation begins to tap his foot, the other is likely to do so as well and that doing so can increase perceived interpersonal affinity (Lakin & Chartrand 2003). This last finding suggests the prosodic entrainment mentioned above may be, at least in part, a *cause* of a positive interaction and not just a sign of one. A recent review summarizes the evidence for both effects.

> When people talk, they coordinate whose turn it is to speak (Sacks, Schegloff, & Jefferson 1974). They also implicitly agree upon names for novel objects (Brennan & Clark 1996; Clark & Brennan 1991), align their spatial reference frames (Schober 1993), and use each other's syntactic structures (Branigan, Pickering, & Cleland 2000). Their accents become more similar (Giles, Coupland, & Coupland 1992), they sway their bodies in synchrony (Condon & Ogston 1971; Shockley et al. 2003), and they even scratch their noses together (Chartrand & Bargh 1999). These acts of co-ordination serve many purposes, such as making sure that conversation flows easily and intelligibly (Garrod & Pickering 2004) and that conversants are well disposed toward each other (Dijksterhuis & Bargh 2001). (Richardson et al. 2007, p. 407)

In addition to this behavioral evidence, it is also worth recalling some of the neural evidence recounted earlier; language emerged out of efforts at social coordination and found its neural niche among regions of the brain implicated in managing action, interaction, and sociality. Even very basic features of language have apparently been shaped by the reuse of existing capacities in new contexts. For instance, the sounds used for speaking appear to be shaped in part by the postural organization of the motor system (Graziano et al. 2002b; MacNeilage 1998), and the processes of speech perception appear likewise to reuse elements of the motor system (Liberman & Mattingly 1985; see Nusbaum 2011 for discussion). Indeed, it has recently been argued that speech processing is in general supported by the establishment of functional connections between regions of the brain involved in various non-speech-related processes (Price, Thierry, & Griffiths 2005).

In addition, the fact that speech involves a serial code may be a feature rooted in the information-processing characteristics (or, if you like, limitations) of brains that evolved for iterative interaction with the world:

> The seriality of vocal output, most obviously, forces a sequential construction of messages. A perceptual and memory system that is typically a "greedy" processor and has very limited capacity for storing "raw" sensory input of any kind (e.g., Haber 1983)

may, moreover, force a code that can be interpreted incrementally (rather than the many practical codes in communication engineering, in which information is stored in large blocks (e.g., Mackay 2003). (Christiansen & Chater 2008, p. 501)

Because our language has this serial character, language acquisition can be accomplished using existing neural capacities for sequential learning, themselves likely evolved originally for behavioral control purposes. It appears, for instance, that regions of the brain including left perisylvian cortex involved in managing action sequences are also involved in parsing grammatical structures (Pulvermüller & Fadiga 2010). Christiansen and Chater (2008) summarize a number of relevant data points:

The close relationship between sequential learning and grammatical ability has been further corroborated by recent neuroimaging studies, showing that people trained on an artificial language have the same event-related potential (ERP) brainwave patterns to ungrammatical artificial-language sentences as to ungrammatical natural language sentences (Christiansen, Conway, & Onnis 2007; Friederici, Steinhauer, & Pfeifer 2002). Moreover, novel incongruent musical sequences elicit ERP patterns that are statistically indistinguishable from syntactic incongruities in language (Patel et al. 1998). Results from a magnetoencephalography (MEG) experiment further suggest that Broca's area plays a crucial role in processing music sequences (Maess et al. 2001). Finally, event-related functional magnetic resonance imaging (fMRI) has shown that the same brain area—Broca's area—is involved in an artificial grammar-learning task and in normal natural language processing (Petersson, Forkstam, & Ingvar 2004). Further evidence comes from behavioral studies with language-impaired populations, showing that aphasia (Christiansen et al. 2010; Hoen et al. 2003), language learning disability (Plante, Gomez, & Gerken 2002), and specific language impairment (Hsu et al. 2006; Tomblin, Mainela-Arnold, & Zhang 2007) are associated with impaired sequential learning. Together, these studies strongly suggest that there is considerable overlap in the neural mechanisms involved in language and sequential learning. (Christiansen & Chater 2008, p. 502)

In fact, with respect to the learnability of language, recent evidence suggests that when one takes into account the social basis of language and language acquisition—such as the actual nature of child-directed speech—the poverty-of-stimulus arguments fail in principle. It turns out that domain general learning algorithms can indeed induce the syntactic structure of language from online experience, given a more accurate understanding of the nature of that experience (Perfors, Tenenbaum, & Regier 2011).

It is worthwhile to consider the fact that language, as, one presumes, with all artifacts and cultural practices, must itself be adapted to the brain. Language is serviceable in part because, like a pair of scissors, it has a two-way fit both to its user and to its task (Clark 1997). In light of this, the fact that language is learnable without special, dedicated neural machinery should cease to be particularly surprising (Chater, Reali, & Christiansen 2009; Christiansen & Chater 2008). No one supposes that long division requires dedicated, specialized neural modules to be humanly possible. Rather, this particular practice must itself be of a sort that facilitates cultural transmission (Tomasello 1999); long division (among many other things, e.g., reading) must be able to find its neural niche (Dehaene 2005) or the practice will die out. The same must be true of spoken language—and no such system or practice needs to initially emerge in perfectly adapted form. Rather, one would expect the trajectory of development to ever better suit such practices both to the capacities of the practitioners and to the purposes of the practice:

> At the outset, it is natural that language will be the outcome of competing selectional forces. On the one hand, as we shall note, there will be a variety of selectional forces that make the language "easier" for speakers/hearers; on the other, it is likely that [expressivity] is a powerful selectional constraint, tending to increase linguistic complexity over evolutionary time. ... From this perspective, the problem of language acquisition is very different from learning, say, some aspect of the physical world. In learning naïve physics, the constraints to be learned (e.g., how rigid bodies move, how fluids flow, and so on) are defined by processes outside the cognitive system. External processes define the "right" answers to which learners converge. But in language acquisition, the structure of the language to be learned is itself determined by the learning of generations of previous learners (see Zuidema 2003). ... [I]n language acquisition, the learner's biases, if shared by other learners, are likely to be helpful in acquiring the language—because the language has been shaped by processes of selection to conform with those biases. ... [L]anguage has been shaped to be learnable from the kind of ... input available to young children. Thus, language acquisition is constrained by substantial biological constraints—but these constraints emerge from cognitive machinery that is not language specific. (Christiansen & Chater 2008, pp. 503, 507)

Thus, it appears that a misunderstanding of the nature of children's linguistic experience, combined with a failure to carefully consider the processes that shape language *development* over time, may have led to a systematic overestimation of the real difficulty faced by language learners. Similarly, it appears that the computational complexity of language may have also been systematically overestimated. To consider just a single instance, the possibility of indefinite recursion, current thinking suggests that (1) not all languages in fact allow for recursion (that is, it is not a universal feature of language), and (2) it is typically severely restricted even in those that do (Evans & Levinson 2009):

> In computational linguistics, recursion without limit (to infinity) is the basis for analyzing the power of different grammars such as [Finite State Grammars] and [Context Free Grammars]. In reality, humans rarely achieve, and only awkwardly, even a recursive level of three when using center embedded sentences in natural speech ('The house that the man I married bought, burned.'). ... We propose that embracing an infinite recursion for linguistics unsupported by corresponding data of actual human performance arises from an unfettered first-principle perspective. (Margoliash & Nusbaum 2009, p. 509)

This is important both because bounded recursion can be handled by mechanisms with considerably less computational power than those required for indefinite recursion and also because it suggests that handling embedded clauses is something that we can *learn* to do but is not necessarily something for which we are innately equipped (Frank, Bod & Christiansen 2012). Naturally, during learning we can avail ourselves of multiple forms of external support for the capacity, including external symbols. As we have already demonstrated in chapter 6, human brains coupled with external symbols can instantiate a Turing complete system, which is sufficiently powerful for indefinite recursion, even when the naked brain is not. It is finally worth noting that one can observe nested action sequences in the manipulation and combination of physical objects and that the neural substrates for *word* combination appear to be largely shared with those supporting *object* combination (Greenfield 1991).

I take such evidence to indicate—consistent with the theoretical perspective on language advocated in section 7.1—that language is indeed largely supported via the repurposing of existing neural machinery developed to support action, interaction, and social relations. This is important in two ways for understanding the evolutionary pathways for language: first, it greatly extends the time over which language-relevant neural systems can be said to have been under selection pressure; repurposed systems could be considered preadaptations for language (Christiansen & Chater 2008; Elman 1999). Second, it makes it probable that neural redeployment, which likely works over shorter timescales than mosaic evolution, has played a significant role in the emergence of our natural language capacities. Both of these facts are crucial given the time constraints placed on language evolution:

> The biological underpinnings of language are so recently evolved that they cannot be remotely compared, for instance, to echolocation in bats (*pace* Jackendoff 2002, p. 79). Echolocation is an ancient adaptation shared by 17 families (the Microchiroptera) with nearly 1,000 species and over 50 million years of evolution (Teeling et al. 2005), whereas language is an ability very recently acquired with spiraling culture

in perhaps the last 200,000 to 300,000 years by a single species. Language therefore must exploit preexisting brain machinery, which continues to do other things to this day. ... The null hypothesis here is that all needed brain mechanisms, outside the vocal tract adaptation for speech, were co-opted from preexisting adaptations not specific to language (though perhaps specific to communication and sociality more generally). (Evans & Levinson 2009, pp. 446–447)

Naturally, none of this evidence proves that there have *not* been any adaptations or genetic mutations resulting in neural "modules" specific to language. What it shows instead is that it is unlikely that we *need* there to have been any such events to explain the emergence of language. As with the hypothesis of reconstructive perception, which was theoretically necessitated by a certain understanding (that turned out to be a *mis*understanding) of the nature of our perceptual systems and their epistemic access to the world, it appears that the theoretical necessity for dedicated neural modules supporting a language faculty was rooted in a misunderstanding of the nature of language.

I should nevertheless hasten to admit that I have *not* here provided (nor even tried to provide) anything like a complete account of the nature, emergence, and evolution of language. For this one would most certainly have to emphasize the general role of culture in evolution (Richerson & Boyd 2005) and the specific interactions between individual brains and the social practices that govern cultural transmission (Boyd, Richerson, & Henrich 2011; Tomasello 1999), among many other things. My goal was only to demonstrate that there is little good reason to suppose that language requires a brain evolved (or otherwise constructed) specifically for symbol processing; on the contrary, language is just the sort of thing that brain evolved for action and interaction can master. More broadly, I hope to suggest that our practical brain is just the sort of brain we need in order to be the sorts of cultural beings we are. The view being propounded here, this is to say, places neuroscience, embodied cognition, and what Bruner (1990) calls "cultural psychology" into a single coherent framework. The aim is not just to provide a more nuanced view of the functional architecture of the brain but to support the (eventual) elucidation of the biological foundations of a culturally situated psychology.

7.3 Language Is Leverage

Lurking behind the notion of the two-way fit mentioned above is an evolutionary and developmental process that should be brought to the fore: niche construction (Kendal, Tehrani, & Odling-Smee 2011), itself an instance of a larger process of gene-culture coevolution that has been remarkably important to the human species (Boyd et al. 2011; Richerson & Boyd 2005). The idea is simple but remarkably powerful. Organisms routinely alter their environments to better adapt the environments to their uses. They build nests and burrows, dam streams, change the pH of the soil, and do many other things besides. These changes modify the selection pressures acting on organisms, which adapt them to their (altered) environments, which organisms then change further, creating a virtuous cycle in which organisms become ever better able to survive and thrive in the world.

In the case of human organisms, the social world is most certainly implicated in this process: the social and cultural environment consists of our conspecifics but also the practices, material culture, and artificial environments bequeathed by previous generations. Humans have adapted not just to the "natural" environment but to the total environment as shaped and constructed by past and ongoing human activity. It is this realization that prompts Tomasello to remark that learning to use language is to "[learn] the ways [one's] forebears in the culture have found it useful to manipulate the attention of others in the past" (Tomasello 1999, p. 126). Insofar as we are like our forebears and live in a similar total environment, we are likely to find the particular tools and cultural practices bequeathed us to be useful for accomplishing our ends, and they will in turn attune us to the kinds of ends we might pursue. We will largely find these practices suited to us and us to them. And, as I have been arguing all through this volume, our brain is especially adapted to interacting with such external supports—maps, diagrams, symbols, and tools of various description—in pursuit of cognitive ends. As Evans and Levinson suggest:

> On this view, cognition is less like the proverbial toolbox of ready-made tools than a machine tool, capable of manufacturing special tools for special jobs. ... Culture provides the impetus for new tools of many different kinds—whether calculating, playing the piano, reading right to left, or speaking Arabic. (Evans & Levinson 2009, p. 447)

One of the most important aspects of our total environment and the tools it offers is language itself, and here I would like to review some of the ways that language helps us to pursue our cognitive ends. In this I am largely following Andy Clark's (1997, 1998b) pioneering efforts to draw our attention to the ways in which language acts as a cognitive scaffold, allowing us to do things we would not otherwise be able to do so easily, if at all. The view I am explicating differs from Clark's in only one way, albeit an important one. Although Clark (1997, 1998b) clearly understood language to be primarily a pragmatic tool for effecting change in the world—that is, he had largely rejected the picture theory of language in favor of a more causal one (Gauker 1990)—he apparently still retained the notion that utterances instantiate a dyadic relationship between speaker and world. That is to say, he did not fully recognize the irreducibly *social* nature of language (as the multiple references in these works to "public" language—as if there were a different sort—perhaps attests; see Love 2004 for a similar complaint). As we have seen above, utterances are in fact *triadic* and intersubjective and function in large measure by effecting phenotypic reorganization (e.g., biasing attention) within frames and footings. This being said, the main effect of this reform in our view of the nature and function of language will actually be to confirm and strengthen Clark's hypotheses, as we will see, below.

7.3.1 Language and Self-Control

One of the more famous observations in child development is that children speak to themselves as part of mastering new tasks (Piaget 1962). Clark, following Vygotsky (1962), suggests that the main function of this self-directed speech may be to facilitate task performance.

> When a child is "talked though" a tricky challenge by a more experienced agent, the child can often succeed at a task that would otherwise prove impossible. (Think of learning to tie your shoelaces.) Later, when the adult is absent, the child can conduct a similar dialogue, but this time with herself. ... In such cases, the role of language is to guide and shape our own behavior—it is a tool for structuring and controlling action, not merely a medium of information transfer between agents. (Clark 1997, p. 195)

From the standpoint of the view of language outlined above, this is indeed the most natural hypothesis to entertain. Far from constituting a highly speculative conjecture, the notion that the main function of language is effecting changes in the world by affecting the actions one's hearers will take is at the heart of current sociolinguistic theory. It is not that language has somehow been extended to this new use as a tool of behavioral control; this is what it was all along. The child in this imagined example is essentially simulating the sociolinguistic environment as a means of biasing her own attention in task-relevant ways that mimic the biases she would have acquired as a result of interaction with an adult. Language here is functioning the way language always does, albeit with one fewer person immediately involved.

Not surprisingly, the evidence bears out the hypothesis that language functions in this way and thus serves as a crucial developmental scaffold for children in learning new skills. For instance, Bivens and Berk (1990) showed that task-relevant, self-directed speech in elementary school students was developmentally related to math achievement in that the amount of task-relevant private speech predicted math performance 1–2 years in the future better than it did concurrently. Similarly, Winsler, Diaz, and Montero (1997) found (1) that such task-relevant speech was associated with task success; (2) that it was more likely to be used following verbal problem-solving help from an adult; and (3) that children were more likely to succeed when using task-relevant, self-directed speech following such interventions than they were when not using such speech.

In addition to the role of *spoken* language in self-control there is evidence that *written* language and other symbols can also have a salutary effect on self-regulation via a rather different pathway. The evidence comes from *reverse-reward contingency tasks* in which subjects are asked to indicate the *lesser* of two rewards (e.g., one cookie rather than three) in order to receive the *greater* reward. It has been shown that both young children and nonhuman primates have difficulty learning this task when the items are desirable in themselves (like cookies) but learn much more easily when Arabic numerals or plastic tokens are substituted for the food items (Boysen 2006; Carlson, Davis, & Leach 2005). Apparently, the external physical symbols (which are meaningless in and of themselves outside of practical context) do not elicit the prepotent response to grab for the larger reward. This allows space and time for associative and other learning mechanisms to do their work.

7.3.2 Objectification

As already detailed in chapter 6, what I am calling objectification refers to the ways in which meaningful external symbols can engage our highly developed capacities for manipulating and combining physical objects and tools. We can push symbols around to help us solve mathematical problems (Landy & Goldstone 2009) and move around chunks of other texts to help us formulate and evaluate arguments—as I am physically doing with books, articles, notebooks, and note cards as I write this chapter. As Clark describes:

> As I (literally, physically) move these things about, interacting first with one, then another, making new notes, annotations and plans, so the intellectual shape of the chapter grows and solidifies. It is a shape which [sic] does not spring fully developed from inner cogitations. Instead, it is the product of a sustained and iterated sequence of *interactions* between my brain and a variety of external props. (Clark 1998b, p. 171)

7.3.3 Memory Augmentation

The most obvious case of using language as a memory augmentation is the case of using inner speech—the so-called phonological loop—to remember something like a phone number or grocery list (Baddeley & Hitch 1974). But we also often write things down (or, perhaps more frequently these days, type them into a device) so as to store them for later. And of course the vast resources of the written word represent an immense memory resource made possible by written language and the social practices of storing volumes of it.

It should also be clear that linguistic memory augmentation serves not just an individual but an important social and cultural purpose (or several of these in fact; see below). Language helps implement what Tomasello (1999) calls the ratchet effect:

> The process of cumulative cultural evolution requires not only creative invention but also, and just as important, faithful social transmission that can work as a ratchet to prevent slippage backward—so that the newly invented artifact or practice preserves its new and improved form at least somewhat faithfully until a further modification or improvement comes along. Perhaps surprisingly, for many animal species it is not the creative component, but rather the stabilizing ratchet component, that is the difficult feat. Thus, many nonhuman primate individuals regularly produce intelligent behavioral innovations and novelties, but then their groupmates do not engage in the kinds of social learning that would enable, over time, the cultural ratchet to do its work (Kummer & Goodall 1985). (Tomasello 1999, p. 5)

Tomasello is largely concerned with the ways in which spoken language facilitates the social learning required to produce the ratchet effect, but of course stored texts can underlie an especially powerful version of this effect that works across time and distance. When combined with the role of language in objectification mentioned above, ideas can be easily compared, combined, and modified, and any (potential) improvements likewise preserved and passed along.

7.3.4 Environmental Simplification

There are myriad ways in which we impose regular structure on our environment so as to simplify the cognitive burdens we bear in negotiating it (McClamrock 1995). An important if simple instance of this strategy using language is signposting—on streets, inside buildings, even in our kitchens, linguistic labels allow "a little learning to go a long way" (Clark 1998b p. 168), lessening the burdens on memory, planning, and procedural knowledge.

7.3.5 Social Coordination

Because social coordination was the subject of much of the exegesis above, I only make the obvious point here: well-coordinated groups can accomplish more than individuals alone and individuals in poorly coordinated groups. From managing complex tasks such as ship navigation to building construction to large-scale scientific experiments (e.g., using the Large Hadron Collider), linguistic coordination immeasurably extends our practical and cognitive reach (Hutchins 1995).

7.3.6 Biasing Attention

Because the effect of language on phenotypic reorganization in general, and attention in particular, was likewise a central part of the treatment of language above, I am confining this section to discussing a suggestive piece of evidence that this picture of language is on the right track.

The Russian language makes an obligatory linguistic distinction between light blue and dark blue—between *goluboy* and *siniy*—that English does not make; it would be linguistically permissible in English to call the whole range of blue shades "blue," but there is no equivalent Russian color-category term. To investigate the effect of language on perception, Winawer and colleagues (2007) employed a collection of color chips that formed a smooth gradient of shades across the color blue, such that the differences between the shades of neighboring chips were equal across the gradient, and the gradient centered on the *goluboy-siniy* distinction. That is, of the 20 chips used, 10 were *goluboy* and 10 were *siniy*, even though the darkest *goluboy* and lightest *siniy* were the same "shade distance" apart as any other two neighboring chips.

Russian speakers and English speakers were shown a target chip along with two others, one of which was the same shade as the target and one that differed by one shade, and asked to identify the chip that matched. Winawer and colleagues found that Russian speakers, but not English speakers, were faster to make this judgment when the difference between the two chips spanned the goluboy-siniy distinction. Earlier work (Witthoft et al. 2003) showed a similar facilitation effect for English speakers when the chips spanned the blue-green boundary. Interestingly, and crucially, the facilitation effect disappears when participants are subject to verbal interference (silent rehearsal of an eight digit number), but it survives spatial interference (remembering a pattern of black-and-white squares), which suggests that it is in fact the ability to (covertly) *use* language during the perceptual task, and not just knowing the language, that accounts for the observed facilitation.

Although such experiments have often been glossed as an empirical test of linguistic determinism—the thesis that the world appears different to speakers of different languages—I would like to suggest that the more natural interpretation of such results is as a demonstration of the way language can bias attention: certain differences, which are perceivable regardless of language, appear more salient to speakers of one language versus another. Language probably does not make the world look different, but it can make some things easier to see (*see* also Treisman 1986).

7.3.7 Mitigating Path Dependence

As reviewed in chapter 2, learning is subject to order effects—also called path dependencies—such that certain learning states are accessible only from a limited number of "places" and by following a finite number of possible learning paths through the learning state space; likewise, the learning state you have currently achieved restricts the number and direction of future paths open to you. We generally think of this issue in terms of discovering the optimally effective paths to certain desirable educational outcomes and avoiding paths known to lead to suboptimal results. But of course this assumes that we already know the desired ends relative to which the learning states in question are to be evaluated (e.g., to gain proficiency in calculus or reading). Naturally, this is not always the case; a given learning state may produce poor math skills but facilitate some unknown competence. Generally, exclusively following known educational paths will of necessity leave most of the state space unexplored. This restriction might make sense given confidence that the best paths to the only ends worth pursuing have already been discovered. Some societies and subcultures do indeed seem to believe this. But a seed of doubt might lead one to embrace a certain amount of diversity in learning and in the pursuit of useful ends, and it is in this sort of mix that language can play a profound and powerful role.

An insight that, given his unique path through learning pace, would only occur to Joe to formulate might have a catalytic effect when heard by Mary, for it will cause her to perceive something new in the personal intellectual landscape she (but not Joe) occupies (I am here paraphrasing Clark 1998b, pp. 169–170). Although there may have been no possible *individual* path to the state that Mary now occupies as a result, there was in fact a collective path made possible by a combination of intellectual diversity and linguistic interaction. This fact is perhaps the origin of the observation that serendipity counts for as much as preparation in the production of scientific and cultural breakthroughs.

What applies to the effects of interaction between individuals within a culture applies also to interactions between individuals from different cultures, and naturally these effects are both individual *and* cultural.

Some years ago I was engaged in a period of moral reflection and reassessment and remarked to a friend that as part of this process I was rereading some Ancient Greek philosophy and literature. "Yes," he replied, "it's nice that they are always there waiting for us." Indeed, it is not just nice, it is arguably crucial to progress to have access to the preserved wisdom of alien cultures. My late-twentieth-century encounter with Ancient Greek writing allowed me to move beyond the (personally) suboptimal state into which I was locked by past experience without actually returning to any truly Greek modes of construing the world—because although they saw things I could not, perhaps the reverse was also true. So too it seems possible that cultures as a whole might, via interactions of just this sort, walk paths through the cultural state space not possible for a single culture alone. After all, what was a personal renaissance for me was a broad cultural one for Medieval Europe.

I wrote above that people largely find that their inherited cultural practices suit them. This is so, but the exceptions to this rule are crucially important as well. For in light of the discussion of this section, it should be clear that no culture can enable all human potentials, that some may enable more or fewer than others, and that some may enable better or worse possibilities (and likely some of both). This cultural diversity can perhaps *also* be leveraged via cultural exchange to mitigate the path dependencies that mark cultural development as surely as they do individual development. Recognizing that insights from one culture can be hard to express in the language of another—that a given culture's construals can be hard to adopt—is an important part of the collective search through the cultural state space, for it highlights precisely those differences in perspective that make such exchange potentially enriching.

7.4 How to Study Language and the Brain

Does any of the foregoing mean that grammar and linguistic form are not a legitimate object of study? Of course not. But the message coming from the body of work discussed above is that any particular grammar is, strictly speaking, inessential to language (Bruner 1990). A language must have structure to function at all, to serve as a reliable system of communication and coordination. But this structure is conventional, created by and preserved from practice: "yesterday's discourse is today's syntax" (Givón 1979;

cited in Christiansen & Chater 2008, p. 503). Linguistic forms are preserved and transmitted *if and only if* they are effective, and thus the structure of a language, far from representing any universal constraints imposed by some inherited language faculty, in fact indexes the practical wisdom of a given linguistic culture. At the deepest possible levels, form follows function.

Grammar, then, is acquired and perceived instrumentally as the form one's linguistic actions must take to achieve their desired ends. In this sense learning the grammar of one's language is little different from learning to use any other complex tool or machine—and as we saw it appears to depend on the same learning mechanisms and neural substrates that govern sequence learning and object manipulation, among other things. As with the doorknob in Andrew Pickering's example (chapter 5), we acquire a performative relationship with a grammar, and whatever explicit models we induce might best be understood in such pragmatic terms.

Finally, we should be ever mindful of the fact that the formal representation of linguistic structure in external symbolic media, such as the diagrams found in grammar primers, may have transformative effects on our linguistic performance and perception—which then no longer, strictly speaking, reflect the nature of our native capacities (Henrich, Heine, & Norenzayan 2010; Linell 2005; Port 2010). To study grammar—especially in Western literate societies—is in fact to reify and quantify a set of *cultural* practices, the ability to master which depends not just on our native cognitive and neural endowments but on the external supports offered by the social and symbolic structures in our environments (Port 2010). Thus, the study of grammar does not, in the end, allow us to uncover any universal and necessary features of language (Evans & Levinson 2009), nor does it offer a special window onto our innate neurocognitive capacities.

How then should we approach the study of the brain bases of language? The most important reform will be to begin to study the neural supports for the act of communication rather than for metalinguistic judgments, which approach dominates current work. We should focus instead on what brain resources are deployed when, for instance, interlocutors establish (or subvert) a conversational frame. When frames are nested, is this relying on neural resources for nested action patterns developed in the service of action? Which brain regions interact to make this possible, and what is the nature of the interaction? How do interlocutors establish joint attention? What are the roles of repetition and lexical entrainment in maintaining a successful linguistic interaction, and how exactly—with what resources deployed in what patterns—does the brain support such aspects of dialogue? When we understand language in relational terms—as, for instance, managing hierarchy and closeness—how does this express itself in terms of the brain's propensities for managing organism-environment relationships? Does language rely relatively more on brain regions with high loads on NRP factors related to social relationships? Or do regions with high loads on factors related to the management of physical relationships dominate? (Does this distinction even strictly hold in the brain?) Possibly we will discover that language is supported by a unique form of interaction *between* regions engaged by primarily social and those engaged by primarily physical tasks. Perhaps the most exciting scientific avenue involves investigating the relation between phenotypic reorganization in action and dynamically adopting footings in discourse. This, it seems to me, will allow researchers to get at the heart of the way the brain's action-oriented resources are deployed in support of linguistic interaction.

There is a wealth of opportunity just waiting to be mined, for these topics have been almost entirely neglected in the study of language and the brain. Here I would like to summarize just one existing empirical paradigm for studying language as interaction that, it seems to me, could be readily applied to the study of language and the brain (the brief discussion offered here owes much to conversations with Rick Dale, Chris Kello, and Anne Warlaumont). Central to the approach is the insight that coordinated activity—whether the intrapersonal coordination needed for walking, throwing, eating, and the like or the interpersonal coordination seen in joint actions including team sports and conversation—always involves constraining one's degrees of freedom. If one is to throw a rock at a target, one's shoulder and elbow and hand cannot move just any which way; they must coordinate to achieve the goal. Similarly, one cannot be perceptually attuned to *all* the information in and properties of one's environment but must instead attend to the few things that matter to the success of the action, such as estimating the effort needed in light of the distance and motion of the target and the heft—the throwability—of the rock. As we have seen already in the discussion of phenotypic reorganization, to constitute oneself as a thrower is to take a particular stance within and toward one's environment that involves a goal-directed coordination of eyes and limbs, attention, and action. Interestingly and crucially, this coordination is not a matter of individually constraining limbs or joints but involves a kind of mutual constraint within and between the coordinated parts. One does not lock one's shoulder into a particular arc; one allows the shoulder to be coordinated with and constrained by the elbow, hand, legs, and eyes. Coordinated action involves, this is to say, the construction of a synergy: "a functionally driven reduction of degrees of freedom, where components do not simply align, but also complement and compensate for each other" (Dale et al. 2013, p. 54)

> A classic example is Bernstein's analysis of chisel and hammer. If we want to strike a chisel with a hammer, this gives direction to and constrains the workings of our body. The exact timing and force of contraction and relaxation of all the individual muscles in our hands, fingers, and arm are locally regulated to comply with that overall goal and the unfolding interaction with the environment. This intuition was tested empirically by measuring the precision of movements at all relevant joints in a blacksmith's hammering of a chisel. The variability of the trajectory of the tip of the hammer across a series of strikes turned out to be smaller than the variability of the trajectories of the individual joints on the hammering arm (Bernstein 1967; Latash 2008). The joints are not acting independently but correcting each other's errors at the relevant timescale ... (Dale et al. 2013, pp. 54–55)

Naturally, the coordination involves more than just muscles and joints but also perception and attention, and the action as a whole crucially involves neural coordination—not as a top-down control process but as part and parcel of the whole-body synergy. Indeed, it is only because the synergy includes the brain that the hammering is coordinated not just in the moment but across many timescales and integrated with the ongoing behavior and overall purposes of the agent.

As noted above, the creation of such synergies is not central just to intrapersonal coordination but applies also to the interpersonal coordination seen in conversation:

> [T]wo people interacting in a joint task come to form their behaviors through compensatory complementary behaviors. These behaviors influence one another locally and incrementally, making the whole conversational performance itself a kind of self-organizing synergy. (Dale et al. 2013, p. 56)

This is to say, one of the effects of conversation—of communication—is the construction of social synergies in which dialogue moves invite and induce phenotypic reorganization in one's dialogue partners. Such stance-taking in dialogue has the effect of reducing one's degrees of freedom, simplifying what might otherwise be an intractable problem of social coordination. If one's interlocutors might really (and sometimes actually) do *anything* moment to moment, communication would be impossible to sustain. As Dale et al. note:

> One intuitive way of reducing the complexity in interpersonal interactions is to diminish the range of possible behaviors via a progressive adaptation to each other. By becoming increasingly similar, the interlocutors greatly simplify the cognitive load needed to interact with the other. (Dale et al. 2013, p. 69)

We have already reviewed a great deal of evidence for such convergence of behavior in dialogue partners. Here we see that this convergence is a sign not just that language is social but also that dialogue both creates and is sustained by interpersonal synergies of mutually constraining parts. Dale et al. (2013) propose that insofar as it is possible, these dynamics should be studied all at once rather than narrowing the focus to measuring one kind of coordination, as has been the common practice in most studies. Figure 7.1, adapted from figure 2.8 in Dale et al. (2013), illustrates both the power and the predictions of this approach.

For a simple case, imagine recording the values of four variables—gesture type (A); gesture speed (B); speech act category (C); and prosodic inflection direction (e.g., up or down) (D)—in two people over the course of a conversation. We know from previous work (reviewed above) that we can expect to see coordination across multiple modalities and both within and between behaviors and persons. We can measure the degree to which these variables are coupled—mutually predictive—over time and represent these findings in the form of graphs like those illustrated in figure 7.1, which shows high degrees of coordination (darker lines) both between some of the behaviors of each individual and between the individuals themselves. Thus, in this hypothetical illustration, we find that the speech act category of person 1 (node *C1*) strongly predicts gesture speed in person 1, *and also* gesture speed in person 2 at time t_1; and in turn, gesture speed in person 1 strongly predicts both speech act category and gesture speed in person 2 and time t_2. This coupling within and between persons—the phenotypic reorganization that dialogic interaction facilitates—serves to restrict the degrees of freedom of the system as a whole, both simplifying and in an important sense *creating* the desired linguistic coordination. As Dale et al. explain:

> We know through extensive explorations … that speech, gesture, and other features of interaction will exhibit coordination (e.g., see Louwerse et al. 2009). This can be identified as clusters in the network that become tightly entrained over time. … As an interaction changes across time, the network structure might change, but the degrees of freedom may stay the same. … This network analysis approach may serve as a powerful means of visualizing and quantifying the "surface configurations" of an interaction and [provide] clues to underlying mechanisms. (Dale et al. 2013, p. 81)

I believe this approach offers immense promise, but it does seem to me that including not just behavioral but also neural measures promises even greater insight into underlying mechanisms. Indeed, it is in part because the *neural* coordination dynamics are often left out of the study of coordinated

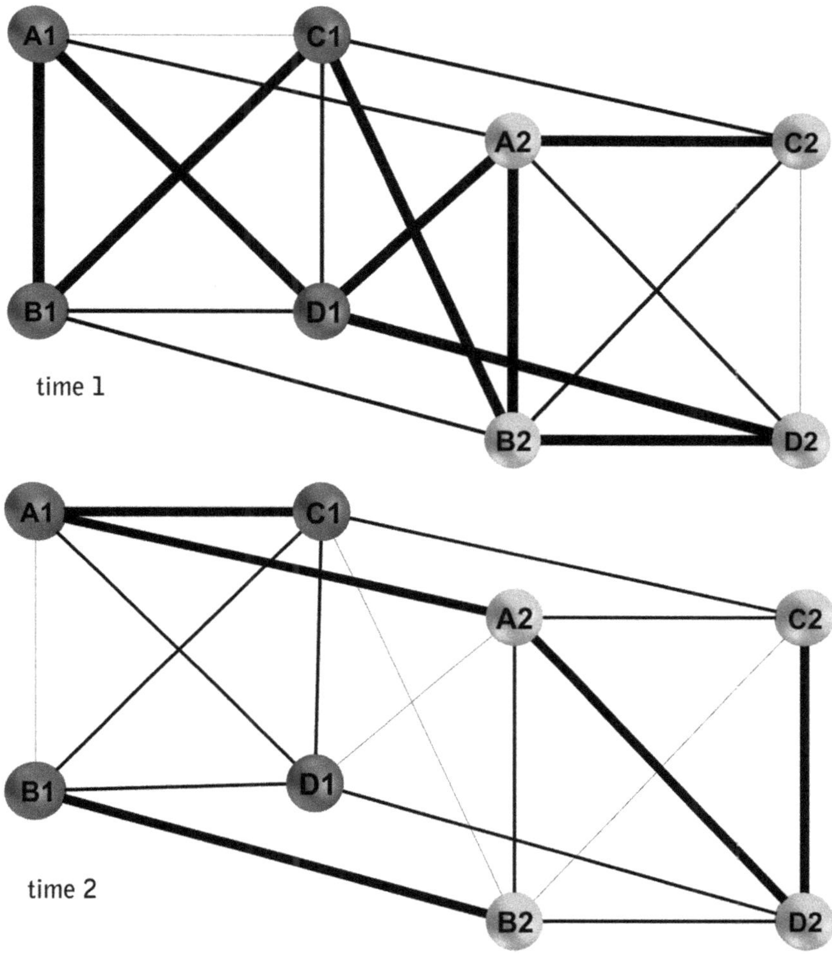

Figure 7.1
Illustration of hypothetical coordination relationships measured among four behaviors (A–D) exhibited by two people (1, 2) at two different times. Width of edge indicates strength of correlation.

action (but see Kelso 1995), and most studies of the brain bases of language do not treat language as a type of coordinated action, that the opportunities here are so enormous.

It is of course no coincidence that figure 7.1 looks rather like figure 1.3, which illustrates neural coordination patterns under different conditions. The nodes in these coordination graphs need not represent just measureable behaviors but can also represent the activity in regions of the brain. If we were to include such nodes in future empirical work of this sort, we could expect to see not just changing cooperation patterns within the brain, within the behaviors, and within the individuals but *between* these, and the particulars of these interactions promise to tell us a great deal about how the brain works to manage and maintain social and linguistic coordination. Which neural clusters, representing which sets of NRP factor loads, are associated with which speech acts, and which sorts of phenotypic reorganization (itself indexed by clusters within and between brain and behavior) do they predict in speaker and listener? How do the dynamics unfold at multiple time scales, as frames and footings are established and maintained, dialogue strategies are implemented, individual speech acts are performed, and various gestural and linguistic moves are made? The questions—and the scientific benefits of answering them—are truly unlimited.

Naturally, there are various technical and theoretical challenges to be met before such possibilities can be realized. Some of the challenges are technological and some methodological. I briefly review a few of these here as a way of outlining part of the research agenda that we must follow going forward.

One important technological innovation that will be required will involve finding ways to measure brain function in people while they interact with one another rather than while they respond to preprocessed computer-presented stimuli. This challenge in fact represents a nexus of different limitations that conspire to make studying discourse and the brain especially difficult. First is the issue of using technologies that can measure brain function in freely behaving individuals. Electroencephalography (EEG) would seem like an initially good candidate because signals can often be collected wirelessly (Debener et al. 2012), and even high-density, wired EEG systems do not impede face-to-face interaction. Moreover, EEG is able to detect and resolve neural events on the scale of milliseconds, giving it temporal resolution at the scale needed for studying rapid, real-time linguistic interaction. On the other hand, the spatial resolution of even high-density EEG systems is generally considered to be poor; although neural events can be rapidly and reliably detected, it is often not possible to know exactly *where* the event took place in the brain. This limitation is due to a general problem in electrical source localization known as the inverse problem (Grech et al. 2008). The technical details of the problem need not concern us here, but we can get the flavor of it by imagining that we wish, by detecting the size and frequency of waves hitting the side of a swimming pool, to determine where a rock (or rocks) had been thrown into the water. Consider trying to solve the problem with a single sensor: it takes but a moment of reflection to see that at a single sensor the difference between a single large rock, and two smaller rocks thrown near each other with the correct timing (such that their wave forms reinforce) would be indistinguishable. Adding sensors in the right place(s) would render these two events distinguishable, but there will always be multiple possible explanations for the set of observations. The problem is insoluble in the general case.

Perhaps, then, it would be better to use a technique with high spatial resolution such as functional magnetic resonance imaging (fMRI). Here we face some complementary limitations. First, because of the strength of the magnetic fields being used, fMRI generally requires the isolation of the test participants, which hinders interaction; and moreover, to get accurate images participants must remain very still, which hinders natural responsivity. Second, although fMRI has excellent spatial resolution, its temporal resolution is fairly poor, for which reason it is generally not used to study fast-moving processes such as sentence comprehension.

Barring breakthroughs in neuroimaging techniques, then, we must try to find ways to use the complementary strengths of these techniques to compensate for their complementary weaknesses. I noted above that the inverse problem is insoluble in the general case. It is, however, possible to rule out competing solutions for the locations of electrical sources with reasonable assumptions about which solutions are realistic. Some general assumptions include limitations on the *strength* and *number* of sources as well as limitations on their possible *locations*—for instance, they must be inside the skull! I think it might be possible, following the program I have outlined here, to make such assumptions about possible or likely locations more specific. For instance, maps of NRP factor loads across the brain, generated during multiple fMRI studies, can be used as constraints on the likely origins of neural oscillations due to language processing. This would be a way of using fMRI data asynchronously with EEG as a prior constraint on the source analysis of the EEG signal so as to allow researchers to locate the neural nodes that could be included in dynamic coherence graphs like that depicted in figure 7.1.

In addition to this sort of asynchronous cooperation between fMRI and EEG, it is also possible that utilizing multiple fMRI sites networked together so to allow for communication between participants could help mitigate some of the social isolation generally imposed by the technologies, which is especially detrimental to the study of language. Recent work by Uri Hasson and collaborators is pointing the way here (Hasson et al. 2012; Hasson & Honey 2012). And performing simultaneous EEG and fMRI in such networked machines might offer a way to mitigate the weaknesses in spatiotemporal resolution of each technology in the course of a single study (Ritter & Villiringer 2006).

The methodological challenges are perhaps even greater. Chief among them is the difficulty of designing *controlled* studies while still allowing for the spontaneity of interactive dialogue. The paradigm outlined above has to date focused on the recording and post-hoc coding of dialogic behavior as the main empirical approach. This is, of course, a perfectly legitimate choice, but it would be beneficial to be able to supplement this with more coder-independent measures. In addition, it seems possible to design interactions that constrain in advance the likely dialogue moves in a naturalistic way (utilizing, for instance, a restaurant "script" [Schank & Abelson 1975, 1977] by having participants interact as if ordering dinner). Such a possibility might allow the interaction to *feel* unconstrained while it still allows for a great deal of reproducibility across trials.

As noted above, the field is open to myriad possibilities. I hope here only to have showcased some of the promise offered by the framework I have been developing and to convince the reader that there is a fruitful path forward, should any be inclined to walk it.

Interlude 7 What Mindedness Is

Recent advances in theoretical cognitive science can be fruitfully characterized as part of the ongoing attempt to come to grips with the very idea of *Homo sapiens*—an entity at once biological and intelligent—and among the more striking developments has been the emergence of a philosophical anthropology that, contra Descartes and his thinking thing (*res cogitans*), instead puts doing at the center of human being (Anderson 2003; Chemero 2009; Wheeler 2005). This shift in our understanding of human nature is owed proximally to the work of Heidegger and Merleau-Ponty but also has clear precursors in such figures as William James and Hegel—and more specifically Marx and Marxist interpreters of Hegel such as Kojève. Naturally, Darwin must be considered as central as any philosopher, and many of the recent developments also echo the Aristotelian sense that being-at-work is the primary way of being anything at all.

The ongoing engagement with these ideas and figures has put (back) into play two competing visions of human being and its place in nature: one is a conventionally Modern scientific view, quintessentially represented by certain schools of neuroscience, that treats the body primarily as a reactive mechanism whose main purpose is to house and feed (sensations and sustenance to) the brain—or, as Chemero (2009, p. 177) puts it, "the problem is that neuroscientists tend to ignore the animals attached to the brains they are interested in studying." The other is more influenced by ecology and evolutionary biology and takes "human being" to be rooted in and by agency and practical activity. Although these two positions in fact suggest fairly complete visions of the nature of human life (not just biological but ethical, social, emotional, and political—something to which the inclusion of Marx and Aristotle in the list of intellectual forebears already alludes), I would like to focus here just on the competing views of mindedness itself: what it is, where it is, and how it might be possible.

As has been indicated already, the central tension concerns the relation of agency and practical activity to mindedness, and it therefore concerns precisely the role of the body in (and for) mind. Whereas in the traditional Cartesian view the body is understood as the source of afferent stimulus and the target of efferent output—as, therefore, neither more nor less than a set of sensory receptors and physical effectors, peripheral devices playing subordinate roles to the brain-as-CPU (where representation and calculation occur, on this view the central hallmarks of intelligence)—in the enactive view, the body and its activity play not a peripheral but a central role in the processes of mind. In fact, the activity of an organism in relation to its environment can be considered not just the most salient *expression* of mindedness (its location, if you will) but also in some sense its *constitution*.

Pretending for a moment that mindedness is composed of perception and cognition (it is not; not only does this leave out such important elements of mindedness as emotion—the recently emerged conception of emotion as an embodied cognitive system is an extremely important development in the overall program of cognitive science—but this distinction between perception and cognition is itself Cartesian in origin) will provide an opportunity for drawing these distinctions somewhat more finely. In considering the issue of perception, the Cartesian asks first how it is that the features and elements of the outside world can be captured and represented inside the organism. Note how this simple question, thus framed by the notions of inner and outer and centrally featuring the idea of representation, points us in the direction of the familiar and intractable anxieties of Modern philosophy: how to relate the accessible inner given to an outer reality, how to determine truth and representational accuracy, how to adjudicate the relation between the well-known, easily accessible self and the social world, how to see meaning in a mechanical world. The assumption is that the end product of perception is an inner world that fully reproduces—that in its elements and their relations is appropriately homologous to—the outer. This approach to perception accords perfectly well with a notion of mind that is contemplative (or, perhaps better, *reflective*, mirror-like) in character; such a mind—withdrawn, narcissistic, engaged only in its own productions—needs inner objects to behold, to alter. Thus, cognition, in this view, is the manipulation of, and the calculation over, such inner objects, a notion that points us in the direction of such harmful abstractions in ethical and social thought as the "rational calculation" of individual and collective utilities.

In contrast, the enactive view treats perception first and foremost as an organism's means for negotiating its environment. This suggests at least two things: first, that perception is a tool of exploration, and second, that it is intimately bound up with and primarily fitted to the service of action. Perception is not the passive reception of abstract qualities from the environment but is itself active, often highly selective, and goal-directed, designed to mine from the world all and only information of importance to the current (or ongoing background) purposes of the agent. The primary task of perception, then, is not the construction of inner objects but the detection of opportunities for action, a notion that recalls the familiar phenomenological claim that the perceptual field is always an action field, that the perceived world is always known in terms directly related to an agent's current behavioral options. To put it in terms of affordances, the perceived availability of things to certain interventions: the world is *seen* as a continuous series of invitations to action.

As with perception, so too with cognition. In the enactive view, cognition first emerged from, and is still organized around, mechanisms to control the behavior and augment the survival of particular agents in particular environments. Given the real-time demands of rapidly changing circumstances, one would expect systems to develop that, rather than rely strictly on "inner" manipulations of abstract constructions, instead utilize and exploit the various features of the environment to drive decision making. Thus, in the simplest sort of case, the frog's vision system is highly attuned to contrast and motion, and prey capture is a hard-wired response to the detection of small, dark, moving dots. The frog does not represent individual insects; it cannot distinguish between them or recognize one in particular. Nor does it model its whole environment and *decide* which objects are tasty. Indeed, the detection of a fly and the eating of that fly are not really separate events; eating is the sign of detection.

This illustrates two important principles of the enactive approach to cognition. First, the classification of an object or situation and the response to it are deeply related. We perceive not weight but throwability, not slope but climbability, and we do so because these are the properties of the world we care about. To *see as* is often to act *as if*, and, more generally, what one is sees is a function of what one does. Indeed, as we will see in more detail below, it is often untenable to speak of a separation of visual and motor systems; seeing is not a single process, the information from which is neutrally specified and centrally available, but is often highly task specialized, such that separate, encapsulated systems have evolved to support the distinct visual needs of different classes of action. Thus, for instance, the frog has distinct visuomotor systems for orienting itself to its prey and for capturing it. These systems are unified and coherent *not* due to the production of some shared representation of the world, an integrated motion picture forever showing in the Cartesian theater, but simply in virtue of being in the same body, experiencing and acting on the same environment.

Second, the organism exploits facts about its environment to turn simple mechanisms to somewhat more sophisticated uses: black dot detection *is*, in the frog's environment, fly detection. Were the world different, the environment different, or the frog's tastes more discriminating, the mechanism would not work. Put differently: typical behavior-guiding mechanisms are built on—and in some cases out of—environmental features. And, indeed, many human perception-action mechanisms are like this, based on satisficing machinery that is good enough for the usual conditions but easily foiled by a change in circumstance.

It is of course open for someone to object that the frog example is misleading—the frog is not thinking but only reacting; this is not an example of cognition but merely of instinct. Whatever the force of such an argument from within a Cartesian world view, it makes little sense as a specific objection to the enactive view. For the frog doesn't eat indiscriminately; it exhibits very specific and appropriate responses to differential aspects of the environment. It eats bugs and avoids predators. It shrewdly negotiates its environment in accord with its limited set of needs and goals. For the enactive view this *is* intelligence—and note the conflation, in the example, of the perceptual, cognitive, and performative elements of the phenomenon in question; this is typical of the enactive approach but looks sloppy to the Cartesian. Indeed, given the Cartesian notion that the cognitive and perceptual can and should be distinguished, one suspects that were the frog also endowed with a limited vocabulary—along the lines of Wittgenstein's builders, so it said "fly!" when it saw a fly and "hawk!" just before diving under cover of water to avoid the bird overhead, and perhaps whistled just so when it saw an attractive potential mate—that its intelligence would be more widely recognized and praised. Herein lies a simple prejudice, which the enactive view does not share.

Besides which, there is evidence that a significant part of the human's visual system is not entirely unlike the frog's (which is the sort of fact one comes to expect when looking at these matters from an evolutionary perspective; solutions to common adaptive problems are often common across species and conserved along lines of descent). For the human visual system is likewise split into (at least) two separate pathways: a "dorsal stream" (also known as the "where" pathway) that tracks the location, size, and shape of objects and a "ventral stream" (the "what" pathway) that facilitates classifying and identifying objects. The dorsal stream is a perception-action system optimized for calculating and directing motor responses aimed at an object in virtue of its location, orientation, and spatial extent. This system guides such things as reaching and grasping and the orientation of sense organs for optimal perception and perceptual tracking. Thus, the natural way to characterize what one knows in virtue of dorsal stream operation is in terms of egocentric spatial coordinates: where something is in relation to one's self and what might be done to get the self-object relation into a preferred state. One might say that the dorsal stream "sees" objects in an egocentric action field; the object is thereby experienced in these terms. Like the dorsal stream, the ventral stream is a perception-action system but in this case optimized for making classifications, generating descriptions, and for other more traditionally "cognitive" activities.

As is often the case in cognitive science, some of the most striking illustrations of this separation and its importance come from studies of individuals with specific neural deficits. Thus, for instance, patient DF, who has widespread lesions in the ventral stream caused by carbon monoxide poisoning, although unable to identify objects by sight (she can neither draw nor describe them), can nevertheless reach for these objects with fluent and appropriately sized and oriented grips. Similarly, although she is unable to perceive and describe the orientation of a letter slot, she is easily able to post a letter through it. In contrast, optic ataxics who have dorsal stream lesions are able to see and describe visual scenes without trouble (for instance, the objects on a table or the letter-slot in a wall) but are unable to fluently grasp those objects or post a letter through the slot despite the apparent clarity of their visual experience.

The same kind of disconnect between conscious perception and perceptually guided action can be seen for *everyone* in the case of a clever variation of the "Titchener circles" illusion (figure 7a.1). In this experiment subjects were presented with poker chips arranged like the discs in the

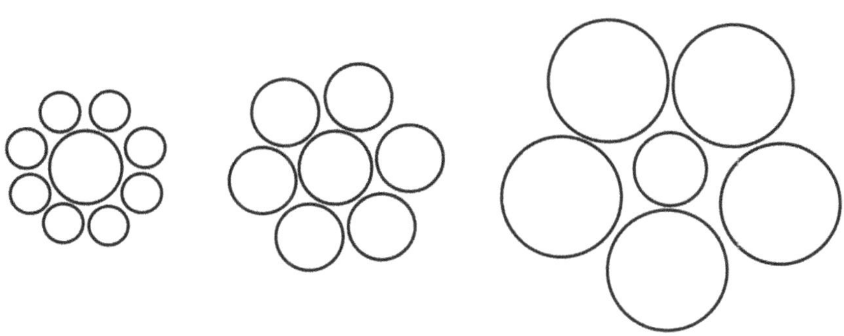

Figure 7a.1
The Titchener circles illusion. The three center circles are the same size.

Titchener circles diagram and were told to pick up the center chip on the left if the center chips appeared to be of the same size, and on the right if they appeared to be different. Although the *choice* of chip showed that the participants were subject to the relevant illusion, in reaching for the chip, they used a grip perfectly suited to its *actual* and not its *perceived* size. Subjective experience to the contrary, it appears that much of our behavior is (still) governed by specific, individual, and sometimes even unconscious visuomotor systems. When it comes to actually acting in the world, we are perhaps more frog-like than we care to admit.

Still, one is certainly entitled to doubt whether this dorsally mediated orientation to the environment can account for the complex tasks routinely faced by the typical parent, who shuttles the children to their various activities at the right time, in the right order, meanwhile figuring out ways to work in time for laundry, dry cleaning, and grocery shopping. Surely he cannot simply let the world unconsciously guide him and expect to accomplish his daily tasks; rather, he must think about, and plan in reference to, the way the world *will be*. Does this not require world modeling, concepts, and representations—cognition as traditionally understood? Possibly. But whenever in our (pre-)history symbols and representations emerged as a cognitive tool, they did so in a context—in an environment—already dominated by effective solutions to perceptual-behavioral coordination. It seems unlikely that new solutions radically broke with the old. Instead, we should ask what existing resources might have been exapted, redeployed, recycled, or otherwise adapted to these new purposes.

From this perspective what emerges as the critical (and fabulously interesting) question is: What are the relations between the lower-level, older, specialized sensory-motor systems and the structure, elements, and rules of operation of any more general, highly flexible, symbolic computational system we may possess? My bet, for what it is worth, is that these are significantly intertwined, with bidirectional feedback and cooperation—that, for instance, some conceptual contents can be traced to specific sensory-motor systems, and some sensory-motor systems have been adapted to utilize some of the resources of (or at least be responsive to) more general conceptual systems. At the very least one must concede that whatever representations emerged in such an environment are likely to be themselves action oriented, built on faculties that govern our ability to move and act in a dynamic environment. And so the moral remains the same: the first work of cognition is to provide for action; as the organism's possibilities for action become more sophisticated, so, too, must the structures that support that activity. But the very nature of perception, of cognition—of mindedness—is not fundamentally transformed by this process.

This returns us to a point we left behind earlier: it isn't just that perception and cognition are action oriented but that they are *interactive*, exploiting properties of their environment to guide and simplify cognitive tasks. Thus, to understand the character of (advanced) cognition one needs to understand not just the basic faculties that support and constrain it but also the nature of the environment within which an organism exercises those faculties. Put somewhat differently, thinking and perceiving are activities of embodied agents in particular circumstances. These processes crucially rely on neural, corporeal, and environmental resources and are thus not easily localized "inside" or "outside" the agent. Because this is true of both the human *and* frog, the massively greater sophistication of human intelligence needs to be explained in a way that does justice *both* to the enhanced (behavioral, categorical, representational, linguistic) faculties of human brains *and* to the richer resources of the human environment (not to mention the structure and behavioral repertoire of the human body). For human-level cognition is marked by the use of and interaction with the environment in myriad ways: using a pencil and paper to store intermediate results in long division or large-number multiplication; arranging a hand of cards or Scrabble tiles to better see relevant patterns, matches, or potential words; rotating puzzle pieces to better discern their fit; making grocery lists, labels, signs, encyclopedias, and otherwise storing information in the world to be consulted later; and using management structures and the constraints imposed by individual roles to accomplish complex tasks such as ship navigation or building construction.

The overall picture that this suggests is of an intelligence that lies less in the individual brain and more in the dynamic interaction of agents with and within the wider world. Mindedness emerges as—*is*—the activity of making the world a home, one that reflects the nature of its occupant. Its primary sign is a kind of adaptive integration with one's environment, including especially the social and cultural worlds that are so important to human cognition. Note this is the precise opposite of mind in the Cartesian view, which shows itself fully only in disengagement and alienation. Thus, in my judgment, discovering and detailing the particular physical characteristics and environmental integrations that shape and support the various aspects of mindedness are among the central projects of cognitive science. What we are (or should be) doing is attempting to understand how the activity of human mindedness emerges from—is related to, shaped by, and influences—the structures and characteristics of human biology and society. This physical grounding project encompasses enterprises ranging from specifying the particular influence of physical or neurological structures on the contents of experience; to modeling the principles guiding the reuse of existing neural, behavioral, and environmental resources for new purposes over evolutionary and developmental time; to understanding the simple interactions with the physical environment that aid in calculation, memory, and decision making (some of which have been mentioned already); to grappling with how we give abstract, linguistic, and mathematical symbols concrete meaning, something that involves supporting integrations not just with the physical environment but also, and perhaps especially, with the social world; all the way to the extremely difficult question of how to understand the very formation, in its social and physical context, of subjectivity and self-hood.

Enactive cognitive science therefore sits at the junction of biology, psychology, philosophy, and the various humanistic sciences, including anthropology, sociology, and economics, with the hope that a vision of mindedness that insists from the outset on *staying* at this critical intersection can help to unify—or at least make consistent—the myriad visions of human being expressed in these various fields.

8 A Functionalist Neuroscience for the Twenty-First Century

In the *Principles of Psychology* (1890), William James advocated for a psychology rooted in the study of the brain and shaped by an understanding of evolution. Modern psychologists, James wrote, must be "cerebralists" (James 1950, p. 5). And yet psychology could not be *only* about the brain.

> [I]t will be safe to lay down the general law that *no mental modification ever occurs which is not accompanied or followed by a bodily change.* The ideas and feelings, e.g., which these present printed characters excite in the reader's mind not only occasion movements of his eyes and nascent movements of articulation in him, but will some day make him speak, or take sides in a discussion, or give advice, or choose a book to read, differently from what would have been the case had they never impressed his retina. Our psychology must therefore take account not only of the conditions antecedent to mental states, but of their resultant consequences as well. (James 1950, p. 5)

To study the brain well one must also study the body and behavior. This claim follows naturally from James's evolutionary perspective. Following Herbert Spencer, James thought the purpose of the mind is to adapt us to the environment. As Spencer put it,

> The fundamental condition of vitality, is, that the internal order shall be continually adjusted to the external order. If the internal order is altogether unrelated to the external order, there can be no adaptation between the actions going on in the organism and those going on in its environment: and life becomes impossible. (Spencer 1855, §173, p. 506)

Such adaptation occurs only via action that adjusts the body so that it fits in with the world, or, as has been emphasized in recent thinking, adjusts the world to fit the body (Odling-Smee, Laland, & Geldman 2005). Thus, for James, the subject matter of psychology has to be *every* aspect of our mental life, understood in the context in which it adapts us to the environment. It is this feature of Jamesian psychology that led him and his followers to be condescendingly called *functionalists* (Titchener 1898) because they believed that the way to do psychology is to understand thoughts, habits, emotions, and so on in terms of their function, that is, in terms of how they contribute to the overall fit of the animal to its environment. As should be clear by now, this volume has embraced this perspective and attempted to spell out a functionalist psychology and neuroscience for the twenty-first century. In this I have explicitly followed the example of the great Spanish neuroanatomist Santiago Ramón y Cajal. In these concluding remarks I would like to remind the reader of Ramón y Cajal's contributions, both to place the current work in historical context and also to help illuminate the road not (yet) taken, in the hope that many will now choose to walk this less-traveled path.

8.1 Ramón y Cajal's Functionalist Neuroscience

Like James, Ramón y Cajal was importantly influenced by Spencer's evolutionary approach to understanding brain and behavior. Spencer argued that one had to approach the investigation of life and mind with some fundamental principles in mind. First was the primacy of adaptation, the continual adjustment of inner to outer conditions. Second was a principle of growth and development, whereby both an organism's repertoire of responses as well as the biological structures supporting them increase in number, diversity, and complexity. And finally, there is the principle of continuity, which states that new developments emerge from, build on, and (partly) preserve what came before.

Organisms evolve and develop by becoming at one and the same time more differentiated and more integrated or coordinated in both structure and behavior. It is from these parallel developments (and not from either acting alone) that the increasing complexity of organisms emerges over time.

> In the progress from an eye that appreciates only the difference between light and darkness, to one which appreciates degrees of difference between them, and afterwards to one which appreciates differences of colour and degrees of colour—in the progress from the power of distinguishing a few strongly contrasted smells or tastes, to the power of distinguishing an infinite variety of slightly contrasted smells or tastes … in all those cases which present merely a greater ability to discriminate between varieties of the same simple phenomenon; there is increase in the speciality of the correspondence without increase in its complexity. … But where the stimulus responded to consists, not of a single sensation but of several; or where

the response is not one action but a group of actions; the increase in speciality of correspondence results from an increase in its complexity. (Spencer 1855, §154, pp. 445–446.)

Such development is made possible in part by the way that existing resources are both modified and preserved over time. This principle of continuity implies not just that organisms can be arrayed on a biological and psychological continuum, with many differences in degree but few sharp fundamental discontinuities to be found between the mental powers of "higher" and "lower" organisms, but also that within each organism higher mental faculties develop from and rest on the foundations of the lower. As Robert Wozniak comments:

> The implications of these evolutionary conceptions ... are clear. The brain is the most highly developed physical system we know, and the cortex is the most developed level of the brain. As such, it must be heterogeneous, differentiated, and complex. Furthermore, if the cortex is a continuous development from subcortical structures, the sensory-motor principles that govern subcortical localization must hold in the cortex as well. Finally, if higher mental processes are the end product of a continuous process of development from the simplest irritation through reflexes and instincts, there is no justification for drawing a sharp distinction between mind and body. The mind/body dichotomy that for two centuries had supported the notion that the cerebrum, functioning as the seat of higher mental processes, must function according to principles radically different from those descriptive of subcerebral nervous function, had to be abandoned. (Wozniak 1992)

Ramón y Cajal took these principles to heart, and clearly saw them reflected in the neural structures that he was so adept at describing. It is perhaps easiest to start with his summary of three trends in the evolution of neural organization that he observed. The first was a "proliferation of neurons and neuronal processes that ... increased the complexity of relationships between various tissues and organs" (Ramón y Cajal 1995, p. 11). Such proliferation was necessitated by the increase in the number and complexity of *other* cells in the organism that is observed over evolutionary time. As Ramón y Cajal points out, an increase in the size and complexity of an organism without an attendant increase in the number of neural cells it possesses would precipitate a decrease in sensory acuity and presumably in agility as well, given the increase in the ratio between body parts and the sensory and motor neurons that would serve them. This already points in the direction of the primary sensorimotor function of the nervous system, but Ramón y Cajal ties neural development especially closely to the motor system.

Once it has appeared, the nervous system comes to direct the muscular system through a series of actions and reactions. Indeed, because of the concurrent specializations that occur in animals, both the nervous system and the muscular system not only appear together, but are also functionally interdependent. (Ramón y Cajal 1995, p. 5)

The second evolutionary trend detailed by Ramón y Cajal was "an adaptive differentiation of neuronal morphology and fine structure," and the third was "a progressive unification of the nervous system, a concentration of its elements into neural masses," that is, the emergence of central ganglia including the brain and spinal cord (Ramón y Cajal 1995, p. 12). The effect of this centralization is crucial to function:

Motor neurons that before were peripheral and isolated from one another are now juxtaposed in a single central nucleus; they are transversely integrated, to use Herbert Spencer's phrase. … [T]he sensory neurons can excite all of the aggregated motor neurons, and only a few additional expansions are necessary for the sensory arborizations to expand their spheres of motor influence. (Ramón y Cajal 1995, p. 14)

In what must appear a paradox to those accustomed to understanding the brain in terms of the localization of psychological faculties, the anatomic consolidation that Ramón y Cajal describes in fact permits function to be *less* localized, even as the supporting tissues become more central. This arrangement, however, makes perfect sense when one expects the brain to be not a collection of organs with distinct local functions but rather a structure whose purpose is to establish the functional relationships between cells that will best perform the job of coordinating the organism's interaction with its environment.

Coordination, control, and complexity are achieved via the emergence of two new classes of neural cells in addition to sensory and motor neurons: association neurons and psychomotor neurons. Association neurons mediate the link between sensory and motor cells, allowing the emergence of complex responses to sensory stimuli.

With the association neuron, multicellular organisms become true animals. Sensory stimuli, even if localized to one point on the integument, are no longer isolated. … The association pathways that interrelate various muscle fields and the areas of the integument with which they are connected are by no means randomly distributed. Evolution and adaptation have determined their organization, and the precision of their distribution is such that each stimulus received by a sensory cell causes the animal to respond with what Exner has called a *combination of movements*, that is to say, with a complex movement that is appropriately coordinated for the animal's self-preservation and procurement of nutritional requirements. (Ramón y Cajal 1995, pp. 5, 7)

Psychomotor neurons were understood by Ramón y Cajal to be exceptionally powerful and centralized association neurons, able to exert their influence over an extraordinarily broad range of circumstances and behaviors. Psychomotor neurons are able to modulate behavior based not just on external stimuli but also on internal conditions, and not just on current stimulation but also past experience.

In the evolution of the nervous system, this element, which underlies the still largely unexplored world of psychological (psychic) phenomena, is a more recent addition than the association neuron. It too is interpolated between sensory and motor neurons, but at a distance, and is generally located in one particular ganglion: the cerebral ganglion of invertebrates and the cerebral cortex of vertebrates. … It becomes so dominant in certain vertebrates (humans), and leaves an imprint so striking, that one can reasonably divide animal evolution into distinct neural eras. Unicellular animals and sponges would constitute the era of irritability; coelenterates, the era of two basic neurons; higher invertebrates, the era of association neurons; and vertebrates, and particularly humans, the era of psychomotor neurons. Of course, the advances of preceding eras are conserved and highly refined in each era … The empire of the psychomotor neuron, together with the various ganglia distributed throughout the body, constitute the organism's newest and most useful weapons in the struggle for survival. … (Ramón y Cajal 1995, p. 8)

Ramón y Cajal's choice of metaphor is striking, for in his view the psychomotor neuron truly does rule over vast swaths of behavior. Two things especially are important to note: the first is that this regulatory capacity is made anatomically possible *only* because of the centralization of neural structures, which permits even single cells to affect a wide range of inputs and outputs; and second is that the power of psychomotor neurons reflects not any *intrinsic* properties of these cells but is rather a product of the functional *relationships* that define them.

Wherein lies the superiority—the supremacy—of the cephalic ganglion? In our view it derives from the inherent superiority of the functional relationships established between the external world and this ganglion. Let us explain. The abdominal ganglia are linked to the sensory nerve cells that relay simple, rather poorly defined and crude tactile and thermal sensations from the integument. The cephalic ganglion, in contrast, is connected to the very specialized neurons that subserve vision, hearing and smell, and this receives preorganized patterns (including more complex temporal and spatial information) that provide the most accurate representations of the external world. This difference in type of connections is mostly responsible for the preeminence of the cerebral ganglion. And the eye and the ear are the major artisans of this preeminence. In essence these organs are *computational devices*, to use Max Nordau's pleasing expression, that select in a very specific way from the middle range of the immensely broad energy spectrum those wavelengths for which they are adapted. (Ramón y Cajal 1995, p. 8)

Interestingly, for Ramón y Cajal the precision and accuracy with which the sense organs represent the external world obviate the need for central structures to do so.

[T]he cerebral cortex of vertebrates, and the cerebral ganglion of invertebrates, do not need to create images; complete images are formed by the sense organs and supplied instead to the cerebral cortex or cerebral ganglion in highly refined ways that actually reflect the intensity and all the subtle nuances inherent in the excitatory stimuli. In the final analysis, the marvelous structural organization of the eye and ear is the primary reason for the dominant position of the cerebral cortex. (Ramón y Cajal 1995, pp. 8–9)

There is much that is striking in Ramón y Cajal's perspective. First is his focus not on intrinsic function or localized faculties in the regions of the brain but rather on the establishment of functional relationships. Indeed, Ramón y Cajal takes this perspective so seriously that he is led to predict the outcome of experiments—different in detail but identical in intent—first performed over 80 years after the time of this writing (e.g., Sur et al. 1988):

Insights provided by the evolution of central neural centers have now so convinced us of the preeminent role played by the nature of their relationship to the external world that we are tempted to propose the following: If by some capricious and seemingly impossible developmental anomaly the optic nerve should end in the spinal cord, visual sensations would be elaborated in the region occupied by motor neurons! (Ramón y Cajal 1995, p. 9)

It is worth emphasizing an important consequence of this focus on neural relationships: differences in neural morphology should not be taken to indicate differences in intrinsic function but rather to reflect abilities to establish different sorts of functional connections or coordination. It is only for this reason that anatomic, morphological differentiation can effect increasing functional—which is to say behavioral—complexity.

Second, we see a recognition of the importance of peripheral structures to cognition not just as input channels but as organs of cognition in their own right. Indeed, it would not be inappropriate to see in Ramón y Cajal's insight that sense organs play a role in selecting and structuring stimuli a precursor to current recognition of the importance of bodily activity and morphology to cognitive processes (Anderson 2003; Barrett 2011; Chemero 2009), including such recently emerging notions as morphological computation (Hochner 2012; Paul, Lungarella, & Iida 2006).

Last but certainly not least is Ramón y Cajal's fundamental orientation toward action.

> What utilitarian goal has nature (which never seems to act in vain) pursued in forcing nervous system differentiation to these lengths? ... [T]he refinement and enhancement of reflex activity, which protects the life of both the individual and the species. ... Such reflexes constitute the fundamental repository of neural adaptations that provide an animal with the necessities of life. ... To the hierarchy of increasingly more complex reflexes—irritability in protozoa, simple reflexes in lower vertebrates, and more complex reflexes in higher invertebrates and vertebrates—one must add the all-powerful psychic reflex of vertebrates, and especially the higher vertebrates. In the latter ... neural and nonneural structures are not simply under the influence of external stimuli; they are also subject to internal stimuli arising from control centers within the organism itself. (Ramón y Cajal 1995, p. 16)

For Ramón y Cajal, the *telos* of cognition is action, and for this reason even "complex and deferred responses ... are true reflexes" (Ramón y Cajal 1995, p. 17). Because in our time we tend to reserve the term "reflex" for those simple, stereotyped (and generally spinally mediated) motor responses to strong, simple stimuli, it would be easy to dismiss Ramón y Cajal's view here as not reflecting the true complexity of the brain's function. In point of fact, given his insistence on a "hierarchy" of reflexes and the more general point that evolution tends to preserve, adapt, and enhance existing structure and function, what he appears to have in mind is a functional arrangement not unlike the subsumption architecture proposed by Rodney Brooks (1991), whereby simpler, specific reflex responses are modulated or suppressed by higher "reflexes" that reflect more general sensory-motor coordinations. It is in any case clear that Ramón y Cajal imagined overall brain function was achieved via the establishment of a hierarchical continuum of sensorimotor control processes, all aimed at "conferring advantage in the struggle for survival" (Ramón y Cajal 1995, p. 17).

8.2 Embodied Cognition and the Brain

Despite the growing interest in and information about the details of neural processing and the development of dynamic models of neural processing that emphasize the establishment of functional relationships among perception, memory, and behavior, much of the cognitive neuroscience of the last two decades has still been largely guided by cognitivist principles and the symbol system model for the brain (see Miller 2003; Posner et al. 1988 for discussion). As we have seen throughout this volume, computational cognitive science treats perceiving as picturing and thinking as the transformation of the pictures constructed out of the stimulations that the senses provide. Such an approach encourages scientists to abstract away from the details of action and perception and focus instead on "central" processes. After all, if cognition is a computational process, it is natural to treat the body as merely a peripheral conduit such as a keyboard or Internet connection that merely provides information about the environment to the central processor, which does the real cognitive work. Indeed, I have argued here that it is *because* the cognitive neurosciences have been guided by these assumptions that work in the field is dominated by the search for the nature of the "representations" and "computational operations" supported by the neural activity observed in particular regions of the brain. The perspective also means that evolution can be mostly ignored or at best studied separately. Of course the brain has an evolutionary and developmental history, but that history is inessential to understanding the basic nature of its functions, which will be primarily dictated, in this model, by informational and computational constraints. This perspective might account for the antipathy that many of the founders of computational cognitive science have shown to evolution by natural selection. Infamously, Chomsky has argued that the human language faculty could not have evolved by natural selection (Chomsky 1988). More recently, Jerry Fodor has gone from arguing that evolution by natural selection cannot explain how thoughts have meaning (Fodor 1990) to arguing that evolution by natural selection cannot explain the nature of cognition (Fodor 2000) to arguing that evolution by natural selection is simply ill conceived (Fodor 2007).

Because of the prevalence of these attitudes in the cognitive sciences generally, and in the cognitive neurosciences particularly, a truly functionalist cognitive neuroscience has yet to fully emerge. I certainly hope that the current volume can help point the way, and so here I review some of the main principles and tenets that I am arguing we need to embrace. In light of the discussion of James and Ramón y Cajal in section 8.1, there would appear to be three principles or commitments that could be said to characterize a functionalist neuroscience. A functionalist believes that (1) the functional architecture of the brain has been established by natural selection and, more particularly, via a process marked by both functional differentiation and continuity; (2) our complex and diverse behavioral repertoire is supported primarily by the brain's ability to dynamically establish multiple different functional coalitions, coordinating both neural partnerships and external resources; and (3) the brain is fundamentally action oriented, with its primary purpose to coordinate the organism's ongoing adjustments to external circumstances.

Consider the first principle, that the functional architecture of the brain should reflect an evolutionary history marked by both functional differentiation and also the incorporation and reuse of existing structures for new purposes. If the last 100 years of neuroscience have established anything, it is that the various regions of the brain are functionally differentiated. But although for most of that time this fact has been taken to indicate that each region of the brain is highly functionally specialized and implements a single cognitive operation (e.g., Kanwisher 2010; Posner et al. 1988), much recent work, reviewed in detail in chapter 4, has turned instead to characterizing the functional differentiation observed in the brain in a multidimensional manner that highlights the functional differences between regions while recognizing that in point of fact each region of the brain appears to be active in multiple diverse circumstances (Anderson, Kinnison, & Pessoa 2013; Barrett & Satpute 2013; Hanson & Schmidt 2011; Lindquist & Barrett 2012; Poldrack et al. 2009).

Indeed, it is at this point well established that individual regions of the brain support many different tasks across multiple task categories, as would be predicted by the principle of continuity. This work was reviewed throughout the present volume and in chapter 1 in particular. That this apparent functional diversity is a reflection of the evolutionary history of the brain is supported by some interesting features of the *pattern* of use and reuse of individual regions of the brain across multiple circumstances. For instance, it appears that, *ceteris paribus*, older regions of the brain tend to be used in more tasks—presumably because they have been around longer and have thus had more opportunity to be incorporated into multiple functional coalitions (Anderson 2007a). In addition, more recently emerging cognitive functions, such as language, appear to be supported by more and more widely scattered brain regions than are evolutionarily older functions such as vision and attention (Anderson 2008a, 2010b; Anderson & Penner-Wilger 2013). Again, this makes sense in light of both differentiation and continuity, for the later a given cognitive process or behavioral competence emerges, the greater the number and diversity of neural structures that will be available to support the new competence, and there is little reason to believe the useful structures will be near one another in the brain.

This brings us to the second principle, that achieving behavioral competence is a matter of establishing the right functional coalitions to support the tasks in question. Although there is little work that specifically investigates the neural supports for the incorporation of external resources into cognitive processing (but see, e.g., Iriki 2005; Iriki & Sakura 2008), there is certainly evidence that *within* the brain cognitive function is a matter of assembling the right coalition of neural partners. Recall, for instance, the demonstration in chapter 1 that, although many of the same regions of the brain were used and reused in multiple tasks across the domains, the regions cooperated with one another in different patterns in each task domain (Anderson & Penner-Wilger 2013).

Experimental work investigating temporal coherence in the brain also points in the direction of large-scale modulation of neural partnerships in support of cognitive function. For instance, changes in the functional connectivity between brain regions (local and long-distance) appear to play a role in sensory binding, in the modulation of attention, and in other cognitive functions as well (Steinmetz et al. 2000; Varela et al. 2001). For instance, Friston (1997) demonstrated that whether a given region of inferotemporal cortex was face selective depended on the level of activity in posterior parietal cortex; and McIntosh et al. (1994) showed that during face processing superior parietal cortex cooperated strongly with a region of inferotemporal cortex; but during attention tasks that same region of parietal cortex cooperated more strongly with a prefrontal area. Similar patterns of changing functional connectivity are observed over developmental time, which suggests that acquiring new skills involves changes to both local and long-distance functional partnerships (Fair et al. 2009; Supekar et al. 2009). In chapter 2 I reviewed a large and growing amount of evidence pointing to this conclusion and furthermore sketched an account of the mechanisms that might be responsible for this changing connectivity. It seems reasonable to predict that such results will continue to emerge, given the increasing interest in network-oriented approaches to the brain (Sporns 2011).

Finally, we consider the third principle, that the brain should be understood as an action-oriented system. It is here, perhaps, that there is the most work left to be done to establish a functionalist neuroscience because the symbol systems model for the brain still dominates the cognitive neurosciences, leading most researchers to interpret the neural activity they observe during their experiments as reflecting information-processing, rather than action coordination, operations. It is nevertheless worth highlighting one recent line of work in the neural bases of decision making to illustrate what a more action-oriented approach to the brain might look like. This work was reviewed in chapters 5 and 6, but it is worth returning to the idea again here.

Prevailing models of decision making generally accept the cognitivist notion that perception involves building an objectively specified world model, which can then be used for multiple purposes, including the generation and selection of action plans and the specification of how the motor system will enact them. But what I have been arguing here is that from the functionalist, embodied perspective, perception should be thought of as the assessment of the values of salient organism-environment relationships and the detection of opportunities for changing those values through action. Given this model, deciding what to do is largely a matter of choosing which perception-action path—which perceived affordances—to follow. As Paul Cisek writes of his particular affordance-based model of decision making:

The proposal made here is that *the process of action selection and specification occur simultaneously* and continue even during overt performance of movements. That is, sensory information arriving from the world is continuously used to specify currently available potential actions. ... From this perspective, behaviour is viewed as a constant competition between internal representations of the potential actions which Gibson (1979) termed "affordances." Hence, the framework presented here is called the "affordance competition hypothesis." (Cisek 2007, p. 1586)

There are two central tenets to his hypothesis, both of which fit in nicely with the functionalist approach to the brain being described here. First, given that the brain evolved to support interactive behavior, and perception is the detection of opportunities for action, the processes of selecting and specifying actions are in fact continuous, ongoing parts of simply perceiving and acting in the world. Second, these processes are not neutrally segregated; action selection and specification occur in the very same regions of the brain as part of the ongoing process of affordance competition.

The model reflects not just the notion of cognitive continuity but also the potential power of action-oriented representations in interpreting neuroimaging data. If the brain is an action-oriented system, then whatever representations it generates and utilizes should reflect this functional inheritance. Cisek argues that the decision-making literature points in precisely this direction:

The affordance competition hypothesis ... differs in several important ways from the cognitive neuroscience frameworks within which models of decision making are usually developed. Importantly, it lacks the traditional emphasis on explicit representations which capture knowledge about the world. For example, the activity in the dorsal stream and the fronto-parietal system is not proposed to encode a representation of objects in space, or a representation of motor plans, or cognitive variables such as expected value. Instead, it implements a particular, functionally motivated mixture of all these variables. From a traditional perspective, such activity appears surprising because it does not have any of the expected properties of a sensory, cognitive or motor representation. It does not capture knowledge about the world in the explicit descriptive sense expected from cognitive theories and has proven difficult to interpret from that perspective. … However, from the perspective of affordance competition, mixtures of sensory information with motor plans and cognitive biases make perfect sense. Their functional role is not to describe the world, but to mediate adaptive interaction with the world. (Cisek 2007, p. 1594)

For a functionalist neuroscience to fully emerge, this action-oriented perspective must be taken up into every corner of the field. Given the apparent challenges facing the status quo and the importance of integrating the study of the mind and brain more fully with the ecological and evolutionary biology, this is a possibility we should both welcome and encourage.

8.3 The Road from Here

To see where we are going it will be helpful first to review where we have been. Over the past seven chapters I have argued that the collective evidence from psychology, neuroscience, ecology, and evolutionary biology necessitates a fairly radical reconsideration of how we should study the brain. Chapter 1, on neural reuse, highlighted the evidence that regions of the brain are functionally differentiated from one another and functionally diverse in and of themselves. Overall function is achieved by establishing dynamic functional partnerships between these regions that change with changing task demands, as modulated by genetic and electrochemical factors at multiple levels of organization. Chapter 2 developed the interactive differentiation and search framework for understanding the development of these partnerships and pointed to some of the biological mechanisms that might be responsible for this particular—and largely neglected—form of neural plasticity. In both chapters 1 and 2 I have argued that because the brain does *not* appear to be composed of nearly decomposable subsystems, we must rethink the mechanisms by and through which the brain adapts over evolutionary time. The functional architecture of the brain is a product of both evolutionary and developmental processes, and more attention needs to be paid to the ways in which these processes interact (Anderson & Finlay 2014). In particular, my suggestion is that evolutionary pressures primarily target the developmental processes that work to ensure a diverse range of initial cortical biases, which are relatively species-typical and heritable both because of the stereotyped projections of major neuronal fiber tracts in early development and because of the stable statistical properties of the environment and thus of the initial input to the system.

This antimodularist thread was further developed in chapter 3, where I introduced the notion of TALoNS, the transiently assembled local neural subsystems in which function is temporary, repeatable, and determined by the interaction of bottom-up and top-down influences. Because TALoNS are put to multiple uses, and local neural circuits form different TALoNS at different times, I argued in chapter 4 that the current approach to function-structure mapping, which assumes that single simple functions can be specified for each small part of the brain, needs to be abandoned in favor of defining local—and even network—function in terms of a multidimensional, dispositional vector. One important effect of adopting this approach is to focus scientific attention on discovering and defining the psychological dimensions—the NRP factors—that are most appropriate for simultaneously illuminating both brain and behavior.

The search for the most useful and explanatory NRP factors will not succeed if we do not take a close, evolutionarily informed look at the nature of the organisms whose brains we study. Beginning in chapter 5, I argued that—following James, Spencer, Ramón y Cajal, and others—we should understand brains as action-control systems and that this means recentering our model of cognition around the notion of affordances, moving away from the reconstructive, representation-centric model of thinking offered by prevailing computational approaches to the mind. It also means recognizing the importance to cognition of continuous iterated interactions with the world; virtually all cognitive problems are solved piecemeal, and each cognitive cycle can entail making changes in the world that serve to cause changes in the mind. As we saw in chapter 6, and again in section 8.2 above, this overall perspective suggests we should model the brain as a dynamic action-control system in which biased affordance competition plays a major functional role, even in the case of such "higher" cognitive functions as mathematics and language. Even equations have affordances and have been (unwittingly) designed to "fit" the native capacities of our action-oriented brains.

All of this work puts us in a position to see, as I argued in chapter 7, that far from being the rock on which evolutionary and embodied approaches to the mind will eventually founder, natural language is in fact a natural showcase of the power of our interactive brains to leverage disparate resources in the service of cognitive ends. Language works with, and because of, our native capacities for iterative interaction with the world and one another by serving both as a locus of affordances and as one source of biasing inputs capable of inducing various forms of phenotypic reorganization in speaker and listener alike.

In a nutshell, then, what I have argued over the last seven chapters can be summarized as follows: the Modern, modular, cognitivist assumptions that have guided research during most of the last 50 years of cognitive neuroscience have not been borne out by the data this research produced. It is time to see our science with fresh eyes. To this end, I have attempted to provide an evolutionarily informed framework that captures and does empirical justice to some fundamental features of the functional organization of the brain and of our embeddedness in our environments: (1) individual regions of the brain are functionally diverse; (2) individual regions of the brain are functionally differentiated; (3) there is frequent functional overlap between different brain networks; (4) the brain is fundamentally action oriented, and specializes in managing the organism's interactions with the world; and (5) the brain achieves its functions by assembling the right functional coalitions between both neural and extraneural partners, including supporting interaction with external artifacts—including symbolic ones—for cognitive ends.

What might neuroscience look like if this framework indeed proves useful? In an appendix to this volume I detail a number of open problems for the cognitive neurosciences as they appear from this perspective and suggest some empirical approaches to addressing them. Here I simply paint with broad strokes a picture of neuroscience after phrenology.

1. We should represent the functional activity of individual regions of the brain in a multidimensional manner that captures their underlying functional and dispositional properties without committing us to the notion that the observed neural responses reflect the engagement of a single unified function.
2. We should expect not just local but distributed contributions to overall function, determined by the interactions between top-down and bottom-up, feed-forward and feedback processes. Structurally, we need to attend especially to the interactions between regions, how these change, and how they map onto changes in behavior. Developmentally, this means working hard to establish the mechanisms whereby potential functional partners are discovered and eventual functional partnerships are established and maintained.
3. We should deeply rethink the vocabulary of cognition, ideally giving the brain its voice in this process. In trying to discern what in the world the brain cares about, we should recognize that it is fundamentally an action-control system, specializing in managing the values of salient organism-environment relationships. Thus, many of the properties to which the brain is attuned are likely to be action relevant and relational; throwability and climbability will likely be more important to the brain than weight and slope.

4. We should recognize that cognition does not take place in the brain alone and that we think with and through artifacts and one another. Although it will sometimes be necessary to experimentally bracket off the natural and social worlds to focus on the brain in isolation, we need at the same time to develop experimental paradigms that *include* robust social and environmental interactions and take measurements that capture the details of these interactions among brain, body, and world.
5. In all of this, we should embrace the empirical tools offered us by machine learning, graph theory, independent component analysis, linear algebra, dynamic systems theory, and the like. Focusing on local, linear correlations between brain activity and stimuli will never be by itself sufficient to capture the complexity of the brain and its interacting parts. Our empirical tools must be sensitive to distributed information, able to disentangle the psychological mixtures of which brain signals are made and to detect interactions both linear and not.

I very much look forward to walking the path that lies ahead. This is, I believe, the most exciting time in the history of the neurosciences, a time when major technological advances are allowing us to measure and analyze brain function in ways undreamt of even just a few years ago. If we can manage to sustain a level of conceptual progress that keeps pace with the rapid march of technology, then the future of our science is astonishingly bright. Shall we get to work?

Appendix: Twenty-Three (Hundred) Open Questions after Phrenology

At various junctures in this book, I highlighted places where important questions are left unanswered or more research on a particular matter is called for. Here I simply gather and organize these questions thematically, offering some reflections on why the questions are important and in some cases making suggestions for how to go about answering them. My hope is that some of these questions are sufficiently well formed for a young scientist to easily take on as a dissertation or postdoctoral project. There are others that are further from that degree of precision but that can nevertheless serve to jump-start the process whereby such precise questions can be formulated. And of course, the reader will have identified many questions that I do not have the room to highlight here.

A.1 Learning, Neural Search, and Neuromodulation

As has been extensively discussed, the brain consists of a neural network, an underlying genetic network, and an overlying chemical gradient field. We are far from understanding the dynamic interactions of these three elements. The interactive differentiation and search (IDS) framework suggests that one of the important functional upshots of the interactions between these elements may be the establishment of novel functional partnerships to meet new behavioral challenges. I propose the following series of questions as a starting point:

1. Can we establish an instance in which a neural partnership initially created via volume transmission becomes consolidated into the wiring network?
2. Assuming it is possible for volume transmission to establish synaptic connections between unconnected cells, this mechanism will be limited to the number of cells within the diffusion cloud of the transmitting cell.

The question, then, is: How much "functional coverage" of the brain do the cells within that diffusion cloud provide? For our purposes, "functional coverage" might be measured in terms of the volume of the brain potentially influenced by the downstream connections of each of the cells inside the diffusion cloud. I have called this number C_f. The emerging field of connectomics is clearly going to be crucial to answering this question—especially cell-level connectomics, although the identification of fiber pathways via such technologies as diffusion spectrum imaging may offer a reasonable first approximation. The general question suggests a few subquestions: Is C_f large enough to render plausible the hypothesized role of volume transmission in setting up neural partnerships between anatomically distant parts of the brain? What is the range of values of C_f for the brain? How does C_f change as a function of the size of the functional unit and of the anatomical differences between different regions of the brain?

3. When, where, and by what mechanisms do we see the concerted modulation of the effective connectivity between large numbers of cells, and not just local but large-scale modulation? Much of the work on neuromodulation to date has focused on the modulation of single or very small numbers of cells, as in work on sensory gain modulation in mammalian retina cells and on genetic modulation of synaptic connections in model organisms such as *C. elegans*. However, it seems pretty clear that for neuromodulation to be a significant factor in human cognition, it must be able to produce concerted, reproducible differences in effective connectivity across large numbers of cells. Here again, the large-scale changes in coherence between brain regions during different tasks and the rapid recruitment of new functional partnerships during Braille reading each suggest that neuromodulation does indeed act on large populations of cells, both locally and long-distance.

4. IDS sees learning to read as an especially good candidate for understanding the nature of the neural search process at the level of large-scale recruitment of neural systems. This is due both to the irreducibly cultural nature of the task and the recency of the emergence of this practice but also because of its crucial importance to learning and development more generally and the devastating impact of the various disorders of reading that we are just beginning to understand. If we hope to understand the neural processes that facilitate normal acquisition of reading skills and also the pathways that lead to each of the many reading disabilities, then we need more fine-grained longitudinal data on the neural steps and stages that lead to reading proficiency or the lack thereof. These data should include both fMRI and diffusion spectrum imaging so that it is possible to track emerging

networks at both the functional and structural levels of description. These data will be crucial to answering the theoretical question of how, precisely, *does* reading find and establish the appropriate neural niche, and also the practical question of why the process sometimes goes awry. Ideally, knowing more about the latter will lead to better, earlier interventions.

A.2 Function-Structure Mapping

The complexities we have reviewed extensively in this volume suggest that we need to revisit our approach to function-structure mapping in the brain. This issue has several different interrelated facets, including those reviewed separately in section A.4 on cognitive ontology.

5. How does the physical configuration of a region determine its causal features? Recall that I am urging us to treat the multidimensional functional biases of individual brain regions as indices of some set of underlying causal dispositions. Thinking about this at the task level offers a top-down approach to this question, and this is crucial, but it can also be approached from a bottom-up perspective by trying to find mappings between network motifs—common configurations of functional relations between neural elements—and causal and functional effects. Naturally, until we have a way of identifying small-scale network configurations in living animals, this latter project will be limited to various neural models.
6. How do individual neural regions contribute to the overall function of a complex network? Why should a given particular mix of underlying functional fingerprints lead to an attention network whereas a different mix leads to an adaptive control network? Here again, this question needs to be approached both at the level of looking for patterns in the mappings between functional fingerprints and network functions and by building models that focus on understanding how the right mix of top-down and bottom-up, feed-forward and feedback, and excitatory and inhibitory connections between functionally complex network elements can lead to the temporary functional specificity we observe in the brain.
7. Methodologically, we need techniques that will allow us to identify the underlying network states—the effective connectivity—of the various interacting parts of the brain. Mathematical techniques such as the Granger causality approach to task-based functional connectivity fMRI are very promising for exploring effective connectivity between large-scale systems of the brain. But we need to be able to achieve a much finer grain, both spatially and temporally, in order to increase our confidence in the reliability of the function-structure relationships we propose.

8. I have been urging a *factor-based* and *dispositional* approach to defining function-structure relationships. More philosophical work needs to be done to understand what kinds of explanations such an approach can provide. What are the strengths and the limitations of this approach to understanding function in the brain sciences?
9. Similarly, one more theoretical question that emerges from reflection on questions 6 and 8 is: Are there alternatives to the componential model of achieving large-scale function from the interaction of individual elements? As I indicated at various junctures in this volume, I believe that there will be clear instances in the brain where one can identify individual TALoNS that interact in a fairly straightforward, componential way to produce complex function. But I also suspect there will be instances in which the interactions between the TALoNS involve not just bottom-up additive contributions to function but also top-down constraints that limit, change, or determine the functional properties of the TALoNS. And there will even be cases where the TALoNS prove difficult if not impossible to identify, although we will still have information on which circumscribed neural *regions* contribute to function, on the functional fingerprints of those regions, and on the functional outcomes of their cooperation. Can we develop noncomponential models of how function emerges in these cases? Or will such cases always resist understanding until it proves possible to identify components?
10. We have been focusing in this volume on the various ways in and conditions under which the reuse of regions of the brain in multiple-task contexts results in the inheritance of function from one task to the other—cases in which the region can be said to offer the same service across multiple uses. But it also appears to be the case that these overlaps *sometimes* result in the inheritance of *semantic structure or content*. These cases have played an especially important role in accounts of concept empiricism and conceptual metaphor theory, but one upshot of the findings reported here is that such cases of inheritance of content between, for example, perceptual and conceptual structures is not in fact explained by the existence of overlaps in the supporting neural substrates of perceiving and conceiving. Thus, more detailed work is called for to determine when and under what conditions such overlaps result in semantic inheritance of the sort that is so clearly important to the human cognitive system.

A.3 The Various Uses of Modeling

Clearly, one of the central claims of this volume has been that the existence of massive overlaps in the implementation bases of cognitive processes has to change the way we think about work in the cognitive neurosciences. Modeling may have an especially important role to play in this effort. Here I mention just a few projects that are ripe for exploration.

11. Altering systems such as ACT-R so that different modules share component parts might offer some interesting opportunities to explore the functional consequences of such an architectural arrangement (I noted in chapter 3 that some such efforts are already under way). For instance, it might enable ACT-R to model some cognitive phenomena that would otherwise prove more difficult or perhaps impossible in the current system, for example, the observation that children who receive pitch training starting in first grade perform better in mathematics longitudinally in third grade (Gardiner 2008). It might also form the basis for modeling both functional inheritances and semantic inheritances (discussed in chapter 1 and above).

12. One consequence of the brand of neural reuse in which a local network is used in different partnerships while retaining the same local connectivity is that the local circuit with be applying the same operation, function, or transform to potentially very different kinds of input. There are a number of relevant questions surrounding that fact, including exploration of the conditions under which useful things can be done by local assemblies working with various inputs. Are there theoretical limits on the diversity of inputs that could be entertained without destabilizing the system? What kinds of implementations increase the chances of functionally beneficial outcomes given the fact of reuse?

13. Given the use and reuse of local networks under multiple circumstances, there are a number of questions surrounding how those uses can be coordinated. I argued in chapter 1 that the coordination problem is perhaps less pressing than it might generally seem because the environment does a great deal of task segregation for us, but the issue is nevertheless important. As part of the exploration of this question, I would advocate for a systematic investigation of various kinds of cognitive interference, the mechanisms governing which are poorly understood. Clearly, interference results *in general* from competition for the same resources for multiple purposes. But the details are elusive. For instance, does cognitive interference result only when a given network is needed by two processes while the network remains *in the same* configuration? That is, does interference result because, when the network is in the appropriate configuration, it is at least *possible* for it to be used in another task, but the neural patterns induced by each process interfere? This might imply that it will be simply *impossible* to do a second task if it requires the same resource as the first, but in a different configuration (an extreme case of interference that is not particularly lab

friendly). Or are there cases where the observed performance effects are a matter of the costs of neural configuration switching? In the case of the latter possibility, we would like to understand the mechanisms that allow for sufficient retention of state across configurations to permit even degraded performance on two tasks. On the other hand, if the performance issues are instead a result of trying to use a network—locked in one configuration in virtue of being cognitively occupied—for an alternate use, then why do we generally see bidirectional performance effects (that is, decrements in performance in both of the interfering tasks)? Are there cases of unidirectional (or nearly unidirectional) cognitive interference? If so, what is the scientific implication of such a finding? Would a performance decrement primarily in one of the two interfering tasks indicate that one is dealing with two tasks that generally use the same local network in different configurations with the network unable to switch configurations to accommodate the second task?

14. The investigation of cognitive interference may put us in a position where we can begin to address question 7 in the absence of a technological/methodological breakthrough. For instance, models of a given network that show how it could contribute to some group of tasks, but not to some others while in the same underlying configuration, could provide evidence leading to the generation of task sets that more clearly correspond to underlying network states.

15. Local neural networks are pretty clearly multifunctional in some (and perhaps many) senses of that term. So how should we pin down the meaning of "multifunctional" in concrete terms? One promising option discussed at length in chapter 3 is the notion of programmable neural networks. Here there are multiple research questions: How should the different network weights be conceived of biologically, and how can they be made to represent biologically realistic aspects of target systems? How would learning happen? When would learning involve the tuning of an existing weight matrix, and when would it necessitate the creation of a new one? When there are multiple matrices, how would ongoing learning target the "correct" matrix? Once there are multiple options available for the network configuration, how do task demands cause the network to switch from one option to another?

A.4 The Cognitive Ontology

In addition to reforming the major methods we use to study cognition and the brain, I have been arguing that we also need to take seriously the need to reform the ontology of cognition, the set of taxonomic concepts we bring to bear in the study of human thinking and acting. Here there are a number of questions, both theoretical and methodological.

16. What is the underlying nature of the psychological primitives that we are after here? Russell Poldrack, as I have noted, is firm in the belief that these will be low-level, domain-general computational operations. Lisa Feldman Barrett argues for domain-general processes, which she sometimes calls "ingredients." And I have been arguing here for a set of (orthogonal?) dispositional properties that can be captured as NRP factors. But quite clearly there is a great deal of room for debate and further conceptual clarification.
17. To what will these primitives map? How complex with the mapping be? Here Russ Poldrack has argued for one-to-one mappings between these operators and regions of the brain; indeed, for him a failure of selectivity is a sign of the need to reform the implicated psychological concepts. In contrast, both Lisa Feldman Barrett and I expect to see many-to-many mappings between psychological primitives and neural elements. Barrett argues that the primitives will map to networks, whereas I think they might be usefully mapped to both networks and local regions. This being said, although Poldrack has argued in favor of selectivity, he has in some work relied instead on the weaker notion of *separability*: so long as distributed patterns of neural activity reliably indicated the engagement of a particular cognitive process, that would be evidence for its psychological reality. So clearly, there is room here also for an investigation of what selectivity in fact requires and indicates.
18. Echoing some issues raised in question 9, there is also the issue of how the psychological primitives relate to one another in the brain. Poldrack appears to expect that the relationship between the computational operators will be best understood in a componential way, whereas I have been questioning whether a componential explanation will always be applicable. Instead, I have been suggesting that the primary mechanisms of interaction between the neural primitives will be indirect: patterns of neural activity, which themselves index and indicate the differential engagement of neural elements in accord with their different loads on the primitive NRP factors, will compete via biased pattern competition to implement iterative control loops responding to the perception of affordances. In contrast to both of us, Barrett has been developing more chemical metaphors for the interactions between the elements, or "ingredients," of psychological mixtures. Clearly there is room for a clarification of these issues.

19. Methodologically, there is also a great deal of room for future development. In chapter 4 I highlighted the many ways in which machine learning, dimensional reduction, multidimensional scaling, and other techniques have been applied to both brain activity data and also behavioral data (including such things as similarity judgments) to try to uncover the underlying structure of our psychological processes. Here there is room for both experimental procedures and large-scale data mining. For instance, one ongoing project in my lab attempts to use various action- and attention- biasing cues to get participants to categorize objects along action- (or affordance-) oriented dimensions and then uses multidimensional scaling to try to get some sense for the structure of the underlying space. As a complementary project, in chapter 5 I reported on efforts to mine the very large OpenMind Common Sense database to illuminate the underlying dimensions structuring the facts we all take largely for granted. I think much more work along these lines is needed.
20. The very attempt to use neuroscience to reform the cognitive ontology raises with new force the old question of how much weight neuroscientific data should have when modeling psychological processes. To what degree is the validity of psychological categories beholden to neuroscientific data? The endeavor clearly raises the specter of reductionism, but might the result instead be an enrichment of our psychological vocabulary? These efforts call for a renewed discussion of these matters.

A.5 Embodied and Interactive Accounts of Math, Language, and "Higher" Cognition

From the standpoint of the general theoretical perspective adopted by work in ecological psychology and embodied cognition, the research program going forward is *not* to discover *what* is represented *where* in the brain, which is a natural question to ask within the CCTM framework. Rather, we must ask: What is the nature and range of environmental and bodily structures that, when coupled with appropriately tuned neural networks, can implement our celebrated capacities for planning, abstract thought, and reasoning? More specifically, we might investigate some of the following questions:

21. What precise roles do perception and action play in human computation, and what are the functional side effects of deploying perception-action loops to accomplish abstract semantic (rather than physical) tasks? When external symbols are not available for perception and manipulation,

what internal resources (e.g., memory, imagination) can be deployed in their stead? Under what circumstances is this substitution possible, and what limits does it impose on human computation? What kinds of networks are there, and what are their functional structures? Which networks are more suited for interacting with the natural environment, which for interacting with the social environment, and which interacting with the artificial symbolic environment offered by language and mathematics? What are the differences between these networks, if any? How do they interact? Are (entire) networks developed for one purpose often reused for other purposes? If so, under what circumstances, and with what functional and structural effects? Or are the networks themselves relatively selective and specialized but composed of parts that are marshaled in the service of multiple networks and operate across these different targets of interaction (natural, social, symbolic)? These and many similar questions await the concerted efforts of the field.

22. In chapter 7 I proposed extending the notion of an affordance from the binary relationship between organism and environment to the ternary relationship between interacting individuals and shared artifacts (including words). But there is much work to be done to clarify this proposal: How should we understand the nature of such cultural affordances? How should we model our *awareness* of cultural affordances? Will it be the same kind of direct perception that obtains in our awareness of environmental affordances? Or would it be better to understand cultural affordances as praxis-based representations of cultural conventions that exert an indirect, modulatory influence on our direct perception of the affordances in our environments?

23. Last but certainly not least, we come to a series of specific questions about how we should approach the study of the brain bases of language. The most important reform will be to begin to study the neural supports for the act of communication rather than for metalinguistic judgments. We should focus our efforts instead on what brain resources are deployed when (for instance) interlocutors establish (or subvert) a conversational frame. More generally, which neural clusters, representing which sets of NRP factor loads, are associated with which speech acts, and which sorts of phenotypic reorganization (itself indexed by dynamic interactions within and between brain and behavior) do they predict in speaker and listener? How do the dynamics unfold at multiple timescales as frames and footings are established and maintained, dialogue strategies are implemented, individual speech acts are performed, and various gestural and linguistic moves are made? Given the central role of language in coordinating social

relationships, does this translate into a reliance on brain regions with high loads on NRP factors related to such relationships? Or do regions with high loads on factors related to the management of physical relationships dominate? (Does this distinction even strictly hold in the brain?)

As should be abundantly clear, there is much work to be done. But it should also be clear that the framework I have outlined here provides a rich research agenda rather different from the one advocated by currently prevailing frameworks for the study of brain and behavior. I do hope that the framework I have offered proves cogent, and the agenda compelling.

References

Achard, S., Salvador, R., Witcher, B., Suckling, J., & Bullmore, E. T. (2006). A resilient, low frequency, small world human brain functional network with highly connected association cortical hubs. *Journal of Neuroscience, 26*, 63–72.

Ackerman, J. M., Nocera, C. C., & Bargh, J. A. (2010). Incidental haptic sensations influence social judgments. *Science, 328*, 1712–1715.

Adret, P. (1993). Operant conditioning, song learning and imprinting to taped song in the zebra finch. *Animal Behaviour, 46*, 149–159.

Agnati, L. F., Guidolin, D., Guescini, M., Genedani, S., & Fuxe, K. (2010). Understanding wiring and volume transmission. *Brain Research. Brain Research Reviews, 64*, 137–159.

Agnati, L. H., & Fuxe, K. (2000). Volume transmission as a key feature of information handling in the central nervous system: Possible new interpretive value of Turing's B-type machine. *Progress in Brain Research, 125*, 3–19.

Agre, P., & Chapman, D. (1987). Pengi: An implementation of a theory of activity. *Proceedings of the Sixth National Conference on Artificial Intelligence* (pp. 268–272). Menlo Park, CA: AAAI Press.

Akil, H., Watson, S. J., Young, E., Lewis, M. E., Khachaturian, H., & Walker, J. M. (1984). Endogenous opioids: Biology and function. *Annual Review of Neuroscience, 7*, 223–255.

Albertson, D. G., & Thomson, J. N. (1984). The pharynx of *C. elegans*. *Philosophical Transactions of the Royal Society of London, 275B*, 299–325.

Allport, G. W. (1937). *Personality: A psychological interpretation*. New York: Holt.

Altun, Z. F., & Hall, D. H. (2011). Nervous system, general description. In Z. F.Altun, L. A. Herndon, C. Crocker, R. Lints, & D. H. Hall (Eds.), *2002–2010 WormAtlas*. Retrieved July 16, 2012.

Amedi, A., Jacobson, G., Hendler, T., Malach, R., & Zohary, E. (2002). Convergence of visual and tactile shape processing in the human lateral occipital complex. *Cerebral Cortex, 12*(11), 1202–1212.

Anderson, D. L., & Stufflebeam, R. (2012). Introduction to the science of vision. Consortium on Cognitive Science Instruction. http://www.mind.ilstu.edu/curriculum/vision_science_intro/vision_science_intro.php?modgui=204&compgui=1940&itemgui=3375.

Anderson, J. R. (1980). *Cognitive psychology and its implications.* San Francisco: Freeman.

Anderson, J. R. (2007). *How can the human mind occur in the physical universe?* Oxford: Oxford University Press.

Anderson, J. R., & Lebiere, C. (1998). *The atomic components of thought.* Hillsdale, NJ: Lawrence Erlbaum Associates.

Anderson, J. R., Qin, Y., Junk, K. J., & Carter, C. S. (2007). Information processing modules and their relative modality specificity. *Cognitive Psychology, 57*, 185–217.

Anderson, M. L. (2003). Embodied cognition: A field guide. *Artificial Intelligence, 149*(1), 91–130.

Anderson, M. L. (2006). Cognitive science and epistemic openness. *Phenomenology and the Cognitive Sciences, 5*(2), 125–154.

Anderson, M. L. (2007a). Evolution of cognitive function via redeployment of brain areas. *Neuroscientist, 13*(1), 13–21.

Anderson, M. L. (2007b). Massive redeployment, exaptation, and the functional integration of cognitive operations. *Synthese, 159*(3), 329–345.

Anderson, M. L. (2007c). The massive redeployment hypothesis and the functional topography of the brain. *Philosophical Psychology, 21*(2), 143–174.

Anderson, M. L. (2008a). Circuit sharing and the implementation of intelligent systems. *Connection Science, 20*(4), 239–251.

Anderson, M. L. (2008b). On the grounds of x-grounded cognition. In P. Calvo & T. Gomila (Eds.), *The Elsevier handbook of cognitive science: An embodied approach* (pp. 423–435). Oxford: Elsevier.

Anderson, M. L. (2009). What mindedness is. *Europe's Journal of Psychology, 5*(4), 1–12.

Anderson, M. L. (2010a). A review of: *Neuroeconomics: Decision making and the brain. Journal of Economic Psychology, 31*, 151–154.

Anderson, M. L. (2010b). Neural reuse: A fundamental organizational principle of the brain. (Target article). *Behavioral and Brain Sciences, 33*(4), 245–266.

Anderson, M. L., Brumbaugh, J., & Şuben, A. (2010). Investigating functional cooperation in the human brain using simple graph-theoretic methods. In A. Chaovalitwongse, P. M. Pardalos, & P. Xanthopoulos (Eds.), *Computational neuroscience* (pp. 31–42). New York: Springer.

Anderson, M. L., & Chemero, T. (2009). Affordances and intentionality: Reply to Roberts. *Journal of Mind and Behavior, 30*(4), 301–312.

Anderson, M. L. & Chemero, T. (in press). Brains evolved to control action. *Handbook of Evolutionary Neuroscience.*

Anderson, M. L., & Finlay, B. L. (2014). Allocating structure to function: the strong links between neuroplasticity and natural selection. *Frontiers in Human Neuroscience, 7,* 918.

Anderson, M. L., Kinnison, J., & Pessoa, L. (2013). Describing functional diversity of brain regions and brain networks. *NeuroImage, 73,* 50–58.

Anderson, M. L., & Penner-Wilger, M. (2013). Neural reuse in the evolution and development of the brain: Evidence for developmental homology? *Developmental Psychobiology, 55*(1), 42–51.

Anderson, M. L., & Pessoa, L. (2011). Quantifying the diversity of neural activations in individual brain regions. In L. Carlson, C. Hölscher, & T. Shipley (Eds.), *Proceedings of the 33rd Annual Conference of the Cognitive Science Society* (pp. 2421–2426). Austin, TX: Cognitive Science Society.

Anderson, M. L., Richardson, M., & Chemero, T. (2012). Eroding the boundaries of cognition: Implications of embodiment. *Topics in Cognitive Science, 4*(4), 717–730.

Anderson, M. L., & Rosenberg, G. (2008). Content and action: The guidance theory of representation. In D. Smith (Ed.), *Evolutionary biology and the central problems of cognitive science,* a special issue of *Journal of Mind and Behavior, 29*(1–2), 55–86.

Andres, M., Seron, X., & Oliver, E. (2007). Contribution of hand motor circuits to counting. *Journal of Cognitive Neuroscience, 19,* 563–576.

Andrews-Hanna, J. R., Reidler, J. S., Sepulcre, J., Poulin, R., & Buckner, R. L. (2010). Functional-anatomic fractionation of the brain's default network. *Neuron, 65,* 550–562.

Anstey, M. L., Rogers, S. M., Ott, S. R., Burrows, M., & Simpson, S. J. (2009). Serotonin mediates behavioral gregarization underlying swarm formation in desert locusts. *Science, 323,* 627.

Arbib, M. A. (2010). Mirror system activity for action and language is embedded in the integration of dorsal and ventral pathways. *Brain and Language, 112*(1). 12–24.

Argyle, M., & Little, B. R. (1972). Do personality traits apply to social behavior. *Journal for the Theory of Social Behaviour, 2,* 1–35.

Aron, A., Fisher, H., Mashek, D., Strong, G., Li, H., & Brown, L. L. (2005). Reward, motivation, and emotion systems associated with early-stage intense romantic love. *Journal of Neurophysiology, 94*, 327–337.

Atallah, H. E., Frank, M. J., & O'Reilly, R. C. (2004). Hippocampus, cortex and basal ganglia: Insights from computational models of complementary learning systems. *Neurobiology of Learning and Memory, 82*(3), 253–267.

Atkinson, J. (1984). Human visual development over the first six years of life: A review and a hypothesis. *Human Neurobiology, 3*, 61–74.

Atmanspacher, H., & Filk, T. (2006). Complexity and non-commutativity of learning operations on graphs. *Bio Systems, 85*, 84–93.

Austin, J. L. (1975). *How to do things with words* (2nd ed.). Cambridge, MA: Harvard University Press.

Avidan, G., Hasson, U., Malach, R., & Behrmann, M. (2005). Detailed exploration of face-related processing in congenital prosopagnosia: 2. Functional neuroimaging findings. *Journal of Cognitive Neuroscience, 17*, 1150–1167.

Awh, E., Jonides, J., Smith, E. E., Schumacher, E. H., Koeppe, R. A., & Katz, S. (1996). Dissociation of storage and rehearsal in verbal working memory: Evidence from positron emission tomography. *Psychological Science, 7*, 25–31.

Bach-y-Rita, P. (1993). Nonsynaptic diffusion neurotransmission (NDN) in the brain. *Neurochemistry International, 23*(4), 297–318.

Bach-y-Rita, P. (1995). *Nonsynaptic diffusion neurotransmission and late brain reorganization*. New York: Demos-Vermande.

Bach-y-Rita, P. (2004). Emerging concepts of brain function. *Journal of Integrative Neuroscience, 4*(2), 183–205.

Bach-y-Rita, P., Collins, C. C., Saunders, F., White, B., & Scadden, L. (1969). Vision substitution by tactile image projection. *Nature, 221*, 963–964.

Bach-y-Rita, P., & Kercel, S. W. (2003). Sensory substitution and the human-machine interface. *Trends in Cognitive Neuroscience, 7*(12), 541–546.

Baddeley, A. D. (1986). *Working memory*. Oxford: Oxford University Press.

Baddeley, A. D. (1995). Working memory. In M. S. Gazzaniga (Ed.), *The cognitive neurosciences* (pp. 755–764). Cambridge, MA: MIT Press.

Baddeley, A. D., & Hitch, G. (1974). Working memory. In G. H. Bower (Ed.), *The psychology of learning and motivation* (pp. 647–667). New York: Lawrence Erlbaum Associates.

Baddeley, A. D., & Hitch, G. (1994). Developments in the concept of working memory. *Neuropsychology, 8*, 485–493.

Barabási, A.-L., & Albert, R. (1999). Emergence of scaling in random networks. *Science, 286*, 509–512.

Barabási, A.-L., Albert, R., & Jeong, H. (2000). Scale-free characteristics of random networks: The topology of the World Wide Web. *Physica A, 281*, 69–77.

Barbey A. K., Colom, R., Solomon, J., Krueger, F., Forbes, C., & Grafman, J. (2012). An integrative architecture for general intelligence and executive function revealed by lesion mapping. *Brain, 135*(4), 1154–1164.

Bargmann, C. I. (2012). Beyond the connectome: How neuromodulators shape neural circuits. *BioEssays, 34*(6), 458–465.

Barkow, J., Cosmides, L., & Tooby, J. (Eds.). (1992). *The adapted mind: Evolutionary psychology and the generation of culture.* New York: Oxford University Press.

Barrett, H. C., & Kurzban, R. (2006). Modularity in cognition: Framing the debate. *Psychological Review, 113*(3), 628–647.

Barrett, L. (2011). *Beyond the brain: How body and environment shape animal and human minds.* Princeton, NJ: Princeton University Press.

Barrett, L. F., & Satpute, A. (2013). Large-scale brain networks in affective and social neuroscience: Towards an integrative architecture of the human brain. *Current Opinion in Neurobiology, 23*, 1–12.

Barsalou, L. W. (1999). Perceptual symbol systems. *Behavioral and Brain Sciences, 22*, 577–660.

Barsalou, L. W. (2008). Grounded cognition. *Annual Review of Psychology, 59*, 617–645.

Barsalou, L. W., Solomon, K. O., & Wu, L. L. (1999). Perceptual simulation in conceptual tasks. In M. K. Hiraga, C. Sinha, & S. Wilcox (Eds.), *Cultural, typological, and psychological perspectives in cognitive linguistics: The proceedings of the 4th conference of the International Cognitive Linguistics Association,* (Vol. 3, pp. 209–228). Amsterdam: John Benjamins.

Barttfeld, P., Wickera, B., Cukier, S., Navarta, S., Lew, S., & Sigma, M. (2011). A big world network in ASD: Dynamical connectivity analysis reflects a deficit in long-range connections and an excess of short-range connections. *Neuropsychologia, 49*, 254–263.

Bates, E., Wilson, S. M., Saygin, A. P., Dick, F., Sereno, M. I., Knight, R. T., et al. (2003). Voxel-based lesion-symptom mapping. *Nature Neuroscience, 6*(5), 448–450.

Bateson, G. (2000). *Steps to an ecology of mind: Collected essays in anthropology, psychiatry, evolution, and epistemology*. Chicago: University of Chicago Press.

Baucom, L. B., Wedell, D. H., Wang, J., Blitzer, D. N., & Shinkareva, S. V. (2012). Decoding the neural representation of affective states. *NeuroImage, 59*(1–2), 718–727.

Bechtel, W. (1998). Representations and cognitive explanations: Assessing the dynamicist's challenge in cognitive science. *Cognitive Science, 22*, 295–318.

Bechtel, W. (2003). Modules, brain parts, and evolutionary psychology. In S. J. Scher & F. Rauscher (Eds.), *Evolutionary psychology: Alternative approaches* (pp. 211–227). New York: Kluwer.

Bechtel, W. (2012). Referring to localized cognitive operations in parts of dynamically active brains. In A. Raftopoulos & P. Machamer (Eds.), *Perception, realism and the problem of reference* (pp. 262–284). Cambridge: Cambridge University Press.

Bechtel, W., & Abrahamsen, A. (2010). Dynamic mechanistic explanation: Computational modeling of circadian rhythms as an exemplar for cognitive science. *Studies in History and Philosophy of Science Part A, 41*(3), 321–333.

Bechtel, W., & Richardson, R. C. (1993). *Discovering complexity: Decomposition and localization as strategies in scientific research.* Princeton, NJ: Princeton University Press.

Bechtel, W., & Richardson, R. C. (2010). *Discovering complexity: Decomposition and localization as strategies in scientific research* (2nd ed.). Cambridge, MA: MIT Press/ Bradford Books.

Becker, A. L. (1988). Language in particular: A lecture. In D. Tannen (Ed.), *Linguistics in context: Connecting observation and understanding* (pp. 17–35). New York: Praeger.

Becker, A. L. (1991). Language and languaging. *Language & Communication, 11*(1), 33–35.

Becker, A. L. (1995). *Beyond translation: Essays towards a modern philology*. Ann Arbor: University of Michigan Press.

Bedny, M., & Thompson-Schill, S. L. (2006). Neuroanatomically separable effects of imageability and grammatical class during single word comprehension. *Brain and Language, 98*, 127–139.

Beer, R. (1996). Toward the evolution of dynamical neural networks for minimally cognitive behavior. In P. Maes, M. Mataric, J. A. Meyer, J. Pollack, & S. Wilson (Eds.), *From animals to animats 4: Proceedings of the Fourth International Conference on Simulation of Adaptive Behavior* (pp. 421–429). Cambridge, MA: MIT Press.

Beer, R. (2003). The dynamics of active categorical perception in an evolved model agent. *Adaptive Behavior, 11*(4), 209–243.

Beevor, C., & Horsley, V. (1890). An experimental investigation into the arrangement of the excitable fibres of the internal capsule of the bonnet monkey (*Macacus sinicus*). *Philosophical Transactions of the Royal Society of London, B, 181*, 49–88.

Bennett, M. V., & Zukin, R. S. (2004). Electrical coupling and neuronal synchronization in the mammalian brain. *Neuron, 41*, 495–511.

Benton, A. L., Hamsher, K. D., Varney, N. R., & Spreen, O. (1983). *Contributions to neuropsychology assessment: A clinical manual*. New York: Oxford University Press.

Bergeron, V. (2007). Anatomical and functional modularity in cognitive science: Shifting the focus. *Philosophical Psychology, 20*(2), 175–195.

Bernstein, N. A. (1967). *Coordination and regulation of movement*. New York: Pergamon Press.

Bhalla, M., & Proffitt, D. R. (1999). Visual-motor recalibration in geographical slant perception. *Journal of Experimental Psychology. Human Perception and Performance, 25*, 1076–1096.

Bickerton, D. (1995). *Language and human behavior*. Seattle: University of Washington Press.

Bickhard, M. H. (2009). Interactivism: A manifesto. *New Ideas in Psychology, 27*(1), 85–95.

Bigelow, A. E. (1998). Infants' sensitivity to familiar contingencies in social interaction. *Infant Behavior and Development, 21*, 149–162.

Bigelow, A. E., & Rochat, P. (2006). Two-month-old infants' sensitivity to social contingency in mother-infant and stranger-infant interaction. *Infancy, 9*(3), 313–325.

Bilder, R. M., Sabb, F. W., Parker, D. S., Kalar, D., Chu, W. W., Fox, J., et al. (2009). Cognitive ontologies for neuropsychiatric phenomics research. *Cognitive Neuropsychiatry, 14*, 419–450.

Bingham, G. P., Schmidt, R. C., & Rosenblum, L. D. (1989). Hefting for a maximum distance throw: A smart perceptual mechanism. *Journal of Experimental Psychology. Human Perception and Performance, 15*, 507–528.

Binkofski, F., Amunts, K., Stephan, K. M., Posse, S., Schormann, T., Freund, H.-J., et al. (2000). Broca's region subserves imagery of motion: A combined cytoarchitectonic and fMRI study. *Human Brain Mapping, 11*, 273–285.

Bisson, G., Bianconi, G., & Torre, V. (2012). The dynamics of group formation among leeches. *Frontiers in Physiology, 3*, 133.

Bitan, T., Cheon, J., Lu, D., Burman, D. D., & Booth, J. R. (2009). Developmental increase in top-down and bottom-up processing in a phonological task: An effective connectivity, fMRI study. *Journal of Cognitive Neuroscience, 21*(6), 1135–1145.

Bivens, J. A., & Berk, L. E. (1990). A longitudinal study of the development of elementary school children's private speech. *Merrill-Palmer Quarterly, 36*(4), 443–463.

Blakemore, S.-J., Boyer, P., Pachot-Clouard, M., Meltzoff, A., Segebarth, C., & Decety, J. (2003). The detection of contingency and animacy from simple animations in the human brain. *Cerebral Cortex, 13*(8), 837–844.

Block, N. (1997). Anti-reductionism slaps back. In J. E. Tomberlin (Ed.), *Mind, causation, and world. Philosophical Perspectives, 11* (pp. 107–133). Atascadero, CA: Ridgeview.

Bloom, K., Russell, A., & Wessenberg, K. (1987). Turn taking affects the quality of infant vocalizations. *Journal of Child Language, 14*, 211–227.

Blum-Kulka, S. (1997). *Dinner talk: Cultural patterns of sociability and socialization in family discourse.* Mahwah, NJ: Lawrence Erlbaum Associates.

Bonaiuto, J., & Arbib, M. A. (2010). Extending the mirror neuron system model, II: What did I just do? A new role for mirror neurons. *Biological Cybernetics, 102*(4), 341–359.

Boroditsky, L., & Ramscar, M. (2002). The roles of body and mind in abstract thought. *Psychological Science, 13*(2), 185–188.

Bowers, K. S. (1973). Situationism in psychology: An analysis and a critique. *Psychological Review, 80*(5), 307–336.

Boyd, R., Richerson, P. J., & Henrich, J. (2011). The cultural niche: Why social learning is essential for human adaptation. *Proceedings of the National Academy of Sciences of the United States of America, 108*(Suppl 2), 10918–10925.

Boyer, D., Miramontes, O., Ramos-Fernández, G., Mateos, J. L., & Cocho, G. (2004). Modeling the searching behavior of social monkeys. *Physica A, 342*, 329–335.

Boysen, S. T. (2006). The impact of symbolic representations on chimpanzee cognition. In S. Hurley & M. Nudds (Eds.), *Rational animals?* (pp. 489–511). New York: Oxford University Press.

Braitenberg, V. (1986). Vehicles: *Experiments in synthetic psychology*. Cambridge, MA: MIT Press.

Branicky, M. S. (1995). Universal computation and other capabilities of hybrid and continuous dynamical systems. *Theoretical Computer Science, 138*, 67–100.

Branigan, H. P., Pickering, M. J., & Cleland, A. A. (2000). Syntactic coordination in dialogue. *Cognition, 75*, B13–B25.

Brennan, S. E. (1996). Lexical entrainment in spontaneous dialog. *Proceedings of ISSD, 96*, 41–44.

Brennan, S. E., & Clark, H. H. (1996). Conceptual pacts and lexical choice in conversation. *Journal of Experimental Psychology. Learning, Memory, and Cognition, 22*, 1482–1493.

Broadbent, D. (1958). *Perception and communication.* London: Pergamon Press.

Broca, P. P. (1861). Loss of speech, chronic softening and partial destruction of the anterior left lobe of the brain. *Bulletin de la Société Anthropologique, 2,* 235–238.

Brooks, R. (1991). Intelligence without representation. *Artificial Intelligence, 47,* 139–159.

Brown, C. T., Liebovitch, L. S., & Glendon, R. (2007). Lévy flights in Dobe Ju/'hoansi foraging patterns. *Human Ecology, 35,* 129–138.

Bruner, J. S. (1990). *Acts of meaning.* Cambridge, MA: Harvard University Press.

Buckner, R. L. (2010). The role of the hippocampus in prediction and imagination. *Annual Review of Psychology, 61,* 27–48.

Buckner, R. L., Sepulcre, J., Talukdar, T., Krienen, F. M., Liu, H., Hedden, T., et al. (2009). Cortical hubs revealed by intrinsic functional connectivity: Mapping, assessment of stability, and relation to Alzheimer's disease. *Journal of Neuroscience, 29*(6), 1860–1873.

Buss, A. R. (1979). The trait-situation controversy and the concept of interaction. *Personality and Social Psychology Bulletin, 5,* 191–195.

Buss, D. M. (2005). *The handbook of evolutionary psychology.* Hoboken, NJ: John Wiley & Sons.

Butterworth, B. (1999). *What counts—How every brain is hardwired for math.* New York: The Free Press.

Byers, L. (1997). *Telling the stories of our lives: Relational maintenance as illustrated through family communication.* Doctoral dissertation, Ohio University.

Byrne, J., & Flaherty, J. (2004). Measuring diversity in Australian residential property. *Proceedings of the 10th Annual Conference of the Pacific Rim Real Estate Society,* Bangkok, Thailand.

Cabeza, R., & Nyberg, L. (2000). Imaging cognition II: An empirical review of 275 PET and fMRI studies. *Journal of Cognitive Neuroscience, 12*(1), 1–47.

Campos, J. J., Anderson, D. I., Barbu-Roth, M. A., Hubbard, E. M., Hertenstein, M. J., & Witherinton, D. (2000). Travel broadens the mind. *Infancy, 1,* 149–219.

Cantero, J. L., & Atienza, M. (2005). The role of neural synchronization in the emergence of cognition across the wake-sleep cycle. *Reviews in the Neurosciences, 16*(1), 69–83.

Capaday, C., van Vreeswijk, C., Ethier, C., Ferkinghoff-Borg, J., & Weber, D. (2011). Neural mechanism of activity spread in the cat motor cortex and its relation to the intrinsic connectivity. *Journal of Physiology, 589*(10), 2515–2528.

Carlson, S. M., Davis, A. C., & Leach, J. G. (2005). Less is more: Executive function and symbolic representation in preschool children. *Americal Psychological Society, 16*, 609–616.

Carmena, J. M., Lebedev, M. A., Crist, R. E., O'Doherty, J. E., Santucci, D. M., Dimitrov, D. F., et al. (2003). Learning to control a brain-machine interface for reaching and grasping by primates. *PLoS Biology, 1*(2), e42.

Carnap, R. (1937). *The logical syntax of language* (A. Smeaton, Trans.). London: Routledge.

Carruthers, P. (1992). *Human knowledge and human nature*. Oxford: Oxford University Press.

Carruthers, P. (2002). The cognitive functions of language. *Behavioral and Brain Sciences, 25*(6), 657–674.

Carruthers, P. (2006). *The architecture of the mind: Massive modularity and the flexibility of thought*. Gloucestershire: Clarendon Press.

Casasanto, D., & Boroditsky, L. (2008). Time in the mind: Using space to think about time. *Cognition, 106*, 579–593.

Casasanto, D., & Dijkstra, K. (2010). Motor action and emotional memory. *Cognition, 115*(1), 179–185.

Casey, B. J., Trainor, R. J., Orendi, J. L., Schubert, A. B., Nystrom, L. E., Giedd, J. N., et al. (1997). A developmental functional MRI study of prefrontal activation during performance of a go–no-go task. *Journal of Cognitive Neuroscience, 9*(6), 835–847.

Caspi, A., & Moffitt, T. E. (2006). Gene-environment interactions in psychiatry: Joining forces with neuroscience. *Nature Reviews Neuroscience, 7*, 583–590.

Cattell, R. B. (1943). The description of personality. I. Foundations of trait measurement. *Psychological Review, 50*, 559–594.

Cattell, R. B. (1946). *Description and measurement of personality*. Yonkers on Hudson, NY: World Book Company.

Cauda, F., Costa, T., Torta, D. M. E., Sacco, K., D'Agata, F., Duca, S., et al. (2012). Meta-analytic clustering of the insular cortex: Characterizing the meta-analytic connectivity of the insula when involved in active tasks. *NeuroImage, 62*(1), 343–355.

Chandler, J., & Schwarz, N. (2009). How extending your middle finger affects your perception of others: Learned movements influence concept accessibility. *Journal of Experimental Social Psychology, 45*, 123–128.

Chang, A. J., & Bargmann, C. I. (2008). Hypoxia and HIF-1 transcriptional pathway reorganize a neuronal circuit for oxygen-dependent behavior in *Caenorhabditis elegans*. *Proceedings of the National Academy of Sciences of the United States of America, 105*, 7321–7326.

Chang, A. J., Chronis, N., Karow, D. S., Marletta, M. A., & Bargmann, C. I. (2006). A distributed chemosensory circuit for oxygen preference in *C. elegans*. *PLoS Biology, 4*, e274.

Chang, L. J., Yarkoni, T., Khaw, M. W., & Sanfey, A. G. (2012). Decoding the role of the insula in human cognition: Functional parcellation and large-scale reverse inference. *Cerebral Cortex, 23*(3), 739–749.

Chang, M. J. (1999). Does racial diversity matter? The educational impact of a racially diverse undergraduate population. *Journal of College Student Development, 40*(4), 377–395.

Changizi, M. A., & Shimojo, S. (2005). Character complexity and redundancy in writing systems over human history. *Proceedings. Biological Sciences, 272*, 267–275.

Changizi, M. A., Zhang, Q., Ye, H., & Shimojo, S. (2006). The structures of letters and symbols throughout human history are selected to match those found in objects in natural scenes. *American Naturalist, 167*, E117–E139.

Chao, L. L., & Martin, A. (2000). Representation of manipulable man-made objects in the dorsal stream. *NeuroImage, 12*, 478–484.

Chapman, S. (1968). Catching a baseball. *American Journal of Physics, 36*, 868–870.

Chartrand, J. L., & Bargh, J. A. (1999). The chameleon effect: The perception-behavior link and social interaction. *Journal of Personality and Social Psychology, 76*, 893–910.

Chater, N., & Christiansen, M. H. (2010). Language acquisition meets language evolution. *Cognitive Science, 34*, 1137–1157.

Chater, N., & Ganis, G. (1991). Double dissociation and isolable cognitive processes. *Proceedings of the 13th Annual Meeting of the Cognitive Science Society* (pp. 668–672). Hillsdale, NJ: Lawrence Erlbaum Associates.

Chater, N., Reali, F., & Christiansen, M. H. (2009). Restrictions on biological adaptation in language evolution. *Proceedings of the National Academy of Sciences of the United States of America, 106*(4), 1015–1020.

Chemero, A. (2003). An outline of a theory of affordances. *Ecological Psychology, 15*(2), 181–195.

Chemero, A. (2009). *Radical embodied cognitive science*. Cambridge, MA: MIT Press.

Cheung, B. H., Cohen, M., Rogers, C., Albayram, O., & DeBono, M. (2005). Experience-dependent modulation of *C. elegans* behavior by ambient oxygen. *Current Biology, 15*, 905–917.

Chiao, J. Y. (2010). Neural basis of social status hierarchy across species. *Current Opinion in Neurobiology, 20*, 803–809.

Chomsky, N. (1956). Three models for the description of language. *I.R.E. Transactions on Information Theory, 2*, 113–124.

Chomsky, N. (1957). *Syntactic structures*. The Hague: Mouton and Co.

Chomsky, N. (1959). Review of *Verbal Behavior* by B. F. Skinner. *Language, 35*, 26–57.

Chomsky, N. (1965). *Aspects of the theory of syntax*. Cambridge, MA: MIT Press.

Chomsky, N. (1966). *Cartesian linguistics*. New York: Harper & Row.

Chomsky, N. (1980). *Rules and representations*. New York: Columbia University Press.

Chomsky, N. (1981). *Lectures on government and binding: The Pisa lectures*. Dordrecht: Foris Publications. Reprinted (7th ed.). Berlin and New York: Mouton de Gruyter, 1993.

Chomsky, N. (1986). *Knowledge and language: Its nature, origin, and use*. New York: Praeger.

Chomsky, N. (1988). *Language and problems of knowledge*. Cambridge, MA: MIT Press.

Chomsky, N. (2005). Three factors in language design. *Linguistic Inquiry, 36*, 1–22.

Christiansen, M. H., & Chater, N. (2008). Language as shaped by the brain. *Behavioral and Brain Sciences, 31*, 489–558.

Christiansen, M. H., Conway, C. M., & Onnis, L. (2007). Overlapping neural responses to structural incongruencies in language and statistical learning point to similar underlying mechanisms. In D. S. MacNamara & J. G. Trafton (Eds.), *Proceedings of the 29th Annual Cognitive Science Society Conference* (pp. 173–178). New York: Lawrence Erlbaum Associates.

Christiansen, M. H., Kelly, L., Shillcock, R., & Greenfield, K. (2010). Impaired artificial grammar learning in agrammatism. *Cognition, 116*(3), 382–393.

Christiansen, M. H., & Müller, R.-A. (in press). Cultural recycling of neural substrates during language evolution and development. In M. S. Gazzaniga & G. R. Mangun (Eds.), *The cognitive neurosciences V*. Cambridge, MA: MIT Press.

Church, A. (1936). An unsolvable problem in elementary number theory. *American Journal of Mathematics, 58*, 345–363.

Churchland, P. S. (2002). *Brain-wise*. Cambridge, MA: MIT Press.

Cisek, P. (1999). Beyond the computer metaphor: Behaviour as interaction. *Journal of Consciousness Studies, 6*(11–12), 125–142.

Cisek, P. (2006). Integrated neural processes for defining potential actions and deciding between them: A computational model. *Journal of Neuroscience, 26*(38), 9761–9770.

Cisek, P. (2007). Cortical mechanisms of action selection: The affordance competition hypothesis. *Philosophical Transactions of the Royal Society of London. Series B, Biological Sciences, 362*, 1585–1599.

Cisek, P., & Kalaska, J. F. (2005). Neural correlates of reaching decisions in dorsal premotor cortex: Specification of multiple direction choices and final selection of action. *Neuron, 45*(5), 801–814.

Cisek, P., & Kalaska, J. F. (2010). Neural mechanisms for interacting with a world full of action choices. *Annual Review of Neuroscience, 33*, 269–298.

Cisek, P., Michaud, N., & Kalaska, J. F. (2004). Integration of motor planning and sensory feedback in area 5. *Society for Neuroscience Abstracts, 30*.

Clark, A. (1993). *Associative engines: Connectionism, concepts, and representational change*. Cambridge, MA: MIT Press.

Clark, A. (1997). *Being there: Putting brain, body, and world together again*. Cambridge, MA: MIT Press.

Clark, A. (1998a). Embodied, situated, and distributed cognition. In W. Bechtel & G. Graham (Eds.), *A companion to cognitive science* (pp. 506–517). Birmingham, AL: Blackwell.

Clark, A. (1998b). Magic words: How language augments human computation. In P. Carruthers & J. Boucher (Eds.), *Language and thought: Interdisciplinary themes* (pp. 162–183). Cambridge: Cambridge University Press.

Clark, A. (1999). An embodied cognitive science? *Trends in Cognitive Sciences, 3*(9), 345–351.

Clark, A. (2010). *Super-sizing the mind: Embodiment, action and cognitive extension*. Oxford: Oxford University Press.

Clark, H. H. (1996). *Using language*. Cambridge: Cambridge University Press.

Clark, H. H., & Brennan, S. E. (1991). Grounding in communication. In L. B. Resnick, J. M. Levine, & S. D. Teasley (Eds.), *Perspectives on socially shared cognition* (pp. 127–149). Washington, DC: American Psychological Association.

Clayton, N. S., Bussey, T. J., & Dickinson, A. (2003). Can animals recall the past and plan for the future? *Nature Reviews Neuroscience, 4*, 685–691.

Cleermans, A. (1993). *Mechanisms of implicit learning: Connectionist models of sequence processing*. Cambridge, MA: MIT Press.

Coates, J. C., & de Bono, M. (2002). Antagonistic pathways in neurons exposed to bodily fluid regulate social feeding in *Caenorhabditis elegans*. *Nature, 419*, 925–929.

Coe, B., Tomihara, K., Matsuzawa, M., & Hikosaka, O. (2002). Visual and anticipatory bias in three cortical eye fields of the monkey during an adaptive decision-making task. *Journal of Neuroscience, 22*(12), 5081–5090.

Cohen, L. G., Celnik, P., Pascual-Leone, A., Corwell, B., Faiz, L., Dambrosia, J., et al. (1997. Functional relevance of cross-modal plasticity in the blind. *Nature, 389*, 180–183.

Cole, M. W., Reynolds, J. R., Power, J. D., Repovs, G., Anticevic, A., & Braver, T. S. (2013). Multi-task connectivity reveals flexible hubs for adaptive task control. *Nature Neuroscience, 16*(9), 1348–1355.

Cole, M. W., & Schneider, W. (2007). The cognitive control network: Integrated cortical regions with dissociable functions. *NeuroImage, 37*(1), 343–360.

Coltheart, M. (2001). Assumptions and methods in cognitive neuropsychology. In B. Rapp (Ed.), *The handbook of cognitive neuropsychology* (pp. 3–21). Oxford: Psychology Press.

Condon, W., & Ogston, W. (1971). Speech and body motion synchrony 1 of the speaker-hearer. In D. Horton & J. J. Jenkins (Eds.), *The perception of language* (pp. 150–184). Columbus, OH: Charles E. Merrill.

Connor, C. E., Gallant, J. L., Preddie, D. C., & Van Essen, D. C. (1996). Responses in area V4 depend on the spatial relationship between stimulus and attention. *Journal of Neurophysiology, 75*, 1306–1308.

Corbetta, M., Miezin, F. M., Dobmeyer, S., Shulman, G. L., & Petersen, S. E. (1990). Attentional modulation of neural processing of shape, color and velocity in humans. *Science, 248*, 1556–1559.

Cowley, S. J. (2011). Taking a language stance. *Ecological Psychology, 23*(3), 185–209.

Crain, S., & Nakayama, M. (1987). Structure dependence in grammar formation. *Language, 24*, 139–186.

Crammond, D. J., & Kalaska, J. F. (2000). Prior information in motor and premotor cortex: Activity during the delay period and effect on premovement activity. *Journal of Neurophysiology, 84*(2), 986–1005.

Craver, C. (2007). *Explaining the brain: Mechanisms and the mosaic unity of neuroscience*. New York: Oxford University Press.

Crowley, S. J. (2011). Taking a language stance. *Ecological Psychology, 23*, 185–209.

Culham, J. C., & Kanwisher, N. G. (2001). Neuroimaging of cognitive functions in human parietal cortex. *Current Opinion in Neurobiology, 11*(2), 157–163.

Culham, J. C., & Valyear, K. F. (2006). Human parietal cortex in action. *Current Opinion in Neurobiology, 16*, 205–212.

Dagher, A., Owen, A., Boecker, H., & Brooks, D. (1999). Mapping the network for planning. *Brain, 122*, 1973–1987.

Dale, R., Fusaroli, R., Duran, N. D., & Richardson, D.C. (2013). The self-organization of human interaction. *The Psychology of Learning and Motivation, 59*, 43–96.

Damasio, A., & Tranel, D. (1993). Nouns and verbs are retrieved with differently distributed neural systems. *Proceedings of the National Academy of Sciences of the United States of America, 90*, 4957–4960.

Damasio, H., Grabowski, T. J., Tranel, D., Hichwa, R. D., & Damasio, A. R. (1996). A neural basis for lexical retrieval. *Nature, 380*, 499–505.

Danziger, K. (1997). *Naming the mind: How psychology found its language*. London: Sage Publications.

Darwin, C. (1862). *On the various contrivances by which British and foreign orchids are fertilised by insects, and on the good effects of intercrossing*. London: John Murray.

Darwin, C. (1872). *The expression of the emotions in man and animals*. London: John Murray.

Dayan, P., & Abbott, L. F. (2001). *Theoretical neuroscience*. Cambridge, MA: MIT Press.

Deacon, T. (1997). *The symbolic species*. New York: Norton.

Debener, S., Minow, F., Emkes, R., Gandras, K., & Vos, M. (2012). How about taking a low-cost, small, and wireless EEG for a walk? *Psychophysiology, 49*(11), 1617–1621.

de Bono, M., & Maricq, A. V. (2005). Neuronal substrates of complex behaviors in *C. elegans*. *Annual Review of Neuroscience, 28*, 451–501.

de Bono, M., Tobin, D. M., Davis, M. W., Avery, L., & Bargmann, C. I. (2002). Social feeding in *Caenorhabditis elegans* is induced by neurons that detect aversive stimuli. *Nature, 419*, 899–903.

Decety, J., & Grèzes, J. (1999). Neural mechanisms subserving the perception of human actions. *Trends in Cognitive Sciences, 3*, 172–178.

Decety, J., Grezes, J., Costes, N., Perani, D., Jeannerod, M., Procyk, E., et al. (1997). Brain activity during observation of actions. Influence of action content and subject's strategy. *Brain, 120*, 1763–1777.

Decety, J., Sjoholm, H., Ryding, E., Stenberg, G., & Ingvar, D. (1990). The cerebellum participates in cognitive activity: Tomographic measurements of regional cerebral blood flow. *Brain Research, 535*, 313–317.

Deco, G., & Rolls, E. T. (2003). Attention and working memory: A dynamical model of neuronal activity in the prefrontal cortex. *European Journal of Neuroscience, 18*(8), 2374–2390.

Deen, B., Pitskel, N. B., & Pelphrey, K. A. (2010). Three systems of insular functional connectivity identified with cluster analysis. *Cerebral Cortex, 21*(7), 1498–1506.

Dehaene, S. (2005). Evolution of human cortical circuits for reading and arithmetic: The "neuronal recycling" hypothesis. In S. Dehaene, J.-R. Duhamel, M. D. Hauser, & G. Rizzolatti (Eds.), *From monkey brain to human brain* (pp. 133–157). Cambridge, MA: MIT Press.

Dehaene, S. (2009). *Reading in the brain.* New York: Penguin Viking.

Dehaene, S., Bossini, S., & Giraux, P. (1993). The mental representation of parity and numerical magnitude. *Journal of Experimental Psychology. General, 122*, 371–396.

Dehaene, S., & Cohen, L. (2007). Cultural recycling of cortical maps. *Neuron, 56*, 384–398.

Dehaene, S., Dehaene-Lambertz, G., & Cohen, L. (1998). Abstract representations of numbers in the animal and human brain. *Trends in Neurosciences, 21*, 355–361.

Dehaene, S., Tzourio, N., Frak, V., Raynaud, L., Cohen, L., Mehler, J., et al. (1996). Cerebral activations during number multiplication and comparison: A PET study. *Neuropsychologia, 34*, 1097–1106.

Dehnhardt, G., & Mauck, B. (2008). Mechanoreception in secondarily aquatic vertebrates. In T. G. M. Thewissen & S. Nummela (Eds.), *Sensory evolution on the threshold: Adaptations in secondarily aquatic vertebrates* (pp. 295–314). Berkeley: University of California Press.

de Jong, B. M., van Zomeren, A. H., Willemsen, A. T. M., & Paans, A. M. J. (1996). Brain activity related to serial cognitive performance resembles circuitry of higher order motor control. *Experimental Brain Research, 109*, 136–140.

De-Miguel, F. F., & Fuxe, K. (2012). Extrasynaptic neurotransmission as a way of modulating neuronal functions. *Frontiers in Physiology, 3*(16), 5–6.

De Saussure, F. (2013). *Course in general linguistics.* New York: Columbia University Press.

Desimone, R. (1998). Visual attention mediated by biased competition in extrastriate visual cortex. *Philosophical Transactions of the Royal Society of London. Series B, Biological Sciences, 353*(1373), 1245–1255.

Desimone, R., & Duncan, J. (1995). Neural mechanisms of selective visual attention. *Annual Review of Neuroscience, 18*(1), 193–222.

Devlin, J. T., Jamison, H. L., Matthews, P. M., & Gonnerman, L. M. (2004). Morphology and the external structure of words. *Proceedings of the National Academy of Sciences of the United States of America, 101*(41), 14984–14988.

Dewey, J. (1896). The reflex arc concept in psychology. *Psychological Review, 3*, 357–370.

Diamond, A. (1990a). The development and neural bases of memory functions as indexed by the AB and delayed response task in human infants and infant monkeys. In A. Diamond (Ed.), *The development and neural bases of higher cognitive functions* (pp. 267–309). New York: New York Academy of Sciences.

Diamond, A. (1990b). Developmental time course in human infants and infant monkeys and the neural basis of inhibitory control in reaching. In A. Diamond (Ed.), *The development and neural bases of higher cognitive functions* (pp. 637–676). New York: New York Academy of Sciences.

Dijksterhuis, A., & Bargh, J. A. (2001). The perception behavior expressway: Automatic effects of social perception on social behavior. In M. P. Zanna (Ed.), Advances in experimental social psychology (Vol. 33). *The perception-behavior expressway: Automatic effects of social perception on social behavior* (pp. 1–40). San Diego, CA: Academic Press.

Donnarumma, F., Prevete, R., & Trautteur, G. (2012). Programming in the brain: A neural network theoretical framework. *Connection Science, 24*(2-3), 71–90. doi:10.1080/09540091.2012.684670.

Dörfler, W. (2002). Formation of mathematical objects as decision making. *Mathematical Thinking and Learning, 4*, 337–350.

Dosenbach, N. U., Fair, D. A., Miezin, F. M., Cohen, A. L., Wenger, K. K., Dosenbach, R. A., et al. (2007). Distinct brain networks for adaptive and stable task control in humans. *Proceedings of the National Academy of Sciences of the United States of America, 104*(26), 11073–11078.

Douglas, P. K., Harris, S., Yuille, A., & Cohen, M. S. (2011). Performance comparison of machine learning algorithms and number of independent components used in fMRI decoding of belief vs. disbelief. *NeuroImage, 56*(2), 544–555.

Douglas, R. J., Koch, C., Mahowald, M., Martin, K. A., & Suarez, H. H. (1995). Recurrent excitation in neocortical circuits. *Science, 269*(5226), 981–985.

Dragoi, G., & Tonegawa, S. (2011). Preplay of future place cell sequences by hippocampal cellular assemblies. *Nature, 469*(7330), 397–401.

Dreyfus, H. (1990). *Being-in-the-world: A commentary on Heidegger's* Being and Time, *division I.* Cambridge, MA: MIT Press.

Duchaine, B. C., & Nakayama, K. (2006). Developmental prosopagnosia: A window to content-specific face processing. *Current Opinion in Neurobiology, 16*, 166–173.

Durbin, R. M. (1987). *Studies on the development and organisation of the nervous system of* C. elegans. Ph.D. thesis. University of Cambridge, Cambridge, UK.

Eales, L. A. (1989). The influence of visual and vocal interaction on song learning in zebra finches. *Animal Behaviour, 37*, 507–508.

Edelman, S. (2008). *Computing the mind: How the mind really works*. New York: Oxford University Press.

Edelman, S., Grill-Spector, K., Kushnir, T., & Malach, R. (1998). Toward direct visualization of the internal shape space by fMRI. *Psychobiology, 26*, 309–321.

Eickhoff, S. B., Laird, A. R., Grefkes, C., Wang, L. E., Zilles, K., & Fox, P. T. (2009). Coordinate-based activation likelihood estimation meta-analysis of neuroimaging data: A random-effects approach based on empirical estimates of spatial uncertainty. *Human Brain Mapping, 30*(9), 2907–2926.

Ekehammar, B. (1974). Interactionism in personality from a historical perspective. *Psychological Bulletin, 8*(1), 1026–1048.

Elman, J. L. (1994). Learning and development in neural networks: The importance of starting small. *Cognition, 48*, 71–99.

Elman, J. L. (1999). Origins of language: A conspiracy theory. In B. MacWhinney (Ed.), *The emergence of language* (pp. 1–28). Hillsdale, NJ: Lawrence Erlbaum Associates.

Endler, N. S., & Hum, J. M. (1966). Sources of behavioral variance as measured by the S-R inventory of anxiousness. *Psychological Bulletin, 65*, 336–346.

Endler, N. S., & Hunt, J. M. (1968). S-R inventories of hostility and comparisons of the proportions of variance from persons, responses, and situations for hostility and anxiousness. *Journal of Personality and Social Psychology, 9*, 309–315.

Endler, N. S., & Hunt, J. M. (1969). Generalizability of contributions from sources of variance in the S-R inventories of anxiousness. *Journal of Personality, 37*, 1–24.

Endler, N. S., Hunt, J. M., & Rosenstein, A. J. (1962). An S-R inventory of anxiousness. *Psychological Monographs, 76,* (17, whole no. 536).

Endler, N. S., & Magnusson, D. (1976). Toward an interactional psychology of personality. *Psychological Bulletin, 83*, 956–974.

Engel, A. K., Maye, A., Kurthen, M., & König, P. (2013). Where's the action? The pragmatic turn in cognitive science. *Trends in Cognitive Sciences, 17*(5), 202–209.

Epstein, R., & Kanwisher, N. (1998). A cortical representation of the local visual environment. *Nature, 392*, 598–601.

Epstein, S., & O'Brien, E. J. (1985). The person-situation debate in historical and current perspective. *Psychological Bulletin, 98*(3), 513–537.

Erikson, F. (1986). Listening and speaking. In D. Tannen (Ed.), *Languages and linguistics* (pp. 294–319). Washington, D.C.: Georgetown University Press.

Erikson, F. (1990). The social construction of discourse coherence in a family dinner table conversation. In B. Dorval (Ed.), *Conversational organization and its development* (pp. 207–238). Norwood, NJ: Ables.

Erlhagen, W., & Schöner, G. (2002). Dynamic field theory of movement preparation. *Psychological Review, 109*(3), 545–572.

Evans, N., & Levinson, S. C. (2009). The myth of language universals: Language diversity and its importance for cognitive science. *Behavioral and Brain Sciences, 32*, 429–492.

Fair, D. A., Cohen, A. L., Power, J. D., Dosenbach, N. U., Church, J. A., Miezin, F. M., et al. (2009). Functional brain networks develop from a "local to distributed" organization. *PLoS Computational Biology, 5*, e1000381.

Fair, D. A., Dosenbach, N. U. F., Church, J. A., Cohen, A. L., Brahmbhatt, S., Miezin, F., et al. (2007). Development of distinct control networks through segregation and integration. *Proceedings of the National Academy of Sciences of the United States of America, 104*, 13507–13512.

Fauconnier, G., & Turner, M. (2002). *The way we think: Conceptual blending and the mind's hidden complexities*. New York: Basic Books.

Fedorenko, E., Patel, A., Casasanto, D., Winawer, J., & Gibson, T. (2009). Structural integration in language and music: Evidence for a shared system. *Memory & Cognition, 37*(1), 1–9.

Feldman, J., & Narayanan, S. (2004). Embodied meaning in a neural theory of language. *Brain and Language, 89*, 385–392.

Fields, R. D. (2009). *The other brain*. New York: Simon & Schuster.

Finger, S. (1994). *Origins of neuroscience: A history of explorations into brain function*. New York: Oxford University Press.

Fink, P., Foo, P., & Warren, W. (2009). Catching fly balls in virtual reality: A critical test of the outfielder problem. *Journal of Vision, 9*(13), 14.

Fischel, H. (1926). Transformationserscheinungen bei Gewichtschebungen. *Zeitschrift fur Psychologie mit Zeitschrift fur Angewandte Psychologie, 98*, 342–365.

Fischl, B., van der Kouwe, A., Destrieux, C., Halgren, E., Segonne, F., Salat, D. H., et al. (2004). Automatically parcellating the human cerebral cortex. *Cerebral Cortex, 14*, 11–22.

Fivush, R., Bohanek, J., Robertson, R., & Duke, M. (2004). Family narratives and the development of children's emotional well-being. In M. W. Pratt & B. H. Fiese (Eds.), *Family stories and the lifecourse: Across time and generations* (pp. 55–76). New York: Routledge.

Flanagan, J. R., & Beltzner, M. A. (2000). Independence of perceptual and sensorimotor predictions in the size-weight illusion. *Nature Neuroscience, 3*(7), 737–741.

Fodor, J. A. (1975). *The language of thought.* Cambridge, MA: Harvard University Press.

Fodor, J. A. (1981). *Representations*. Cambridge, MA: MIT Press.

Fodor, J. A. (1983). *The modularity of mind: An essay on faculty psychology*. Cambridge, MA: MIT Press.

Fodor, J. A. (1987). *Psychosemantics*. Cambridge, MA: Bradford Books.

Fodor, J. A. (1990). *A theory of content and other essays*. Cambridge, MA: MIT Press.

Fodor, J. A. (1997). Special sciences: Still autonomous after all these years. *Noûs, 31*(s11), 149–163.

Fodor, J. A. (2000). *The mind doesn't work that way*. Cambridge, MA: MIT Press.

Fodor, J. A. (2007). Why pigs don't have wings. *London Review of Books, 29*(20), 19–22.

Fodor, J. A., & Pylyshyn, Z. W. (1988). Connectionaism and cognitive architecture: A critical analysis. *Cognition, 28*, 3–71.

Fowler, C. A., Rubin, P., Remez, R. E., & Turvey, M. T. (1980). Implications for speech production of a general theory of action. In B. Buttersworth (Ed.), *Language production, vol. 1: Speech and talk* (pp. 373–420). London: Academic Press.

Fox, D. (2011). The limits of intelligence. *Scientific American, 305*(1), 36–43.

Fox, M. D., Snyder, A. Z., Vincent, J. L., Corbetta, M., Van Essen, D. C., & Raichle, M. E. (2005). The human brain is intrinsically organized into dynamic, anticorrelated functional networks. *Proceedings of the National Academy of Sciences of the United States of America, 102*, 9673–9678.

Fox, P. T., & Lancaster, J. L. (2002). Mapping context and content: The BrainMap model. *Nature Reviews Neuroscience, 3*, 319–321.

Fox, P. T., Parsons, L. M., & Lancaster, J. L. (1998). Beyond the single study: Function-location meta-analysis in cognitive neuroimaging. *Current Opinion in Neurobiology, 8*, 178–187.

Frank, S. L., Bod, R., & Christiansen, M. H. (2012). How hierarchical is language use? *Proceedings of the Royal Society B: Biological Sciences, 279*, 4522–4531.

Freeman, W. (2005). NDN, volume transmission, and self-organization in brain dynamics. *Journal of Integrative Neuroscience, 4*(4), 407–421.

Fries, R. C. (2006). *Reliable design of medical devices*. Boca Raton, FL: CRC Press.

Friston, K. J. (1997). Imaging cognitive anatomy. *Trends in Cognitive Sciences, 1*, 21–27.

Fukuda, T. (2007). Structural organization of the gap junction network in the cerebral cortex. *Neuroscientist, 13*, 199–207.

Fuxe, K., Hedlund, P., von Euler, G., Lundgren, K., Martire, M., Ögren, S. O., et al. (1991). Galanin/5-HT interactions in the rat central nervous system. Relevance for depression. In T. Hökfelt, T. Bartfai, D. Jacobowitz, & D. Ottoson (Eds.), *Galanin. A new multifunctional peptide in the neuroendocrine system* Wenner-Gren International Series (Vol. 58, pp. 221–235). London: Macmillan Press.

Gall, F. J. (1798). Schreiben über seinen bereits geendigten Prodromus über die Verichtungen des Gehirns der Menschen und der Thiere an Herrn Jos. Fr. von Retzer. Christoph Martin Wieland (ed.). *Der neue Teutsche Merkur, 3*, 311–332.

Gall, F. J. (1857). Letter from Dr. F. J. Gall, to Joseph Fr[eiherr] von Retzer, upon the Functions of the Brain in Man and Animals. In D. G. Goyder (Trans.) *My Battle for Life: The Autobiography of a Phrenologist* (pp. 143–152). London: Simpkin, Marshall, and Co.

Gallagher, S., & Hutto, D. D. (2008). Understanding others through primary interaction and narrative practice. In J. Zlatev, T. Racine, C. Sinha, & E. Itkonen (Eds.), *The shared mind: Perspectives on intersubjectivity* (pp. 17–38). Amsterdam: John Benjamins.

Gallese, V. (2008). Mirror neurons and the social nature of language: The neural exploitation hypothesis. *Social Neuroscience, 3*(3–4), 317–333.

Gallese, V., Fadiga, L., Fogassi, L., & Rizzolatti, G. (1996). Action recognition in the premotor cortex. *Brain, 119*, 593–609.

Gallese, V., & Goldman, A. (1998). Mirror neurons and the simulation theory of mind-reading. *Trends in Cognitive Sciences, 2*(12), 493–501.

Gallese, V., & Lakoff, G. (2005). The brain's concepts: The role of the sensory-motor system in conceptual knowledge. *Cognitive Neuropsychology, 22*(3–4), 455–479.

Gallistel, C. R., & King, A. P. (2009). *Memory and the computational brain: Why cognitive science will transform neuroscience*. New York: Wiley/Blackwell.

Gardiner, M. F. (2008). Music training, engagement with sequence, and the development of the natural number concept in young learners. *Behavioral and Brain Sciences, 31*(6), 652–653.

Garrod, S., & Pickering, M. J. (2004). Why is conversation so easy? *Trends in Cognitive Sciences, 8*, 8–11.

Gathers, A. D., Bhatt, R., Corbly, C. R., Farley, A. B., & Joseph, J. E. (2004). Developmental shifts in cortical loci for face and object recognition. *Neuroreport, 15*, 1549–1553.

Gauker, C. (1990). How to learn language like a chimpanzee. *Philosophical Psychology, 3*(1), 31–53.

Gauthier, I., & Nelson, C. (2001). The development of face expertise. *Current Opinion in Neurobiology, 11*, 219–224.

Gauthier, I., Skudlarski, P., Gore, J. C., & Anderson, A. W. (2000). Expertise for cars and birds recruits brain areas involved in face recognition. *Nature Neuroscience, 3*(2), 191–197.

Gauthier, I., Tarr, M. J., Anderson, A. W., Skudlarski, P., & Gore, J. C. (1999). Activation of the middle fusiform "face area" increases with expertise in recognizing novel objects. *Nature Neuroscience, 2*, 568–573.

Gazzaley, A., Rissman, J., & D'Esposito, M. (2004). Functional connectivity during working memory maintenance. *Cognitive, Affective & Behavioral Neuroscience, 4*(4), 580–599.

Gazzaniga, M. S., Ivry, R. B., & Magnun, G. R. (2008). *Cognitive neuroscience: The biology of the mind*. New York: W.W. Norton & Company.

Gentner, D., & Stevens, A. L. (Eds.). (1983). *Mental models*. Hillsdale, NJ: Lawrence Erlbaum Associates.

Gerlach, J. (1872). The spinal cord. In S. Streiker (Ed.), *Manual of human and comparative histology* (pp. 327–366). London: The New Sydenham Society.

Getting, P. A., & Denkin, M. S. (1985). Tritonia swimming: A model system for integration within rhythmic motor systems. In A. I. Selverston (Ed.), *Model neural networks and behavior* (pp. 3–20). New York: Plenum Press.

Geuss, M. N., & Stefanucci, J. K. (2010). Arousal and imbalance influence size perception. *Journal of Vision, 10*(7), 59.

Ghez, C., Favilla, M., Ghilardi, M. F., Gordon, J., Bermejo, R., & Pullman, S. (1997). Discrete and continuous planning of hand movements and isometric force trajectories. *Experimental Brain Research, 15*(2), 217–233.

Gibson, E. J., & Walk, R. D. (1960). The "visual cliff." *Scientific American, 202*(4), 67–71.

Gibson, J. J. (1966). *The senses considered as perceptual systems*. Boston: Houghton -Mifflin.

Gibson, J. J. (1979). *The ecological approach to visual perception*. Hillsdale, NJ: Lawrence Erlbaum Associates.

Gigerenzer, G., & Selten, R. (Eds.). (2002). *Bounded rationality: The adaptive toolbox.* Cambridge, MA: MIT Press.

Gigerenzer, G., Todd, P. M., & The ABC Research Group. (1999). *Simple heuristics that make us smart.* Oxford: Oxford University Press.

Giles, H., Coupland, N., & Coupland, J. (1992). Accommodation theory: Communication, context and consequences. In H. Giles, J. Coupland, & N. Coupland (Eds.), *Contexts of accommodation* (pp. 1–68). Cambridge: Cambridge University Press.

Gilovich, T., Griffin, D., & Kahneman, D. (Eds.). (2002). *Heuristics and biases: The psychology of intuitive judgment.* Cambridge: Cambridge University Press.

Givón, T. (1979). *On understanding grammar (Perspectives in neurolinguistics and psycholinguistics).* Salt Lake City, UT: Academic Press.

Glenberg, A. M. (1997). What memory is for. *Behavioral and Brain Sciences, 20,* 1–55.

Glenberg, A. M., Becker, R., Klötzer, S., Kolanko, L., Müller, S., & Rinck, M. (2009). Episodic affordances contribute to language comprehension. *Language and Cognition, 1,* 113–135.

Glenberg, A. M., Brown, M., & Levin, J. R. (2007). Enhancing comprehension in small reading groups using a manipulation strategy. *Contemporary Educational Psychology, 32,* 389–399.

Glenberg, A. M., & Kaschak, M. P. (2002). Grounding language in action. *Psychonomic Bulletin & Review, 9,* 558–565.

Glenberg, A. M., Sato, M., & Cattaneo, L. (2008a). Use-induced motor plasticity affects the processing of abstract and concrete language. *Current Biology, 18,* R290–R291.

Glenberg, A. M., Sato, M., Cattaneo, L., Riggio, L., Palumbo, D., & Buccino, G. (2008b). Processing abstract language modulates motor system activity. *Quarterly Journal of Experimental Psychology, 61,* 905–919.

Glimcher, P. W. (2003). The neurobiology of visual-saccadic decision making. *Annual Review of Neuroscience, 26,* 133–179.

Göbel, S. M., Johansen-Berg, H., Behrens, T., & Rushwort, M. F. S. (2004). Response-selection-related parietal activation during number comparison. *Journal of Cognitive Neuroscience, 16,* 1536–1551.

Goffman, E. (1974). *Frame analysis: An essay on the organization of experience.* London: Harper & Row.

Goffman, E. (1981). *Forms of talk.* Philadelphia: University of Pennsylvania Press.

Gold, J. I., & Shadlen, M. N. (2007). The neural basis of decision making. *Annual Review of Neuroscience, 30,* 535–574.

Gold, K., Havasi, C., Anderson, M., & Arnold, K. (2011). Comparing matrix decomposition methods for meta-analysis and reconstruction of cognitive neuroscience results. *Proceedings of the 24th Annual Conference of the Florida Artificial Intelligence Research Society (FLAIRS-24)*, pp. 21–26.

Goldberg, L. R. (1982). From ace to zombie: Some explorations in the language of personality. In C. D. Spielberger & J. N. Butcher (Eds.), *Advances in personality assessment* (Vol. 1, pp. 203–234). Hillsdale, NJ: Lawrence Erlbaum Associates.

Goldin-Meadow, S. (2003). *Hearing gesture: How our hands help us think*. Cambridge, MA: Belknap Press.

Goldman, A. I. (2012). A moderate approach to embodied cognitive science. *Review of Philosophy and Psychology, 3*(1), 71–88.

Goldman-Rakic, P. S. (1988a). Specification of cerebral cortical areas. *Science, 241*, 170–176.

Goldman-Rakic, P. S. (1988b). Topography of cognition: parallel distributed networks in primate association cortex. *Annual Review of Neuroscience, 11*, 137–156.

Goldstein, E. R. (2012, July 16). The strange neuroscience of immortality. *Chronicle of Higher Education*, (July), 16. https://chronicle.com/article/The-Strange-Neuroscience-of/132819/.

Goldstein, M. H., & Schwade, J. A. (2008). Social feedback to infants' babbling facilitates rapid phonological learning. *Psychological Science, 19*(5), 515–523.

Goodwin, C. (2010). Constructing meaning through prosody in aphasia. In D. Barth-Weingarten, E. Reber, & M. Selting (Eds.), *Prosody in interaction* (Vol. 23, pp. 373–394). Philadelphia: John Benjamins Publishing.

Gopnik, A. (2012). Scientific thinking in young children: Theoretical advances, empirical research and policy implications. *Science, 337*, 1623–1627.

Gopnik, A., & Schulz, L. (2007). *Causal learning: Psychology, philosophy, and computation*. New York: Oxford University Press.

Gordon, C. (2009). *Making meanings, creating family: Intertextuality and framing in family interaction*. Oxford: Oxford University Press.

Gordon, C. (2011). Gumperz and interactional sociolinguistics. In R. Wodak, B. Johnstone, & P. Kerswill (Eds.), *Sage handbook of sociolinguistics* (pp. 67–84). London: Sage.

Gordon, C. (2013). Tannen, Deborah. In C. A. Chapelle (Ed.), *The encyclopedia of applied linguistics*. New York: Blackwell Publishing.

Gordon, C. (in press). Framing and positioning. In D. Schriffrin, D. Tannen, & H. E. Hamilton (Eds.), *The handbook of discourse analysis* (2nd ed.). New York: Wiley Blackwell.

Gould, S. J. (1996). *The mismeasure of man*. New York: W. W. Norton & Company. (Original work published 1981)

Gracia-Bafalluy, M., & Noël, M. P. (2008). Does finger training increase young children's numerical performance? *Cortex, 44*(4), 368–375.

Graziano, M. S. A. (2011). Cables vs. networks: Old and new views on the function of motor cortex. *Journal of Physiology, 589*(10), 2439.

Graziano, M. S. A., Taylor, C. S. R., & Moore, T. (2002a). Complex movements evoked by microstimulation of precentral cortex. *Neuron, 34*, 841–851.

Graziano, M. S. A., Taylor, C. S. R., Moore, T., & Cooke, D. F. (2002b). The cortical control of movement revisited. *Neuron, 36*, 349–362.

Grech, R., Cassar, T., Muscat, J., Camilleri, K., Fabri, S., Zervakis, M., et al. (2008). Review on solving the inverse problem in EEG source analysis. *Journal of Neuroengineering and Rehabilitation, 5*(1), 25.

Grecius, M. D., Krasnow, B., Reiss, A. L., & Menon, V. (2003). Functional connectivity in the resting brain: A network analysis of the default mode hypothesis. *Proceedings of the National Academy of Sciences of the United States of America, 100*(1), 253–258.

Greenfield, P. M. (1991). Language, tools and brain: The ontogeny and phylogeny of hierarchically organized sequential behavior. *Behavioral and Brain Sciences, 14*, 531–595.

Gregory, R. (1998). Brainy mind. *British Medical Journal, 317*, 1693–1695.

Gregory, R. (2002). Perceptions as hypotheses. In A. Nöe & E. Thompson (Eds.), *Vision and mind: Selected writings in the philosophy of perception* (pp. 111–134). Cambridge, MA: MIT Press.

Grice, H. P. (1969). Utterer's meaning and intentions. *The Philosophical Review, 78*. Reprinted as chapter 5 of Grice 1989, pp. 86–116.

Grice, H. P. (1989). *Studies in the way of words*. Cambridge, MA: Harvard University Press.

Grill-Spector, K., Sayres, R., & Ress, D. (2006). High-resolution imaging reveals highly selective nonface clusters in the fusiform face area. *Nature Neuroscience, 9*(9), 1177–1185.

Grodzinsky, Y., & Santi, A. (2008). The battle for Broca's region. *Trends in Cognitive Sciences, 12*, 474–480.

Grush, R. (1997). The architecture of representation. *Philosophical Psychology, 10*, 5–24.

Grush, R. (2004). The emulation theory of representation: Motor control, imagery, and perception. *Behavioral and Brain Sciences, 27*, 377–442.

Guan, C. Q., Meng, W., Yao, R., & Glenberg, A. M. (2013). The motor system contributes to comprehension of abstract language. *PLoS ONE*, *8*(9), e75183.

Guignon, C. (1983). *Heidegger and the problem of knowledge*. Indianapolis: Hackett Publishing Co.

Gumperz, J. J. (1982). *Discourse strategies*. Cambridge: Cambridge University Press.

Guthrie, R. V. (2003). *Even the rat was white: A historical view of psychology*. New York: Allyn & Bacon.

Haber, R. N. (1983). The impending demise of the icon: The role of iconic processes in information processing theories of perception. *Behavioral and Brain Sciences*, *6*, 1–55.

Hacking, I. (1983). *Representing and intervening: Introductory topics in the philosophy of natural science*. Cambridge: Cambridge University Press.

Hagoort, P. (2005). On Broca, brain and binding. *Trends in Cognitive Sciences*, *9*(9), 416–423.

Hamzei, F., Rijntjes, M., Dettmers, C., Glauche, V., Weiller, C., & Büchel, C.. (2003). The human action recognition system and its relationship to Broca's area: An fMRI study. *NeuroImage*, *19*, 637–644.

Hanakawa, T., Honda, M., Sawamoto, N., Okada, T., Yonekura, Y., Fukuyama, H., et al. (2002). The role of rostral Brodmann area 6 in mental-operation tasks: An integrative neuroimaging approach. *Cerebral Cortex*, *12*, 1157–1170.

Hanson, S. J., & Schmidt, A. (2011). High-resolution imaging of the fusiform face area (FFA) using multivariate non-linear classifiers shows diagnosticity for non-face categories. *NeuroImage*, *54*, 1715–1734.

Harnad, S. (1990). The symbol grounding problem. *Physica D. Nonlinear Phenomena*, *42*, 335–346.

Harris, C. M., & Wolpert, D. M. (1998). Signal-dependent noise determines motor planning. *Nature*, *394*, 780–784.

Hart, A., Sims, S., & Kaplan, J. (1995). Synaptic code for sensory modalities revealed by *C. elegans* GLR-1 glutamate receptor. *Nature*, *378*, 82–85.

Hasson, U., Ghazanfar, A. A., Galantucci, B., Garrod, S., & Keysers, C. (2012). Brain-to-brain coupling: Mechanism for creating and sharing a social world. *Trends in Cognitive Sciences*, *16*(2), 114–121.

Hasson, U., & Honey, C. J. (2012). Future trends in neuroimaging: Neural processes as expressed within real-life contexts. *NeuroImage*, *62*, 1272–1278.

Haugeland, J. (1978). The nature and plausibility of cognitivism. *Behavioral and Brain Sciences*, *2*, 215–226.

Haugeland, J. (Ed.). (1981). *Mind design*. Cambridge, MA: MIT Press.

Haxby, J. V. (2012). Multivariate pattern analysis of fMRI: The early beginnings. *NeuroImage, 62*(2), 852–855.

Haxby, J. V., Guntupalli, J. S., Connolly, A. C., Halchenko, Y. O., Conroy, B. R., Gobbini, M. I., et al. (2011). A common, high-dimensional model of the representational space in human ventral temporal cortex. *Neuron, 72*(2), 404–416.

Haynes, J.-D., & Rees, G. (2006). Decoding mental states from brain activity in humans. *Nature Reviews Neuroscience, 7*(7), 523–534.

Haynes, J.-D., Sakai, K., Rees, G., Gilbert, S., Frith, C., & Passingham, R. E. (2007). Reading hidden intentions in the human brain. *Current Biology, 17*, 323–328.

Heidegger, M. (1962). *Being and time* (J. Macquarrie & E. Robinson, Trans.). London: SCM Press. (Original work published 1927)

Henley, N. M. (1969). A psychological study of the semantics of animal terms. *Journal of Verbal Learning and Verbal Behavior, 8*, 176–184.

Henrich, J., Heine, S. J., & Norenzayan, A. (2010). The weirdest people in the world. *Behavioral and Brain Sciences, 33*(2–3), 61–83.

Hermans, E. J., van Marle, H. J. F., Ossewaarde, L., Hencken, M. J. A. G., Qin, S., van Kesteren, M.T. R., et al. (2011). Stress-related noradrenergic activity prompts large-scale neural network reconfiguration. *Science, 25*, 334 (6059), 1151–1153.

Hernández-Lemus, E. (2012). Extrasynaptic release of serotonin affects the social dynamics of leeches. *Frontiers in Physiology, 3*, 158.

Hill, M. O. (1973). Diversity and evenness: A unifying notation and its consequences. *Ecology, 54*, 427–432.

Hinton, G. E., McClelland, J. L., & Rumelhart, D. E. (1986). *Parallel distributed processing: Explorations of the microstructure of cognition*, Vol. 1: *Foundations*. Cambridge, MA: MIT Press.

Hochner, B. (2012). An embodied view of octopus neurobiology. *Current Biology, 22*, R887–R892.

Hoen, M., Golembiowski, M., Guyot, E., Deprez, V., Caplan, D., & Dominey, P. F. (2003). Training with cognitive sequences improves syntactic comprehension in agrammatic aphasics. *Neuroreport, 14*, 495–499.

Holtmaat, A., & Svoboda, K. (2009). Experience dependent structural synaptic plasticity in the mammalian brain. *Nature Reviews Neuroscience, 10*, 647–658.

Honey, C. J., Kötter, R., Breakspear, M., & Sporns, O. (2007). Network structure of cerebral cortex shapes functional connectivity on multiple time scales. *Proceedings of the National Academy of Sciences of the United States of America, 104*, 10240–10245.

Honey, C. J., Sporns, O., Cammoun, L., Gigandet, X., Thiran, J. P., Meuli, R., et al. (2009). Predicting human resting-state functional connectivity from structural connectivity. *Proceedings of the National Academy of Sciences of the United States of America, 106*, 2035–2040.

Hopkin, V. D. (1995). *Human factors in air traffic control.* Boca Raton, FL: CRC Press.

Howard, L. A., & Tipper, S. P. (1997). Hand deviations away from visual cues: Indirect evidence for inhibition. *Experimental Brain Research, 113*(1), 144–152.

Howard-Jones, P. A., Blakemore, S.-J., Samuel, E. A., Summers, I. R., & Claxton, G. (2005). Semantic divergence and creative story generation: An fMRI investigation. *Brain Research. Cognitive Brain Research, 25*, 240–250.

Hsu, H.-J., Christiansen, M. H., Tomblin, J. B., Zhang, X., & Gomez, R. L. (2006). Statistical learning of nonadjacent dependencies in adolescents with and without language impairment. *Poster presented at the 2006 Symposium on Research in Child Language Disorders, Madison, WI.*

Hubbard, E. M., Arman, A. C., Ramachandran, V. S., & Boynton, G. M. (2005a). Individual differences among grapheme-color synesthetes: Brain-behavior correlations. *Neuron, 45*, 975–985.

Hubbard, E. M., Piazza, M., Pinel, P., & Dehaene, S. (2005b). Interactions between number and space in parietal cortex. *Nature Reviews Neuroscience, 6*, 435–448.

Huneman, P. (2010). Topological explanations and robustness in biological sciences. *Synthese, 177*, 213–245.

Hurley, S. (2005). The shared circuits hypothesis: A unified functional architecture for control, imitation and simulation. In S. Hurley & N. Chater (Eds.), *Perspectives on imitation: From neuroscience to social science* (pp. 76–95). Cambridge, MA: MIT Press.

Hurley, S. (2008). The shared circuits model (SCM): How control, mirroring, and simulation can enable imitation, deliberation, and mindreading. *Behavioral and Brain Sciences, 31*, 1–58.

Hutchins, E. (1995). *Cognition in the wild.* Cambridge, MA: MIT Press.

Iriki, A. (2005). A prototype of homo-faber: A silent precursor of human intelligence in the tool-using monkey brain. In S. Dehaene, J. R. Duhamel, M. Hauser, & G. Rizzolati (Eds.), *From monkey brain to human brain* (pp. 133–157). Cambridge, MA: MIT Press.

Iriki, A., & Sakura, O. (2008). Neuroscience of primate intellectual evolution: Natural selection and passive and intentional niche construction. *Philosophical Transactions of the Royal Society of London. Series B, Biological Sciences, 363*, 2229–2241.

Jablonka, E., & Lamb, M. J. (2006). *Evolution in four dimensions.* Cambridge, MA: MIT Press.

Jackendoff, R. (2002). *Foundations of language: Brain, meaning, grammar, evolution.* Oxford: Oxford University Press.

Jacob, F. (1982). *The possible and the actual.* New York: Pantheon.

Jacobs, L. F. (2012). From chemotaxis to the cognitive map: The function of olfaction. *Proceedings of the National Academy of Sciences of the United States of America, 109*(Suppl 1), 10693–10700.

Jacoby, W. G. & Armstrong, D. A. (2014). Bootstrap confidence regions for multidimensional scaling solutions. *American Journal of Political Science, 58*(1), 264–278.

James, W. (1950). *The Principles of psychology.* New York: Dover Publications. (Original work published 1890)

Jancke, L., Loose, R., Lutz, K., Specht, K., & Shah, N. J. (2000). Cortical activations during paced finger-tapping applying visual and auditory pacing stimuli. *Brain Research. Cognitive Brain Research, 10,* 51–66.

Jeannerod, M. (1994). The representing brain: Neural correlates of motor intention and imagery. *Behavioral and Brain Sciences, 17,* 187–245.

Jeong, H., Tombor, B., Albert, R., Oltvai, Z. N., & Barabási, A.-L. (2000). The large-scale organization of metabolic networks. *Nature, 407,* 651–654.

Jilk, D. J., Lebiere, C., O'Reilly, R. C., & Anderson, J. R. (2008). SAL: An explicitly pluralistic cognitive architecture. *Journal of Experimental & Theoretical Artificial Intelligence, 20,* 197–218.

John, O. P., Angleitner, A., & Ostendorf, F. (1988). The lexical approach to personality: A historical review of trait taxonomic research. *European Journal of Personality, 2,* 171–203.

Johnson, M. H. (2001). Functional brain development in humans. *Nature Reviews Neuroscience, 2,* 475–483.

Johnson, M. H. (2011). Interactive specialization: A domain-general framework for human functional brain development? *Developmental Cognitive Neuroscience, 1*(1), 7–21.

Johnson-Laird, P. N. (1983). *Mental models: Towards a cognitive science of language, inference, and consciousness.* Cambridge, MA: Harvard University Press.

Jost, L., & Chao, A. (2008). *Diversity analysis.* New York: Taylor & Francis.

Just, M. A., Cherkassky, V. L., Keller, T. A., & Minshew, N. J. (2004). Cortical activation and synchronization during sentence comprehension in high-functioning autism: Evidence of underconnectivity. *Brain, 127,* 1811–1821.

Just, M. A., & Varma, S. (2007). The organization of thinking: What functional brain imaging reveals about the neuroarchitecture of complex cognition. *Carnegie*

Mellon University, Department of Psychology. Paper 240. http://repository.cmu.edu/psychology/240.

Kaan, E., & Swaab, T. Y. (2002). The neural circuitry of syntactic comprehension. *Trends in Cognitive Sciences, 6*, 350–356.

Kandel, E. (2006). *In search of memory: The emergence of a new science of mind*. New York: Norton.

Kanwisher, N. (2010). Functional specificity in the human brain: A window into the functional architecture of the mind. *Proceedings of the National Academy of Sciences of the United States of America, 107*(25), 11163–11170.

Kanwisher, N., McDermott, J., & Chun, M. M. (1997). The fusiform face area: A module in human extrastriate cortex specialized for face perception. *Journal of Neuroscience, 17*(11), 4302–4311.

Katz, D. (1935). *The world of colour*. London: Kegan Paul.

Katz, P. S. (1999). *Beyond neurotransmission: Neuromodulation and its importance for information processing*. Oxford: Oxford University Press.

Katz, P. S., & Calin-Jageman, R. (2008). Neuromodulation. In L. R. Squire (Ed.), *New encyclopedia of neuroscience* (pp. 497–503). Salt Lake City, UT: Academic Press.

Kay, K. N., Naselaris, T., Prenger, R. J., & Gallant, J. L. (2008). Identifying natural images from human brain activity. *Nature, 452*(7185), 352–355.

Kelly, C., Toro, R., Di Martino, A., Cox, C. L., Bellec, P., Castellanos, F. X., et al. (2012). A convergent functional architecture of the insula emerges across imaging modalities. *NeuroImage, 61*(4), 1129–1142.

Kelso, J. A. S. (1995). *Dynamic patterns: The self-organization of brain and behavior*. Cambridge, MA: MIT Press.

Kelso, J. A. S. (2009). Synergies: Atoms of brain and behavior. In D. Sternad (Ed.), *Progress in motor control* (pp. 83–91). New York: Springer.

Kendal, J., Tehrani, J. J., & Odling-Smee, J. (2011). Human niche construction in interdisciplinary focus. *Philosophical Transactions of the Royal Society of London. Series B, Biological Sciences, 366*, 785–792.

Kim, B., & Basso, M. A. (2008). Saccade target selection in the superior colliculus: A signal detection theory approach. *Journal of Neuroscience, 28*(12), 2991–3007.

Kirschner, M. W., & Gerhart, J. C. (2005). *The plausibility of life: Resolving Darwin's dilemma*. New Haven, CT: Yale University Press.

Kiss, J., & Vizi, E. S. (2001). Nitric oxide: A novel link between synaptic and nonsynaptic transmission. *Trends in Neurosciences, 24*(4), 211–215.

Kitcher, P. (1989). Explanatory unification and the causal structure of the world. In P. Kitcher & W. Salmon (Eds.), *Scientific explanation* (pp. 410–505). Minneapolis: University of Minnesota Press.

Klages, L. (1932). *The Science of Character)*. London: Allen & Unwin. (Original work published 1926)

Klein, C. (2010). Redeployed functions versus spreading activation: A potential confound. *Behavioral and Brain Sciences, 33*, 280–281.

Klein, C. (2012). Cognitive ontology and region- versus network-oriented analyses. *Philosophy of Science, 79*(5), 952–960.

Kloos, H., & Amazeen, E. L. (2002). Perceiving heaviness by dynamic touch: An investigation of the size-weight illusion in preschoolers. *British Journal of Developmental Psychology, 20*, 171–183.

Koch, C., & Segev, I. (2000). The role of single neurons in information processing. *Nature Neuroscience, 3*, 1171–1177.

Komuniecki, R., Harris, G., Hapiak, V., Wragg, R., & Bamber, B. (2012). Monoamines activate neuropeptide signaling cascades to modulate nociception in *C. elegans*: A useful model for the modulation of chronic pain? *Invertebrate Neuroscience, 12*(1), 53-61.

Koshland, D. E. (1980). *Behavioral chemotaxis as a model behavioral system*. New York: Raven Press.

Kosslyn, S. M. (1999). If neuroimaging is the answer, what is the question? *Proceedings. Biological Sciences, 354*, 1283–1294.

Kosslyn, S. M., & Koenig, O. (1995). *Wet mind: The new cognitive neuroscience*. New York: Free Press.

Kosslyn, S. M., Thompson, W. L., & Ganis, G. (2006). *The case for mental imagery*. Oxford: Oxford University Press.

Kosslyn, S. M., Thompson, W. L., Kin, L. J., Rauch, S. L., & Alpert, N. M. (1996). Individual differences in cerebral blood flow in area 17 predict the time to evaluate visualized letters. *Journal of Cognitive Neuroscience, 8*, 78–82.

Kriegeskorte, N., Mur, M., & Bandettini, P. A. (2008a). Representational similarity analysis—Connecting the branches of systems neuroscience. *Frontiers in Systems Neuroscience, 2*, 4.

Kriegeskorte, N., Mur, M., Ruff, D. A., Kiani, R., Bodurka, J., Esteky, H., et al. (2008b). Matching categorical object representations in inferior temporal cortex of man and monkey. *Neuron, 60*(6), 1126–1141.

Kristeva, J. (1980). Word, dialogue and novel. In L. S. Roudiez (Ed.), *Desire in language: A semiotic approach to literature and art* (pp. 64–91). New York: Columbia University Press.

Kuhl, P. K. (2007). Is speech learning "gated" by the social brain? *Developmental Science, 10,* 110–120.

Kuhtz-Buschbeck, J. P., Mahnkopf, C., Holzknecht, C., Siebner, H. R., Ulmer, S., & Jansen, O. (2003). Effector-independent representations of simple and complex imagined finger movements: A combined fMRI and TMS study. *European Journal of Neuroscience, 18,* 3375–3387.

Kummer, H., & Goodall, J. (1985). Conditions of innovative behavior in primates. *Philosophical Transactions of the Royal Society of London. Series B, Biological Sciences, 308,* 203–214.

Kupers, R., Papperns, M., de Noordhout, A. M., Schoenen, J., Ptito, M., & Fumal, A. (2007). rTMS of the occipital cortex abolishes Braille reading and repetition priming in blind subjects. *Neurology, 69,* 691–693.

Kurth, F., Zilles, K., Fox, P. T., Laird, A. R., & Eickhoff, S. B. (2010). A link between the systems: Functional differentiation and integration within the human insula revealed by meta-analysis. *Brain Structure & Function, 214*(5–6), 519–534.

Laird, A. R., Eickhoff, S. B., Li, K., Robin, D. A., Glahn, D. C., & Fox, P. T. (2009). Investigating the functional heterogeneity of the default mode network using coordinate-based meta-analytic modeling. *Journal of Neuroscience, 29*(46), 14496–14505.

Laird, A. R., Lancaster, J. L., & Fox, P. T. (2005). BrainMap: The social evolution of a functional neuroimaging database. *Neuroinformatics, 3,* 65–78.

Lakin, J., & Chartrand, T. L. (2003). Using nonconscious behavioral mimicry to create affiliation and rapport. *Psychological Science, 14,* 334–339.

Lakoff, G., & Johnson, M. (1980). *Metaphors we live by.* Chicago: University of Chicago Press.

Lakoff, G., & Johnson, M. (1999). *Philosophy in the flesh: The embodied mind and its challenge to Western thought.* New York: Basic Books.

Lakoff, G., & Núñez, R. (2000). *Where mathematics comes from: How the embodied mind brings mathematics into being.* New York: Basic Books.

Lancaster, J. L., Laird, A. R., Eickhoff, S. B., Martinez, M. J., Fox, P. M., & Fox, P. T. (2012). Automated regional behavioral analysis for human brain images. *Frontiers in Neuroinformatics, 6,* 23.

Landy, D., & Goldstone, R. L. (2007a). Formal notations are diagrams: Evidence from a production task. *Memory & Cognition, 35*(8), 2033–2040.

Landy, D., & Goldstone, R. L. (2007b). How abstract is symbolic thought? *Journal of Experimental Psychology. Learning, Memory, and Cognition, 33*(4), 720–733.

Landy, D., & Goldstone, R. L. (2009). How much of symbolic manipulation is just symbol pushing? In N. A. Taatgen & H. van Rijn (Eds.), *Proceedings of the 31st Annual Conference of the Cognitive Science Society* (pp. 1072–1077). Austin, TX: Cognitive Science Society.

Landy, D., & Goldstone, R. L. (2010). Proximity and precedence in arithmetic. *Quarterly Journal of Experimental Psychology, 63*(10), 1953–1968.

Landy, D., & Linkenauger, S. (2010). Arithmetic notation ... now in 3D! In S. Ohlsson & R. Catrambone (Eds.), *Proceedings of the 32nd Annual Conference of the Cognitive Science Society* (pp. 2164–2169). Austin, TX: Cognitive Science Society.

Langellier, K. M., & Peterson, E. E. (1993). Family storytelling as a strategy of social control. In D. K. Mumby (Ed.), *Narrative and social control. Critical Perspectives* (pp. 49–76). Newbury Park, CA: Sage.

Lanier, J. (2011). *You are not a gadget: A manifesto.* New York: Vintage.

Lashley, K. S. (1929). *Brain mechanisms and intelligence: A quantitative study of injuries to the brain.* Chicago: University of Chicago Press.

Lashley, K. S. (1950). In search of the engram. *Society of Experimental Biology Symposium, 4*, 454–482.

Lashley, K. S. (1951). The problem of serial order in behavior. In L. A. Jeffress (Ed.), *Cerebral mechanisms in behavior* (pp. 112–136). New York: John Wiley & Sons.

Latash, M. L. (2008). *Synergy.* Oxford, New York: Oxford University Press.

Lebedev, M. A., Carmena, J. M., O'Doherty, J. E., Zacksenhouse, M., Henriquez, C. S., Principe, J. C., et al. (2005). Cortical ensemble adaptation to represent velocity of an artificial actuator controlled by a brain-machine interface. *Journal of Neuroscience, 25*, 4681–4693.

Lebedev, M. A., & Nicoleilis, M. A. L. (2006). Brain machine interfaces: past, present and future. *Trends in Neurosciences, 29*(9), 536–546.

LeCun, Y., Boser, B., Denker, J. S., Henderson, D., Howard, R. E., Hubbard, W., et al. (1989). Backpropagation applied to handwritten zip code recognition. *Neural Computation, 1*(4), 541–551.

Lee, C. C., Black, M., Katsamanis, A., Lammert, A., Baucom, B., Christensen, A., et al. (2010). Quantification of prosodic entrainment in affective spontaneous spoken interactions of married couples. *Proceedings of Interspeech, Makuhari, Japan* (pp. 793–796).

Lee, S. W. S., & Schwarz, N. (2010). Of dirty hands and dirty mouths: Embodiment of the moral purity metaphor is specific to the motor modality involved in moral transgression. *Psychological Science, 21*, 1423–1425.

Lee, S., & Zhou, Z. J. (2006). The synaptic mechanism of direction selectivity in distal processes of starburst amacrine cells. *Neuron, 51*, 787–799.

Lenartowicz, A., Kalar, D., Congdon, E., & Poldrack, R. A. (2010). Towards an ontology of cognitive control. *Topics in Cognitive Science, 2*, 678–692.

Lenneberg, E. H. (1964). A biological perspective on language. In E. H. Lenneberg (Ed.), *New directions in the study of language*, (pp. 65–88). Cambridge, MA: MIT Press.

Lenneberg, E. (1967). *Biological foundations of language*. Indianapolis, IN: Wiley.

Lessard, D. A., Linkenauger, S. A., & Proffitt, D. R. (2009). Look before you leap: Jumping ability affects distance perception. *Perception, 38*, 1863–1866.

Levin, D. M. (1993). *Modernity and the hegemony of vision*. Berkeley: University of California Press.

Lewicki, M. S. (1998). A review of methods for spike sorting: The detection and classification of neural action potentials. *Network (Bristol, England), 9*(4), 53–78.

Lewontin, R. C. (1998). The evolution of cognition: Questions we will never answer. In D. Scarborough & S. Sternberg (Eds.), *An invitation to cognitive science*, vol. 4: *Methods, models and conceptual issues*. Cambridge, MA: MIT Press.

Liberman, A. M., Cooper, F. S., Shankweiler, D. P., & Studdert-Kennedy, M. (1967). Perception of the speech code. *Psychological Review, 74*, 431–461.

Liberman, A. M., & Mattingly, I. (1985). The motor theory of speech perception revised. *Cognition, 51*, 1–36.

Liberman, A. M., & Mattingly, I. G. (1989). A specialization for speech perception. *Science, 243*, 489–496.

Lindquist, K. A., & Barrett, L. F. (2012). A functional architecture of the human brain: Emerging insights from the science of emotion. *Trends in Cognitive Sciences, 16*, 533–540.

Lindquist, K. A., Wager, T. D., Kober, H., Bliss-Moreau, E., & Barrett, L. F. (2012). The brain basis of emotion: A meta-analytic review. *Behavioral and Brain Sciences, 35*(3), 121–143.

Linell, P. (2005). *The written language bias in linguistics: Its nature, origins and transformations*. London: Routledge.

Linell, P. (2009). *Rethinking language, mind and world dialogically: Interactional and contextual theories of sense making*. Charlotte, NC: Information Age Publishing.

Linkenauger, S. A., Witt, J. K., & Proffitt, D. R. (2011). Taking a hands-on approach: Apparent hand size affects the perception of size. *Journal of Experimental Psychology. Human Perception and Performance, 37*(5), 1432-41.

Liu, X., Wang, H., Corbly, C. R., Zhang, J., & Joseph, J. E. (2006). The involvement of the inferior parietal cortex in the numerical Stroop effect and the distance effect in a two-digit number comparison task. *Journal of Cognitive Neuroscience, 18,* 1518–1530.

Lloyd, D. (1989). *Simple minds.* Cambridge, MA: MIT Press.

Lloyd, D. (2000). Terra cognita: From functional neuroimaging to the map of the mind. *Brain and Mind, 1*(1), 93–116.

Louwerse, M. M., Benesh, N., Watanabe, S., Zhang, B., Jeuniaux, P., & Vargheese, D. (2009). The multimodal nature of embodied conversational agents. In N. A. Taatgen, & H. van Rijn (Eds.), *Proceedings of the 31st Annual Conference of the Cognitive Science Society* (pp. 1459–1463). Austin, TX: Cognitive Science Society.

Love, N. (2004). Cognition and the language myth. *Language Sciences, 26,* 525–544.

Maass, W., & Markram, H. (2004). On the computational power of recurrent circuits of spiking neurons. *Journal of Computer and System Sciences, 69*(4), 593–616.

Maass, W., Natschläger, T., & Markram, H. (2002). Real-time computing without stable states: A new framework for neural computation based on perturbations. *Neural Computation, 14*(11), 2531–2560.

Machery, E. (2007). Concept empiricism: A methodological critique. *Cognition, 104,* 19–46.

Mackay, D. J. C. (2003). *Information theory, inference and learning algorithms.* Cambridge: Cambridge University Press.

MacNeilage, P. F. (1998). The frame/content theory of evolution of speech production. *Behavioral and Brain Sciences, 21,* 499–511.

Macosko, E. Z., Pokala, N., Feinberg, E. H., Chalasani, S. H., Butcher, R. A., Clardy, J., et al. (2009). A hub-and-spoke circuit drives pheromone attraction and social behavior in *C. elegans. Nature, 458,* 1171–1175.

Maess, B., Koelsch, S., Gunter, T. C., & Friederici, A. D. (2001). Musical syntax is processed in Broca's area: An MEG study. *Nature Neuroscience, 4,* 540–545.

Mahon, B., & Caramazza, A. (2008). A critical look at the embodied cognition hypothesis and a new proposal for grounding conceptual content. *Journal of Physiology, Paris, 102,* 59–70.

Mann, N. I., & Slater, P. J. B. (1995). Song tutor choice by zebra finches in aviaries. *Animal Behaviour, 49*(3), 811–820.

Marcus, G. (2004). *The birth of the mind: How a tiny number of genes creates the complexities of human thought*. New York: Basic Books.

Marcus, G. (2008). *Kluge: The haphazard construction of the human mind*. Boston, MA: Houghton Mifflin.

Marder, E., & Thirumalai, V. (2002). Cellular, synaptic and network effects of neuromodulation. *Neural Networks, 15*, 479–493.

Margoliash, D., & Nusbaum, H. C. (2009). Language: The perspective from organismal biology. *Trends in Cognitive Sciences, 13*, 505–510.

Maricq, A. V., Peckol, E., Driscoll, M., & Bargmann, C. I. (1995). Mechanosensory signalling in *C. elegans* mediated by the GLR-1 glutamate receptor. *Nature, 378*, 78–81.

Marr, D. (1982). *Vision*. San Francisco: W. H. Freeman.

Martin, A., Haxby, J. V., Lalonde, F. M., Wiggs, C. L., & Ungerleider, L. G. (1995). Discrete cortical regions associated with knowledge of color and knowledge of action. *Science, 270*, 102–105.

Martin, A., Ungerleider, L. G., & Haxby, J. V. (2000). Category-specificity and the brain: The sensorymotor model of semantic representations of objects. In M. S. Gazzaniga (Ed.), *The new cognitive neurosciences* (2nd ed.) (pp. 1023–1036). Cambridge, MA: MIT Press.

Martin, A., Wiggs, C. L., Ungerleider, L. G., & Haxby, J. V. (1996). Neural correlates of category-specific knowledge. *Nature, 379*, 649–652.

Masland, R. H. (2005). The many roles of starburst amacrine cells. *Trends in Neurosciences, 28*(8), 395–396.

Mather, M., & Sutherland, M. R. (2011). Arousal-biased competition in perception and memory. *Perspectives on Psychological Science, 6*(2), 114–133.

McBeath, M. K., Shaffer, D. M., & Kaiser, M. K. (1995). How baseball outfielders determine where to run to catch fly balls. *Science, 268*(5210), 569–573.

McClamrock, R. (1995). *Existential cognition*. Chicago: University of Chicago Press.

McClelland, J. (2009). The place of modeling in cognitive science. *Topics in Cognitive Science, 1*, 11–38.

McClelland, J. (2013). *Explorations in parallel distributed processing: A handbook of models, programs, and exercises*. http://www.stanford.edu/group/pdplab/pdphandbook/

McClelland, J. L., McNaughton, B. L., & O'Reilly, R. C. (1995). Why there are complementary learning systems in the hippocampus and neocortex: Insights from the successes and failures of connectionist models of learning and memory. *Psychological Review, 102*, 419–457.

McClelland, J. L., & Rumelhart, D. E. (Eds.). (1986). *Parallel distributed processing* (Vol. 2). Cambridge, MA: MIT Press.

McIntosh, A. R., Grady, C. L., Ungerleider, L. G., Haxby, J. V., Rapoport, S. I., & Horwitz, B. (1994). Network analysis of cortical visual pathways mapped with PET. *Journal of Neuroscience, 14*, 655–666.

McKeown, M. J., Makeig, S., Brown, G. G., Jung, T. P., Kinderman, S. S., & Sejnowski, T. J. (1998). Spatially independent activity patterns in functional magnetic resonance imaging data during the Stroop color naming task. *Proceedings of the National Academy of Sciences of the United States of America, 95*, 803–810.

McKinstry, C., Dale, R., & Spivey, M. J. (2008). Action dynamics reveal parallel competition in decision making. *Psychological Science, 19*(1), 22–24.

Menon, V., & Uddin, L. Q. (2010). Saliency, switching, attention and control: A network model of insula function. *Brain Structure & Function, 214*(5–6), 655–667.

Menzel, R. (2009). Conditioning: Simple neural circuits in the honeybee. In L. R. Squire (Ed.), *Encyclopedia of neuroscience* (Vol. 3, pp. 43–47). San Diego, CA: Academic Press.

Merabet, L. B., Hamilton, R., Schlaug, G., Swisher, J. D., Kiriakapoulos, E. T., Pitskel, N. B., et al. (2008). Rapid and reversible recruitment of early visual cortex for touch. *PLoS One, 3*(8), e3046, 1–12.

Merabet, L., Maquire, D., Warde, A., Altruesco, K., Stickold, R., & Pascual-Leone, A. (2004). Visual hallucinations during prolonged blindfolding in sighted subjects. *Journal of Neuro-Ophthalmology, 24*, 109–113.

Mesulam, M.-M. (1981). A cortical network for directed attention and unilateral neglect. *Annals of Neurology, 10*, 309–325.

Mesulam, M.-M. (1990). Large-scale neurocognitive networks and distributed processing for attention, language and memory. *Annals of Neurology, 28*, 597–613.

Mesulam, M.-M. (1994). Neurocognitive networks and selectively distributed processing. *Revue Neurologique, 150*, 564–569.

Mesulam, M. M. (1998). From sensation to cognition. *Brain. Journal of Neurology, 121*(6), 1013–1052.

Meunier, D., Lambiotte, R., Fornito, A., Ersche, K. D., & Bullmore, E. T. (2009). Hierarchical modularity in human brain functional networks. *Frontiers in Neuroinformatics, 3*, 37.

Middleton, F. A., & Strick, P. L. (2000). Basal ganglia and cerebellar loops: Motor and cognitive circuits. *Brain Research. Brain Research Reviews, 31*, 236–250.

Miller, E. K. (2000). The prefrontal cortex and cognitive control. *Nature Reviews Neuroscience, 1*, 59–65.

Miller, E. K., & Cohen, J. D. (2001). An integrative theory of prefrontal cortex function. *Annual Review of Neuroscience, 24*, 167–202.

Miller, E., Seppa, C., Kittur, A., Sabb, F., & Poldrack, R. A. (2010). The cognitive atlas: Employing interaction design processes to facilitate collaborative ontology creation. *Nature Precedings*. doi:10.1038/npre.2010.4532.1.

Miller, G. A. (1956). The magical number seven, plus or minus two: Some limits on our capacity for processing information. *Psychological Review, 63*(2), 81–97.

Miller, G. A. (2003). The cognitive revolution: A historical perspective. *Trends in Cognitive Sciences, 7*(3), 141–144.

Millikan, R. (1995). Pushmi-pullyu representations. In J. Tomberlin (Ed.), *Philosophical perspectives, 9* (pp. 185–200). Atascadero, CA: Ridgeview.

Milner, A. D., & Goodale, M. A. (1995). *The visual brain in action*. Oxford: Oxford University Press.

Minshew, N. J., & Keller, T. A. (2010). The nature of brain dysfunction in autism: Functional brain imaging studies. *Current Opinion in Neurology, 23*, 124–130.

Minsky, M. (1975). A framework for representing knowledge. In P. Winston (Ed.), *The psychology of computer vision* (pp. 211–277). New York: McGraw-Hill.

Mischel, W. (1968). *Personality and assessment*. New York: Wiley.

Mischel, W. (1979). On the interface of cognition and personality. *American Psychologist, 34*(9), 740–754.

Mitchell, M. (2006). Complex systems: Network thinking. *Artificial Intelligence, 170*, 1194–1212.

Mitchell, M. (2009). *Complexity: A guided tour*. Oxford: Oxford University Press.

Moran, J., & Desimone, R. (1985). Selective attention gates visual processing in the extrastriate cortex. *Science, 229*, 782–784.

Müller, R.-A., & Basho, S. (2004). Are nonlinguistic functions in "Broca's area" prerequisites for language acquisition? fMRI findings from an ontogenetic viewpoint. *Brain and Language, 89*(2), 329–336.

Murray, D. J., Ellis, R. R., Bandomir, C. A., & Ross, H. E. (1999). Charpentier (1891) on the size-weight illusion. *Perception & Psychophysics, 61*, 1681–1685.

Nedergaard, M., Ransom, B., & Goldman, S. A. (2003). New roles for astrocytes: Redefining the functional architecture of the brain. *Trends in Neurosciences, 26*(10), 523–530.

Neisser, U. (1967). *Cognitive psychology*. New York: Appleton-Century-Crofts.

Newell, A., Shaw, J. C., & Simon, H. A. (1958). Elements of a theory of human problem solving. *Psychological Review, 23*, 342–343.

Newell, A., & Simon, H. A. (1976). Computer science as empirical enquiry. *Communications of the ACM, 19*(3), 113–126.

Newman, M., Barabási, A.-L., & Watts, D. J. (2006). *The structure and dynamics of networks*. Princeton: Princeton University Press.

Newmann, S. W. (1999). The medial extended amygdala in male reproductive behavior. *Annals of the New York Academy of Sciences, 877*, 242–257.

Nielsen, F. Å., Hansen, L. K., & Balslev, D. (2004). Mining for associations between text and brain activation in a functional neuroimaging database. *Neuroinformatics, 2*(4), 369–379.

Nishitani, N., Schürmann, M., Amunts, K., & Hari, R. (2005). Broca's region: From action to language. *Physiology (Bethesda, MD), 20*, 60–69.

Niven, J. E., & Chittka, L. (2010). Reuse of identified neurons in multiple neural circuits. *Behavioral and Brain Sciences, 33*(4), 285.

Noë, A. (2004). *Action in perception*. Cambridge, MA: MIT Press.

Norman, K. A., Polyn, S. M., Detre, G. J., & Haxby, J. V. (2006). Beyond mind-reading: Multi-voxel pattern analysis of fMRI data. *Trends in Cognitive Sciences, 10*(9), 424–430.

Norman, W. T. (1963). Toward an adequate taxonomy of personality attributes: Replicated factor structure in peer nomination personality ratings. *Journal of Abnormal and Social Psychology, 66*, 574–583.

Numminen, J., Schurmann, M., Hiltunen, J., Joensuu, R., Jousmaki, V., Koskinen, S. K., et al. (2004). Cortical activation during a spatiotemporal tactile comparison task. *NeuroImage, 22*, 815–821.

Nusbaum, H. C. (2011). Language and communication. In J. Decety & J. T. Cacioppo (Eds.), *Handbook of social neuroscience* (pp. 668–679). Oxford: Oxford University Press.

Ochs, E., Schegloff, E. A., & Thompson, S. A. (Eds.). (1996). *Interaction and grammar* (Vol. 13). Cambridge: Cambridge University Press.

Ochs, E., Smith, R., & Taylor, C. (1996). Detective stories at dinnertime: Problem-solving through co-narration. In C. L. Briggs (Ed.), *Disorderly discourse: Narrative, conflict, and inequality* (pp. 95–113). Oxford: Oxford University Press.

Ochs, E., & Taylor, C. (1992a). Family narrative as political activity. *Discourse & Society, 3*(3), 301–340.

Ochs, E., & Taylor, C. (1992b). Mother's role in the everyday reconstruction of "father knows best." In K. Hall, M. Bucholtz, & B. Moonwoman (Eds.), *Locating power: Proceedings of the Second Berkeley Women and Language Conference* (pp. 447–463). Berkeley CA: Berkeley Women and Language Group.

Ochs, E., & Taylor, C. (1995). The "father knows best" dynamic in family dinner narratives. In K. Hill & M. Bucholtz (Eds.), *Gender articulated* (pp. 97–120). New York: Routledge.

Ochsner, K. N., Beer, J. S., Robertson, E. R., Cooper, J. C., Gabrieli, J. D. E., Kihsltrom, J. F., et al. (2005). The neural correlates of direct and reflected self-knowledge. *NeuroImage, 28*, 797–814.

Odling-Smee, F. J., Laland, K. N., & Geldman, M. W. (2005). *Niche construction: The neglected process in evolution*. Princeton: Princeton University Press.

O'Donovan-Anderson, M. (Ed.) (1996). *The incorporated self: Interdisciplinary perspectives on embodiment*. Lanham, MD: Rowman & Littlefield.

O'Donovan-Anderson, M. (1997). *Content and comportment: Embodiment and the epistemic availability of the world*. Lanham, MD: Rowman & Littlefield.

O'Reilly, R. C. (1998). Six principles for biologically-based computational models of cortical cognition. *Trends in Cognitive Sciences, 2*, 455–462.

O'Reilly, R. C. (2006). Biologically based computational models of high-level cognition. *Science, 314*(5796), 91–94.

O'Reilly, R. C., Braver, T. S., & Cohen, J. D. (1999). A biologically based computational model of working memory. In A. Miyake & P. Shah (Eds.), *Models of working memory: Mechanisms of active maintenance and executive control* (pp. 375–411). Cambridge: Cambridge University Press.

O'Reilly, R. C., & Frank, M. J. (2006). Making working memory work: A computational model of learning in the prefrontal cortex and basal ganglia. *Neural Computation, 18*, 283–328.

O'Reilly, R. C., & Munakata, Y. (2000). *Computational explorations in cognitive neuroscience: Understanding the mind by simulating the brain*. Cambridge, MA: MIT Press.

O'Toole, A. J., Jiang, F., Abdi, H., & Haxby, J. V. (2005). Partially distributed representations of objects and faces in ventral temporal cortex. *Journal of Cognitive Neuroscience, 17*(4), 580–590.

O'Toole, A. J., Jiang, F., Abdi, H., Penard, N., Dunlop, J. P., & Parent, M. A. (2007). Theoretical, statistical, and practical perspectives on pattern-based classification approaches to the analysis of functional neuroimaging data. *Journal of Cognitive Neuroscience, 19*, 1735–1752.

Overton, W. F., & Reese, H. W. (1973). Models of development: Methodological implications. In J. R. Nesselroade & H. W. Reese (Eds.), *Lifespan developmental psychology: Methodological issues* (pp. 65–86). New York: Academic Press.

Ovsepian, S. V., & Dolly, J. O. (2011). Dendritic SNAREs add a new twist to the old neuron theory. *Proceedings of the National Academy of Sciences of the United States of America, 108(*48), 19113–19120.

Pallas, S. L. (2001). Intrinsic and extrinsic factors that shape neocortical specification. *Trends in Neurosciences, 24*(7), 417–423.

Papousek, M., Papousek, H., & Bornstein, M. H. (1985). The naturalistic vocal environment of young infants: On the significance of homogeneity and variability in parental speech. In T. Field & N. Fox (Eds.), *Social perception in infants* (pp. 269–297). Norwood, NJ: Ablex.

Pardo, J. S. (2006). On phonetic convergence during conversational interaction. *Journal of the Acoustical Society of America, 119,* 2382–2393.

Pascual-Leone, A., & Hamilton, R. (2001). The metamodal organization of the brain. *Progress in Brain Research, 134,* 427–445.

Passarotti, A. M., Paul, B. M., Bussiere, J. R., Buxton, R. B., Wong, E. C., & Stiles, J. (2003). The development of face and location processing: An fMRI study. *Developmental Science, 6,* 100–117.

Passarotti, A. M., Smith, J., DeLano, M., & Huang, J. (2007). Developmental differences in the neural bases of the face inversion effect show progressive tuning of face-selective regions to the upright orientation. *NeuroImage, 34,* 1708–1722.

Passingham, R. E., Stephan, K. E., & Kotter, R. (2002). The anatomical basis for functional localization in the cortex. *Nature Reviews Neuroscience, 3,* 606–616.

Patel, A. D., Gibson, E., Ratner, J., Besson, M., & Holcomb, P. J. (1998). Processing syntactic relations in language and music: An event-related potential study. *Journal of Cognitive Neuroscience, 10,* 717–733.

Paul, C., Lungarella, M., & Iida, F. (2006). Morphology, control and passive dynamics. *Robotics and Autonomous Systems, 54*(8), 617–618.

Pecher, D., Zeelenberg, R., & Barsalou, L. W. (2003). Verifying different-modality properties for concepts produces switching costs. *Psychological Science, 14*(2), 119–124.

Penfield, W., & Rasmussen, T. (1950). *The cerebral cortex of man: A clinical study of localization of function.* New York: Macmillan.

Penner-Wilger, M., & Anderson, M. L. (2008). An alternative view of the relation between finger gnosis and math ability: Redeployment of finger representations for the representation of number. In B. C. Love, K. McRae, & V. M. Sloutsky (Eds.),

Proceedings of the 30th Annual Meeting of the Cognitive Science Society (pp. 1647–1652). Austin, TX: Cognitive Science Society.

Penner-Wilger, M., & Anderson, M. L. (2013). The relation between finger gnosis and mathematical ability: Why redeployment of neural circuits best explains the finding. *Frontiers in Psychology, 4*, 877.

Perfors, A., Tenenbaum, J. B., & Regier, T. (2011). The learnability of abstract syntactic principles. *Cognition, 118*(3), 306–338.

Pesenti, M., Thioux, M., Seron, X., & De Volder, A. (2000). Neuroanatomical substrate of Arabic number processing, numerical comparison and simple addition: A PET study. *Journal of Cognitive Neuroscience, 12*, 461–479.

Pessoa, L. (2008). On the relationship between emotion and cognition. *Nature Reviews Neuroscience, 9*, 148–158.

Pessoa, L. (2012). Beyond brain regions: Network perspective of cognition-emotion interactions. *Behavioral and Brain Sciences, 35*, 158–159.

Pessoa, L. (2013). *The cognitive-emotional brain: From interactions to integration*. Cambridge, MA: MIT Press.

Petersen, S. E., van Mier, H., Fiez, J. A., & Raichle, M. E. (1998). The effects of practice on the functional anatomy of task-performance. *Proceedings of the National Academy of Sciences of the United States of America, 95*(3), 853–860.

Petersson, K. M., Elfgren, C., & Ingvar, M. (1997). A dynamic role of the medial temporal lobe during retrieval of declarative memory in man. *NeuroImage, 6*, 1–11.

Petersson, K. M., Forkstam, C., & Ingvar, M. (2004). Artificial syntactic violations activate Broca's region. *Cognitive Science, 28*, 383–407.

Petrov, A. A., Jilk, D. J., & O'Reilly, R. C. (2010). The leabra architecture: Specialization without modularity. *Behavioral and Brain Sciences, 33*(4), 286–287.

Piaget, J. (1962). *The language and thought of the child.* London: Routledge & Kegan Paul. (Original work, *Le langage et la pensée chez l'enfant,* published 1923)

Pickering, A. (2010). *The cybernetic brain: Sketches of another future.* Chicago: University of Chicago Press.

Pietrini, P., Furey, M. L., Ricciardi, E., Gobbini, M. I., Wu, W. H., Cohen, L., et al. (2004). Beyond sensory images: Object-based representation in the human ventral pathway. *Proceedings of the National Academy of Sciences of the United States of America, 101*(15), 5658–5663.

Pinel, P., Piazza, M., LeBihan, D., & Dehaene, S. (2004). Distributed and overlapping cerebral representations of number size and luminance during comparative judgements. *Neuron, 41*, 983–993.

Pinker, S. (1997). *How the mind works*. New York: Norton.

Pinker, S., & Bloom, P. (1990). Natural language and natural selection. *Behavioral and Brain Sciences, 13*, 707–784.

Pitskel, N. B., Merabet, L. B., Ramos-Estabanez, C., Kauffman, T., & Pascual-Leone, A. (2007). Time-dependent changes in cortical excitability after prolonged visual deprivation. *Neuroreport, 18*, 1703–1707.

Plante, E., Gomez, R. L., & Gerken, L. A. (2002). Sensitivity to word order cues by normal and language/learning disabled adults. *Journal of Communication Disorders, 35*, 453–462.

Platt, M. L. (2002). Neural correlates of decisions. *Current Opinion in Neurobiology, 12*, 141–148.

Platt, M. L., & Glimcher, P. W. (1997). Responses of intraparietal neurons to saccadic targets and visual distractors. *Journal of Neurophysiology, 78*, 1574–1589.

Plaut, D. C. (1995). Double dissociation without modularity: Evidence from connectionist neuropsychology. *Journal of Clinical and Experimental Neuropsychology, 17*, 291–321.

Poldrack, R. A. (2006). Can cognitive processes be inferred from neuroimaging data? *Trends in Cognitive Sciences, 10*, 59–63.

Poldrack, R. A. (2010). Mapping mental function to brain structure: How can cognitive neuroimaging succeed? *Perspectives on Psychological Science, 5*(6), 753–761.

Poldrack, R. A., Desmond, J. E., Glover, G. H., & Gabrieli, J. D. (1998). The neural basis of visual skill learning: An fMRI study of mirror reading. *Cerebral Cortex, 8*, 1–10.

Poldrack, R. A., Halchenko, Y., & Hanson, S. J. (2009). Decoding the large-scale structure of brain function by classifying mental states across individuals. *Psychological Science, 20*, 1364–1372.

Poldrack, R. A., Kittur, A., Kalar, D., Miller, E., Seppa, C., Gil, Y., et al. (2011). The cognitive atlas: Towards a knowledge foundation for cognitive neuroscience. *Frontiers in Neuroinformatics, 5*, 17.

Port, R. F. (2010). Language as a social institution: Why phonemes do not live in the brain. *Ecological Psychology, 22*, 304–326.

Posner, M. I., Petersen, S. E., Fox, P. T., & Raichle, M. E. (1988). Localization of cognitive operations in the human brain. *Science, 240*(4859), 1627–1631.

Potter, J., Wetherell, M., Gill, R., & Edwards, D. (1990). Discourse: Noun, verb or social practice? *Philosophical Psychology, 3*, 205–217.

Prescott, T. J., Redgrave, P., & Gurney, K. (1999). Layered control architectures in robots and vertebrates. *Adaptive Behavior, 7*, 99–127.

Price, C., & Friston, K. (2005). Functional ontologies for cognition: The systematic definition of structure and function. *Cognitive Neuropsychology, 22*, 262–275.

Price, C., Thierry, G., & Griffiths, T. (2005). Speech-specific auditory processing: Where is it? *Trends in Cognitive Sciences, 9*, 271–276.

Prinz, J. (2002). *Furnishing the mind: Concepts and their perceptual basis*. Cambridge, MA: MIT Press.

Prinz, J. (2006). Is the mind really modular? In R. J. Stainton (Ed.), *Contemporary debates in cognitive science* (pp. 22–36). Birmingham, AL: Blackwell.

Prinz, J. J., & Barsalou, L. W. (2000). Steering a course for embodied representation. In E. Dietrich & A. Markman (Eds.), *Cognitive dynamics: Conceptual change in humans and machines* (pp. 51–77). Cambridge, MA: MIT Press.

Proffitt, D. R., Bhalla, M., Grossweiller, R., & Midgett, J. (1995). Perceiving geographical slant. *Psychonomic Bulletin & Review, 16*, 970–972.

Proffitt, D. R., & Linkenauger, S. A. (2013). Perception viewed as a phenotypic expression. In W. Prinz, M. Beisert, & A. Herwig (Eds.), *Action science: Foundations of an Emerging Discipline* (pp. 171–197). Cambridge, MA: MIT Press.

Proffitt, D. R., Stefanucci, J., Banton, T., & Epstein, W. (2003). The role of effort in distance perception. *Psychological Science, 14*, 106–112.

Pulvermüller, F. (2005). Brain mechanisms linking language and action. *Nature Reviews Neuroscience, 6*, 576–582.

Pulvermüller, F., & Fadiga, L. (2010). Active perception: Sensorimotor circuits as a cortical basis for language. *Nature Reviews Neuroscience, 11*(5), 351–360.

Pylyshyn, Z. (1980). Computation and cognition: Issues in the foundation of cognitive science. *Behavioral and Brain Sciences, 3*, 111–132.

Pylyshyn, Z. (1984). *Computation and cognition: Toward a foundation for cognitive science*. Cambridge, MA: MIT Press.

Quartz, S. R., & Sejnowski, T. J. (1997). The neural basis of cognitive development: A constructivist manifesto. *Behavioral and Brain Sciences, 20*, 537–556.

Quine, W. V. O. (1969). *Ontological relativity, and other essays*. New York: Columbia University Press.

Radford, L. (2010). Algebraic thinking from a cultural semiotic perspective. *Research in Mathematics Education, 12*(1), 1–19.

Raichle, M. E. (1983). Positron emission tomography. *Annual Review of Neuroscience, 6*, 249–267.

Raichle, M. (2010). The brain's dark energy. *Scientific American, 302*(3), 44–49.

Raichle, M. E., MacLeod, A. M., Snyder, A. Z., Powers, W. J., Gusnard, D. A., & Shulman, G. L. (2001). A default mode of brain function. *Proceedings of the National Academy of Sciences of the United States of America, 98*, 676–682.

Ramón y Cajal, S. (1995). Histology of the nervous system (Vol. 1). N. Swanson & L. W. Swanson (Trans.). New York: Oxford University Press. (Original work published 1904)

Ramsey, W. (1997). Do connectionist representations earn their explanatory keep? *Mind & Language, 12*, 34–66.

Rasmussen, J., & Vicente, K. J. (1989). Coping with human errors through system design: Implications for ecological interface design. *International Journal of Man-Machine Studies, 31*, 517–534.

Redgrave, P., Prescott, T. J., & Gurney, K. (1999). The basal ganglia: A vertebrate solution to the selection problem? *Neuroscience, 89*, 1009–1023.

Reeve, R., & Webb, B. (2002). New neural networks for robot phonotaxis. *Philosophical Transactions of the Royal Society A, 361*, 2245–2266.

Reid, T. (2002). *Essays on the intellectual powers of man*. University Park, PA: Pennsylvania State University Press. (Original work published 1785)

Rhodes, G., Byatt, G., Michie, P. T., & Puce, A. (2004). Is the fusiform face area specialized for faces, individuation, or expert individuation? *Journal of Cognitive Neuroscience, 16*(2), 189–203.

Richardson, D. C., & Dale, R. (2005). Looking to understand: The coupling between speakers' and listeners' eye movements and its relationship to discourse comprehension. *Cognitive Science, 29*(6), 1045–1060.

Richardson, D. C., Dale, R., & Kirkham, N. Z. (2007). The art of conversation is coordination. *Psychological Science, 18*(5), 407–413.

Richardson, D., Spivey, M., Barsalou, L., & McRae, K. (2003). Spatial representations activated during real-time comprehension of verbs. *Cognitive Science, 27*, 767–780.

Richerson, P. J., & Boyd, R. (2005). *Not by genes alone: How culture transformed human evolution*. Chicago: University of Chicago Press.

Rigotti, M., Barak, O., Warden, M. R., Wang, X. J., Daw, N. D., Miller, E. K., et al. (in press). The importance of mixed selectivity in complex cognitive tasks. *Nature*. doi:10.1038/nature12160

Ritchie, J. B., & Carruthers, P. (2010). Massive modularity is consistent with most forms of neural reuse. *Behavioral and Brain Sciences, 33*(4), 289–290.

Ritter, P., & Villringer, A. (2006). Simultaneous EEG-fMRI. *Neuroscience and Biobehavioral Reviews, 30*(6), 823–838.

Rizzolatti, G., & Arbib, M. A. (1998). Language within our grasp. *Trends in Neurosciences, 21*(5), 188–194.

Rizzolatti, G., Camarda, R., Fogassi, L., Gentilucci, M., Luppino, G., & Matelli, M. (1988). Functional organization of inferior area 6 in the macaque monkey. *Experimental Brain Research, 71*(3), 491–507.

Rizzolatti, G., Fadiga, L., Gallese, V., & Fogassi, L. (1996). Premotor cortex and the recognition of motor actions. *Brain Research. Cognitive Brain Research, 3*, 131–141.

Robinson, H. B. (1964). An experimental examination of the size-weight illusion in young children. *Child Development, 35*, 91–107.

Rogers, C., Persson, A., Cheung, B., & de Bono, M. (2006). Behavioral motifs and neural pathways coordinating O_2 responses and aggregation in *C. elegans*. *Current Biology, 16*, 649–659.

Roitman, J. D., & Shadlen, M. N. (2002). Response of neurons in the lateral intraparietal area during a combined visual discrimination reaction time task. *Journal of Neuroscience, 22*(21), 9475–9489.

Rolls, E. T. (1987). Information representation, processing, and storage in the brain: Analysis at the single neuron level. In J.-P. Changeux & M. Konishi (Eds.), *The neural and molecular bases of learning* (pp. 503–539). Chichester: John Wiley.

Rorty, R. (1980). *Philosophy and the mirror of nature*. Princeton, NJ: Princeton University Press.

Ross, H. E. (1969). When is a weight not illusory? *Quarterly Journal of Experimental Psychology, 21*, 346–355.

Roux, F.-E., Boetto, S., Sacko, O., Chollet, F., & Tremoulet, M. (2003). Writing, calculating, and finger recognition in the region of the angular gyrus: A cortical stimulation study of Gerstmann syndrome. *Journal of Neurosurgery, 99*, 716–727.

Rumelhart, D. E., Hinton, G. E., & Williams, R. J. (1986a). Learning representations by back-propagating errors. *Nature, 323*(6088), 533–536.

Rumelhart, D. E., & McClelland, J. L. (1986). *Parallel distributed processing*. Cambridge, MA: MIT Press.

Rumelhart, D. E., Smolensky, P., McClelland, J. L., & Hinton, G. E. (1986b). Schemata and sequential thought processes in PDP models. In J. L. McClelland & D. E. Rumelhart (Eds.), *Parallel distributed processing: Explorations in the microstructure of cognition,* Vol. 2: *Psychological and biological models* (pp. 7–57). Cambridge, MA: MIT Press.

Rusconi, E., Walsh, V., & Butterworth, B. (2005). Dexterity with numbers: rTMS over left angular gyrus disrupts finger gnosis and number processing. *Neuropsychologia, 43*, 1609–1624.

Russell, B. (1918). The philosophy of logical atomism. *The Monist, 28*, 495–527; *29*, 32–63, 190–222, 345–380.

Russell, S., & Norvig, P. (2009). *Artificial intelligence: A modern approach* (3rd ed.). New York: Prentice Hall.

Ryle, G. (1984). *The concept of mind.* Chicago: Chicago University Press.

Sacks, H., Schegloff, E. A., & Jefferson, G. (1974). A simplest systematics for the organization of turn-taking in conversation. *Language, 50*, 696–735.

Sadato, N., Pascual-Leone, A., Grafmani, J., Ibañez, V., Deiber, M.-P., Dold, G., et al. (1996). Activation of the primary visual cortex by Braille reading in blind subjects. *Nature, 380*, 526–528.

Sakai, K., Hikosaka, O., Miyauchi, S., Takino, R., Sasaki, Y., & Pütz, B. (1998). Transition of brain activation from frontal to parietal areas in visuomotor sequence learning. *Journal of Neuroscience, 18*(5), 1827–1840.

Salmon, W. (Ed.). (1971). *Statistical explanation and statistical relevance.* Pittsburgh: University of Pittsburgh Press.

Salmon, W. (1989). *Four decades of scientific explanation.* Minneapolis: University of Minnesota Press.

Salvucci, D. D. (2005). A multitasking general executive for compound continuous tasks. *Cognitive Science, 29*, 457–492.

Sangha, S., Scheibenstock, A., & Lukowiak, K. (2003). Reconsolidation of a long-term memory in Lymnaea requires new protein and RNA synthesis and the soma of right pedal dorsal 1. *Journal of Neuroscience, 23*, 8034–8040.

Saxberg, B. V. (1987). Projected free-fall trajectories. 1. Theory and simulation. *Biological Cybernetics, 56*, 159–175.

Schall, J. D. (2004). On building a bridge between brain and behavior. *Annual Review of Psychology, 55*, 23–50.

Schall, J. D., & Bichot, N. P. (1998). Neural correlates of visual and motor decision processes. *Current Opinion in Neurobiology, 8*, 211–217.

Schank, R. C., & Abelson, R. P. (1975). *Scripts, plans, and knowledge.* New Haven, CT: Yale University.

Schank, R., & Abelson, R. P. (1977). *Scripts, plans, goals and understanding: An inquiry into human knowledge structures.* Hillsdale, NJ: Lawrence Erlbaum Associates.

Schegloff, E. A. (2007). Sequence organization in interaction (Vol. 1). *A primer in conversation analysis*. Cambridge: Cambridge University Press.

Scher, S. J. (2004). A Lego model of the modularity of the mind. *Journal of Cultural and Evolutionary Psychology, 2*(21), 248–259.

Scherf, K. S., Behrmann, M., Humphreys, K., & Luna, B. (2007). Visual category selectivity for faces, places, and objects emerges along different developmental trajectories. *Developmental Science, 10*, 15–30.

Schieber, M. H. (2001). Constraints on somatotopic organization in the primary motor cortex. *Journal of Neurophysiology, 86*, 2125–2143.

Schieber, M. H., & Hibbard, L. S. (1993). How somatotopic is the motor cortex hand area? *Science, 261*, 489–492.

Schiller, P. (1995). Effect of lesions in visual cortical area V4 on the recognition of transformed objects. *Nature, 376*, 342–344.

Schiller, P. H. (1996). On the specificity of neurons and visual areas. *Behavioural Brain Research, 76*, 21–35.

Schlaggar, B. L., & McCandliss, B. D. (2007). Development of neural systems for reading. *Annual Review of Neuroscience, 30*, 475–503.

Schleuter, D., Daufresne, M., Massol, F., & Argiller, C. (2010). A user's guide to functional diversity indices. *Ecological Monographs, 8*(3), 469–484.

Schmolesky, M. T., Wang, Y., Hanes, D. P., Thompson, K. G., Leutgeb, S., Schall, J. D., et al. (1998). Signal timing across the macaque visual system. *Journal of Neurophysiology, 79*(6), 3272–3278.

Schnall, S., Zadra, J. R., & Proffitt, D. R. (2010). Direct evidence for the economy of actions: Glucose and the perception of geographical slant. *Perception, 39*, 464–482.

Schober, M. F. (1993). Spatial perspective-taking in conversation. *Cognition, 47*, 1–24.

Schulte-Ruether, M., Markowitsch, H. J., Fink, G. R., & Piefke, M. (2007). Mirror neuron and theory of mind mechanisms involved in face-to-face interactions: A functional magnetic resonance imaging approach to empathy. *Journal of Cognitive Neuroscience, 19*(8), 1354–1372.

Schultz, W., Tremblay, L., & Hollerman, J. R. (2000). Reward processing in primate orbitofrontal cortex and basal ganglia. *Cerebral Cortex, 10*, 272–284.

Scollon, R. (2010). Aristotle fails to persuade the donkey: Conflicting logics in narrative social analysis. *eVox, 4*(1), 6–22.

Scollon, R., & Scollon, S. (2001). *Mediated discourse: The nexus of practice*. London: Routledge.

Scollon, R., & Scollon, S. (2003). *Discourses in place: Language in the material world*. London: Routledge.

Scott, S. H. (2004). Optimal feedback control and the neural basis of volitional motor control. *Nature Reviews Neuroscience, 5*, 532–546.

Searle, J. R. (1980). Minds, brains, and programs. *Behavioral and Brain Sciences, 3*(3), 417–457.

Sebanz, N., Bekkering, H., & Knoblich, G. (2006). Joint action: Bodies and minds moving together. *Trends in Cognitive Sciences, 10*(2), 70–76.

Seeck, M., Schomer, D., Mainwaring, N., Ives, J., Dubuissson, D., Blume, H., et al. (1995). Selectively distributed processing of visual object recognition in the temporal and frontal lobes of the human brain. *Annals of Neurology, 37*, 538–45.

Seeley, W. W., Menon, V., Schatzberg, A. F., Keller, J., Glover, G. H., Kenna, H., et al. (2007). Dissociable intrinsic connectivity networks for salience processing and executive control. *Journal of Neuroscience, 27*(9), 2349–2356.

Sejnowski, T. J., & Rosenberg, C. R. (1987). Parallel networks that learn to pronounce English text. *Complex Systems, 1*, 145–168.

Sellars, W. (1997). *Empiricism and the philosophy of mind*. Cambridge, MA: Harvard University Press.

Seth, A. K. (2005). Causal connectivity analysis of evolved neural networks during behavior. *Network (Bristol, England), 16*, 35–55.

Seth, A. K., & Edelman, G. M. (2007). Distinguishing causal interactions in neural populations. *Neural Computation, 19*(4), 910–933.

Seung, S. (2012). *Connectome: How the brain's wiring makes us who we are*. New York: Houghton Mifflin.

Shaffer, D. M., & McBeath, M. K. (2005). Naïve beliefs in baseball: Systematic distortions in perceived time of apex for fly balls. *Journal of Experimental Psychology. Human Perception and Performance, 28*, 335–348.

Shapiro, L. (2011). *Embodied cognition*. New York: Routledge.

Shepard, R. N. (1980). Multidimensional scaling, tree-fitting, and clustering. *Science, 210*, 390–398.

Shepard, R., & Metzler, J. (1971). Mental rotation of three dimensional objects. *Science, 171*(972), 701–703.

Shockley, K., Richardson, D. C., & Dale, R. (2009). Conversation and coordinative structures. *Topics in Cognitive Science, 1*(2), 305–319.

Shockley, K., Santana, M. V., & Fowler, C. A. (2003). Mutual interpersonal postural constraints are involved in cooperative conversation. *Journal of Experimental Psychology. Human Perception and Performance, 29*(2), 326–332.

Simmons, W. K., Ramjee, V., Beauchamp, M. S., McRae, K., Martin, A., & Barsalou, L. W. (2007). A common neural substrate for perceiving and knowing about color. *Neuropsychologia, 45*(12), 2802–2810.

Simon, H. A. (1962). The architecture of complexity. *Proceedings of the American Philosophical Association* 106, 467–82. [Reprinted 1969 in H. Simon, *The sciences of the artificial* (pp. 192–229). Cambridge, MA: MIT Press.]

Singh, P., Lin, T., Mueller, E., Lim, G., Perkins, T., & Zhu, W. L. (2002). Open mind common sense: Knowledge acquisition from the general public. In R. Meersman & Z. Tari (Eds.), *On the move to meaningful internet systems 2002* (pp. 1223–1237). London: Springer-Verlag.

Skudlarski, P., Jagannathan, K., Anderson, K., Stevens, M. C., Calhoun, V. D., Skudlarska, B. A., et al. (2010). Brain connectivity is not only lower but different in schizophrenia: A combined anatomical and functional approach. *Biological Psychiatry, 68*(1), 61–69.

Smith, L. B. (2005). Cognition as a dynamic system: Principles from embodiment. *Developmental Review, 25*, 278–298.

Smolensky, P. (1986). Neural and conceptual interpretation of PDP models. In J. McClelland & D. E. Rumelhart (Eds.), *Parallel distributed processing: Explorations in the microstructure of cognition, Vol. 2: Applications* (pp. 390–431). Cambridge, MA: MIT Press.

Sobel, D. M., & Kirkham, N. Z. (2006). Blickets and babies: The development of causal reasoning in toddlers and infants. *Developmental Psychology, 42*(6), 1103.

Solomon, K. O., & Barsalou, L. W. (2001). Representing properties locally. *Cognitive Psychology, 43*, 129–169.

Speer, R., Havasi, C., & Lieberman, H. (2008). AnalogySpace: Reducing the dimensionality of common sense knowledge. In D. Fox & C. P. Gomes (Eds.), *Proceedings of the 23rd AAAI conference on artificial intelligence,* Chicago, IL, pp. 548–553.

Spelke, E. S. (1982). Perceptual knowledge of objects in infancy. In J. Mehler, M. Garrett, & E. Walker (Eds.), *Perspectives on mental representation.* Hillsdale, NJ: Lawrence Erlbaum Associates.

Spencer, H. (1855). *Principles of psychology*. London: Williams and Norgate.

Spencer, J. P., Thomas, S. C., & McClelland, J. L. (Eds.). (2009). *Toward a unified theory of development: Connectionism and dynamic systems theory re-considered.* Oxford: Oxford University Press.

Spivey, M. (2007). *The continuity of mind*. Oxford: Oxford University Press.

Sporns, O. (2011). *Networks in the brain*. Cambridge, MA: MIT Press.

Sporns, O., Tononi, G., & Edelman, G. M. (2000). Theoretical neuroanatomy: Relating anatomical and functional connectivity in graphs and cortical connection matrices. *Cerebral Cortex, 10*, 127–141.

Stam, C. F., Jones, B. F., Nolte, G., Breakspear, M., & Scheltens, P. H. (2007). Small-world networks and functional connectivity in Alzheimer's disease. *Cerebral Cortex, 17*(1), 92–99.

Stefanucci, J. K., & Geuss, M. N. (2009). Big people, little world: The body influences size perception. *Perception, 38*, 1782–1795.

Stefanucci, J. K., & Geuss, M. N. (2010). Duck! Scaling the height of a horizontal barrier to body height. *Attention, Perception & Psychophysics, 72*, 1338–1349.

Steinmetz, P. N., Roy, A., Fitzgerald, P. J., Hsiao, S. S., Johnson, K. O., & Niebur, E. (2000). Attention modulates synchronized neuronal firing in primate somatosensory cortex. *Nature, 404*, 187–190.

Sternberg, S. (1969). The discovery of processing stages: Extensions of Donders' method. *Acta Psychologica, 30*, 276–315.

Stewart, T. C., & West, R. L. (2007). Cognitive redeployment in ACT-R: Salience, vision, and memory. In *8th International Conference on Cognitive Modelling, Ann Arbor, MI, July*.

Stone, J. V. (2004). *Independent component analysis: A tutorial introduction*. Cambridge, MA: MIT Press.

Sugiura, M., Watanabe, J., Maeda, Y., Matsue, Y., Fukuda, H., & Kawashima, R. (2005). Cortical mechanisms of visual self-recognition. *NeuroImage, 24*, 143–149.

Sulston, J. E., Albertson, D. G., & Thomson, J. N. (1980). The *Caenorhabditis elegans* male: Postembryonic development of nongonadal structures. *Developmental Biology, 78*, 542–576.

Supekar, K. S., Musen, M. A., & Menon, V. (2009). Development of large-scale functional brain networks in children. *PLoS biology, 7*(7), e1000157.

Sur, M., Garraghty, P. E., & Roe, A. W. (1988). Experimentally induced visual projections into auditory thalamus and cortex. *Science, 242*, 1437–1441.

Sur, M., Pallas, S. L., & Roe, A. W. (1990). Cross-modal plasticity in cortical development: Differentiation and specification of sensory neocortex. *Trends in Neurosciences, 13*, 227–233.

Syal, S., & Finlay, B. (2011). Thinking outside the cortex: Social motivation in the evolution and development of language. *Developmental Science, 14*(2), 417–430.

Syková, E. (2004). Extrasynaptic volume transmission and diffusion parameters of the extracellular space. *Neuroscience, 129*, 861–876.

Syková, E., Mazel, T., Hasenöhrl, R. U., Harvey, A. R., Šimonová, Z., Mulders, W. H. A. M., et al. (2002). Learning defecits in aged rats related to decrease in extracellular volume and loss of diffucion anisotropy in hippocampus. *Hippocampus, 12*, 469–479.

Takikawa, Y., Kawagoe, R., & Hikosaka, O. (2002). Reward-dependent spatial selectivity of anticipatory activity in monkey caudate neurons. *Journal of Neurophysiology, 87*(1), 508–515.

Tamás, G., Buhl, E. H., Lörincz, A., & Somogyi, P. (2000). Proximally targeted GABAergic synapses and gap junctions synchronize cortical interneurons. *Nature Neuroscience, 3*, 366–371.

Tanji, J., & Hoshi, E. (2001). Behavioral planning in the prefrontal cortex. *Current Opinion in Neurobiology, 11*, 164–170.

Tannen, D. (Ed.). (1982). *Analyzing discourse: Text and talk*. Washington, DC: Georgetown University Press.

Tannen, D. (Ed.). (1993). *Framing in discourse*. Oxford: Oxford University Press.

Tannen, D. (2007). *Talking voices: Repetition, dialogue and imagery in conversational discourse*. Cambridge: Cambridge University Press.

Tannen, D., & Wallat, C. (1987). Interactive frames and knowledge schemas in interaction: Examples from a medical examination/interview. *Social Psychology Quarterly, 50*(2), 205–216.

Tattersall, I. (1998). *The origins of the human capacity*. New York: American Museum of Natural History.

Taylor, D. M., Tillery, S. I. H., & Schwartz, A. B. (2002). Direct cortical control of 3D neuroprosthetic devices. *Science, 296*(5574), 1829–1832.

Taylor, K. S., Seminowicz, D. A., & Davis, K. D. (2009). Two systems of resting state connectivity between the insula and cingulate cortex. *Human Brain Mapping, 30*(9), 2731–2745.

Teeling, E. C., Springer, M. S., Madsen, O., Bates, P., O'Brien, S. J., & Murphy, W. (2005). A molecular phylogeny for bats illuminates biogeography and the fossil record. *Science, 307*, 580–584.

Tettamanti, M., & Weniger, D. (2006). Broca's area: A supramodal hierarchical processor? *Cortex, 42*, 491–494.

Thagard, P. (1992). *Conceptual revolutions*. Princeton, NJ: Princeton University Press.

Thelen, E. (1995). Time-scale dynamics and the embodiment of embodied cognition. In R. Port & T. van Gelder (Eds.), *Mind as motion* (pp. 69–100). Cambridge, MA: MIT Press.

Thelen, E., & Smith, L. B. (1994). *A dynamic systems approach to the development of cognition and action*. Cambridge, MA: MIT Press.

Thewissen, J. G. M., & Nummela, S. (2008). *Sensory evolution on the threshold: Adaptations in secondarily aquatic vertebrates*. Berkeley, CA: University of California Press.

Thoenissen, D., Zilles, K., & Toni, I. (2002). Differential involvement of parietal and precentral regions in movement preparation and motor intention. *Journal of Neuroscience, 22*, 9024–9034.

Thomas, K. M., King, S. W., Franzen, P. L., Welsh, T. F., Berkowitz, A. L., Noll, D. C., et al. (1999). A developmental functional MRI study of spatial working memory. *NeuroImage, 10*(3), 327–338.

Titchener, E. B. (1898). The postulates of a structural psychology. *Philosophical Review, 7*, 449–465.

Titzer, R., Thelen, E., & Smith, L. B. (2003). Learning about transparency. Unpublished manuscript. Cited in Smith (2005).

Todorov, E., & Jordan, M. I. (2002). Optimal feedback control as a theory of motor coordination. *Nature Neuroscience, 5*, 1226–1235.

Tomasello, M. (1999). *The cultural origins of human cognition*. Cambridge, MA: Harvard University Press.

Tomblin, J. B., Mainela-Arnold, M. E., & Zhang, X. (2007). Procedural learning in adolescents with and without specific language impairment. *Language Learning and Development, 3*, 269–293.

Toro, R., Fox, P. T., & Paus, T. (2008). Functional coactivation map of the human brain. *Cerebral Cortex, 18*(11), 2553–2559.

Torta, D. M., & Cauda, F. (2011). Different functions in the cingulate cortex: A meta-analytic connectivity modeling study. *NeuroImage, 56*(4), 2157–2172.

Treisman, A. (1986). Features and objects in visual processing. *Scientific American, 255*(5), 114–125.

Tsunozaki, M., Chalasani, S. H., & Bargmann, C. I. (2008). A behavioral switch: cGMP and PKC signaling in olfactory neurons reverses odor preference in *C. elegans*. *Neuron, 59*(6), 959–971.

Turing, A. (1936). On computable numbers, with an application to the Entscheidungsproblem. *Proceedings of the London Mathematical Society, ser 2, 42*, 230–265.

Turner, P. (2005). Affordances as context. *Interacting with Computers, 17*(6), 787–800.

Uddin, L. Q., Kinnison, J., Pessoa, L., & Anderson, M. L. (2014). Beyond the tripartite cognition-emotion-interoception model of the human insular cortex. *Journal of Cognitive Neuroscience 26*(1), 16-27.

Uttal, W. (2001). *The new phrenology: The limits of localizing cognitive processes in the brain.* Cambridge, MA: MIT Press.

Valencia, M., Pastor, M. A., Fernández-Seara, M. A., Artieda, J., Martinerie, J., & Chavez, M. (2009). Complex modular structure of large-scale brain networks. *Chaos: An Interdisciplinary Journal of Nonlinear Science, 19*(2), 023119.

Vanderbeeken, R., & Weber, E. (2002). Dispositional explanations of behavior. *Behavior and Philosophy, 30,* 43–60.

Van Dijk, K. R., Hedden, T., Venkataraman, A., Evans, K. C., Lazar, S. W., & Buckner, R. L. (2010). Intrinsic functional connectivity as a tool for human connectomics: Theory, properties, and optimization. *Journal of Neurophysiology, 103*(1), 297–321.

van Fraassen, B. (2002). *The empirical stance.* New Haven, CT: Yale University Press.

van Fraassen, B. (2008). *Scientific representation: Paradoxes of perspective.* Oxford: Oxford University Press.

van Gelder, T. (1990). Compositionality: A connectionist variation on a classical theme. *Cognitive Science, 14,* 355–384.

van Gelder, T. (1991). What is the 'D'in 'PDP': A survey of the concept of distribution. In W. Ramsey (Ed.), *Philosophy and connectionist theory* (pp. 33–59). New York: Routledge.

van Gelder, T. (1995). What might cognition be, if not computation? *Journal of Philosophy, 92*(7), 345–381.

van Gelder, T. (1998). The dynamical hypothesis in cognitive science. *Behavioral and Brain Sciences, 21,* 615–665.

van Orden, G. C., Jansen op de Haar, M., & Bosman, A. (1997). Complex dynamic systems also predict dissociations, but they do not reduce to autonomous components. *Cognitive Neuropsychology, 14,* 131–165.

van Orden, G. C., Pennington, B. F., & Stone, G. O. (2001). What do double dissociations really prove? *Cognitive Science, 25,* 111–172.

Varela, F., Lachaux, J. P., Rodriguez, E., & Martinerie, J. (2001). The brainweb: Phase synchronization and largescale integration. *Nature Reviews Neuroscience, 2*(4), 229–239.

Varela, F. J., Thompson, E., & Rosch, E. (1990). *The embodied mind: Cognitive science and human experience.* Cambridge, MA: MIT Press.

Varshney, L. R., Chen, B. L., Paniagua, E., Hall, D. H., & Chklovskii, D. (2011). Structural properties of the *Caenorhabditis elegans* neuronal network. *PLoS Computational Biology, 7*, e1001066.

Velazquez-Ulloa, N., Blackshaw, S. E., Szczupak, L., Trueta, C., Garcia, E., & De-Miguel, F. F. (2003). Convergence of mechanosensory inputs onto neuromodulatory serotonergic neurons in the leech. *Journal of Neurobiology, 54*, 604–617.

Venkatraman, V., Ansari, D., & Chee, M. W. L. (2005). Neural correlates of symbolic and non-symbolic arithmetic. *Neuropsychologia, 43*, 744–753.

Vera, A. H., & Simon, H. A. (1993). Situated action: A symbolic interpretation. *Cognitive Science, 17*, 7–48.

Voytek, J. B., & Voytek, B. (2012). Automated cognome construction and semi-automated hypothesis generation. *Journal of Neuroscience Methods, 208*(1), 92–100.

Vygotsky, L. S. (1962). *Thought and language* (Hanfmann, E., & Vakar, G., Trans., Eds.). Cambridge, MA: MIT Press. (Original work published 1934)

Wager, T. D., Lindquist, M., & Kaplan, L. (2007). Meta-analysis of functional neuroimaging data: current and future directions. *Social Cognitive and Affective Neuroscience, 2*(2), 150–158.

Wallace, A. R. (1870). *The limits of natural selection as applied to man. Contributions to the theory of natural selection: A series of essays*. New York: Macmillan.

Warren, W. (2005). Direct perception: The view from here. *Philosophical Topics, 33*(1), 335–361.

Weiskopf, D. (2007). Concept empiricism and the vehicles of thought. *Journal of Consciousness Studies, 14*, 156–183.

Wells, A. (2006). *Rethinking cognitive computation: Turing and the science of the mind*. London: Palgrave.

Wheeler, M. (2005). *Reconstructing the cognitive world: The next step*. Cambridge, MA: MIT Press.

White, J. G., Southgate, E., Thomson, J. N., & Brenner, S. (1986). The structure of the nervous system of the nematode *C. elegans*. *Philosophical Transactions of the Royal Society of London. Series B, Biological Sciences, 314*, 1–340.

Wills, T. J., Cacucci, F., Burgess, N., & O'Keefe, J. (2010). Development of hippocampal cognitive map in preweaning rats. *Science, 328*, 1573–1576.

Wilson, M. (2001). The case for sensorimotor coding in working memory. *Psychonomic Bulletin & Review, 8*, 44–57.

Winawer, J., Witthoft, N., Frank, M. C., Wu, L., Wade, A. R., & Boroditsky, L. (2007). Russian blues reveal effects of language on color discrimination. *Proceedings of the National Academy of Sciences of the United States of America, 104*(19), 7780–7785.

Winsler, A., Diaz, R. M., & Montero, I. (1997). The role of private speech in the transition from collaborative to independent task performance in young children. *Early Childhood Research Quarterly, 12*(1), 59–79.

Witt, K. L., Linkenaugher, S. A., Bakdash, J. Z., & Proffitt, D. R. (2008). Putting to a bigger hole: Golf performance relates to perceived size. *Psychonomic Bulletin & Review, 15*, 581–585.

Witt, J. K., & Proffitt, D. R. (2005). See the ball, hit the ball: Apparent ball size is correlated with batting average. *Psychological Science, 16*, 937–938.

Witt, J. K., & Proffitt, D. R. (2008). Action-specific influences on distance perception: A role for motor simulation. *Journal of Experimental Psychology. Human Perception and Performance, 34*, 1479–1492.

Witt, J. K., Proffitt, D. R., & Epstein, W. (2004). Perceiving distance: A role of effort and intent. *Perception, 33*, 570–590.

Witt, J. K., Proffitt, D. R., & Epstein, W. (2005). Tool use affects perceived distance but only when you intend to use it. *Journal of Experimental Humam Perception and Performance, 31*, 80–88.

Wittgenstein, L. (1995). *Tractatus logico-philosophicus* (D. F. Pears, Trans.). London: Routledge. (Original work published 1921)

Wittgenstein, L. (2001). *Philosophical investigations*. Oxford: Blackwell Publishing. (Original work published 1953)

Witthoft, N., Winawer, J., Wu, L., Frank, M., Wade, A., & Boroditsky, L. (2003). Effects of language on color discriminability. In *Proceedings of the 25th annual meeting of the Cognitive Science Society* (pp. 1247–1252). Mahwah, NJ: Lawrence Erlbaum Associates.

Woodward, J. (2007). Interventionist theories of causation in psychological perspective. In A. E. Gopnik & L. E. Schulz (Eds.), *Causal learning: Psychology, philosophy, and computation* (pp. 19–36). Oxford: Oxford University Press.

Wozniak, R. H. (1992). *Mind and body: Rene Déscartes to William James*. Bethesda, MD: National Library of Medicine/American Psychological Association.

Yang, T., & Shadlen, M. N. (2007). Probabilistic reasoning by neurons. *Nature, 447*(7148), 1075–1080.

Yarkoni, T., Poldrack, R. A., Nichols, T. E., Van Essen, D. C., & Wager, T. D. (2011). Large-scale automated synthesis of human functional neuroimaging data. *Nature Methods, 8*(8), 665–670.

Yarkoni, T., Poldrack, R. A., Van Essen, D. C., & Wager, T. D. (2010). Cognitive neuroscience 2.0: Building a cumulative science of human brain function. *Trends in Cognitive Sciences, 14*(11), 489–496.

Yeo, B. T., Krienen, F. M., Sepulcre, J., Sabuncu, M. R., Lashkari, D., Hollinshead, M., et al. (2011). The organization of the human cerebral cortex estimated by intrinsic functional connectivity. *Journal of Neurophysiology, 106*(3), 1125–1165.

Yoshida, K., Watanbe, D., Ishikane, H., Tachibana, M., Pastan, I., & Nakanishi, S. (2001). A key role of startburst amacrine cells in originating retinal directional selectivity and optokinetic eye movement. *Neuron, 30*, 771–780.

Young, L., Dodell-Feder, D., & Saxe, R. (2010). What gets the attention of the temporo-parietal junction? An fMRI investigation of attention and theory of mind. *Neuropsychologia, 48*(9), 2658–2664.

Zacksenhouse, M., Lebedev, M. A., Carmena, J. M., O'Doherty, J. E., Henriquez, C. S., Principe, J. C., et al. (2007). Cortical modulations increase during early sessions with brain-machine interfaces. *PLoS ONE, 2*(7), e619.

Zadra, J., Schnall, S., Weltman, A., & Proffitt, D. R. (2010). Direct physiological evidence for the economy of action: Bioenergetics and the perception of spatial layout. *Journal of Vision, 10*(7), 54.

Zago, L., Pesenti, M., Mellet, E., Crivello, F., Mazoyer, B., & Tzourio-Mazoyer, N. (2001). Neural correlates of simple and complex mental calculation. *NeuroImage, 13*, 314–327.

Zangaladze, A., Epstein, C. M., Grafton, S. T., & Sathian, K. (1999). Involvement of visual cortex in tactile discrimination of orientation. *Nature, 401*(6753), 587–590.

Zebian, S. (2005). Linkages between number concepts, spatial thinking, and directionality of writing: The SNARC effect and the reverse SNARC effect in English and Arabic monoliterates, biliterates, and illiterate Arabic speakers. *Journal of Cognition and Culture, 5*(1–2), 1–2.

Zeruvabel, E. (1991). *The fine line: Making distinctions in everyday life*. New York: The Free Press.

Zhu, Q., & Bingham, G. P. (2008). Is hefting to perceive the affordance for throwing a smart perceptual mechanism? *Journal of Experimental Psychology. Human Perception and Performance, 34*, 929–943.

Zhu, Q., & Bingham, G. P. (2011). Human readiness to throw: The size-weight illusion is not an illusion when picking the best objects to throw. *Evolution and Human Behavior, 32*(4), 288–293.

Zuidema, W. (2003). How the poverty of the stimulus solves the poverty of the stimulus. In S. Becker, S. Thrun, & K. Obermayer (Eds.), *Advances in neural information processing systems 15* (pp. 51–58). Cambridge, MA: MIT Press.

Index

www.ingramcontent.com/pod-product-compliance
Lightning Source LLC
LaVergne TN
LVHW052112210926
848482LV00025B/548
* 9 7 8 0 2 6 2 5 4 4 7 3 3 *